SALMONID FISHES

POPULATION BIOLOGY, GENETICS AND MANAGEMENT

Fish and Aquatic Resources Series

Series Editor: Professor Tony. J. Pitcher
Director, Fisheries Centre, University of British Columbia, Canada

The *Blackwell Science Fish and Aquatic Resources Series* is an initiative aimed at providing key books in this fast-moving field, published to a high international standard.

The Series includes books that review major themes and issues in the science of fishes and the interdisciplinary study of their exploitation in human fisheries. Volumes in the Series combine a broad geographical scope with in-depth focus on concepts, research frontiers and analytical frameworks. These books will be of interest to research workers in the biology, zoology, ichthyology, ecology, physiology of fish and the economics, anthropology, sociology and all aspects of fisheries. They will also appeal to non-specialists such as those with a commercial or industrial stake in fisheries.

It is the aim of the editorial team that books in the *Blackwell Science Fish and Aquatic Resources Series* should adhere to the highest academic standards through being fully peer reviewed and edited by specialists in the field. The Series books are produced by Blackwell Science in a prestigious and distinctive format. The Series Editor, Professor Tony J. Pitcher is an experienced international author, and founding editor of the leading journal in the field of fish and fisheries.

The Series Editor and Publisher at Blackwell Science, Nigel Balmforth, will be pleased to discuss suggestions, advise on scope, and provide evaluations of proposals for books intended for the Series. Please see contact details listed below.

Titles currently included in the Series

1. *Effects of Fishing on Marine Ecosystems and Communities* (S. Hall) 1999
2. *Salmonid Fishes* (Edited by Y. Altukhov *et al.*) 2000
3. *Percid Fishes* (J. Craig) 2000
4. *Fisheries Oceanography* (Edited by P. Harrison & T. Parsons) 2000
5. *Sustainable Fishery Systems* (A. Charles) 2000
6. *Krill* (Edited by I. Everson) 2000

For further information concerning books in the series, please contact:
Nigel Balmforth, Professional Division, Blackwell Science, Osney Mead, Oxford OX2 0EL, UK
Tel: +44 (0) 1865 206206; Fax +44 (0) 1865 721205
e-mail: nigel.balmforth@blacksci.co.uk

SALMONID FISHES

POPULATION BIOLOGY, GENETICS AND MANAGEMENT

Yuri P. Altukhov
Elena A. Salmenkova
Vladimir T. Omelchenko

Vavilov General Genetics Institute,
Russian Academy of Sciences, Moscow

English translation edited by

G.R. Carvalho & J.E. Thorpe

Blackwell
Science

Editorial Offices:
Osney Mead, Oxford OX2 0EL
25 John Street, London WC1N 2BL
23 Ainslie Place, Edinburgh EH3 6AJ
350 Main Street, Malden
MA 02148 5018, USA
54 University Street, Carlton
Victoria 3053, Australia
10, rue Casimir Delavigne
75006 Paris, France

Other Editorial Offices:

Blackwell Wissenschafts-Verlag GmbH
Kurfürstendamm 57
10707 Berlin, Germany

Blackwell Science KK
MG Kodenmacho Building
7–10 Kodenmacho Nihombashi
Chuo-ku, Tokyo 104, Japan

First published 2000

DISTRIBUTORS

Marston Book Services Ltd
PO Box 269
Abingdon
Oxon OX14 4YN
(*Orders:* Tel: 01235 465500
Fax: 01235 465555)

USA
Blackwell Science, Inc.
Commerce Place
350 Main Street
Malden, MA 02148 5018
(*Orders:* Tel: 800 759 6102
781 388 8250
Fax: 781 388 8255)

Canada
Login Brothers Book Company
324 Saulteaux Crescent
Winnipeg, Manitoba R3J 3T2
(*Orders:* Tel: 204 837 2987
Fax: 204 837 3116)

Australia
Blackwell Science Pty Ltd
54 University Street
Carlton, Victoria 3053
(*Orders:* Tel: 03 9347 0300
Fax: 03 9347 5001)

A catalogue record for this title is available from the British Library

ISBN 978-0-632-05587-6

Library of Congress
Cataloging-in-Publication Data
Altukhov, IU. P.
[Populiatsionnaia genetika lososevykh ryb. English]
Salmonid fishes : population biology, genetics, and management / by Yuri P. Altukhov, Elena A. Salmenkova, Vladimir T. Omelchenko. ; English translation edited by J. Thorpe and G. Carvalho.
p. cm.
Includes bibliographical references (p.).
ISBN 0-632-05587-1 (pbk.)
1. Salmonidae–Genetics. 2. Fish populations.
I. Salmenkova, E. A. (Elena Aleksandrovna)
II. Omel'chenko, V. T. (Vladimir Timofeevich)
III. Title.
QL638.S2 A4813 2000
597.5′61788–dc21
99-059650

For further information on
Blackwell Science, visit our website:
www.blackwell-science.com

Contents

Series Foreword

Fish researchers (a.k.a. fish freaks) like to explain, to the bemused bystander, how fish have evolved an astonishing array of adaptations, so much so that it can be difficult for them to comprehend why anyone would study anything else. Yet, at the same time, fish are among the last wild creatures on our planet that are hunted by humans for food. As a consequence, few today would fail to recognize that the reconciliation of exploitation with the conservation of biodiversity provides a major challenge to our current knowledge and expertise. Even evaluating the trade-offs that are needed is a difficult task. Moreover, solving this pivotal issue calls for a multidisciplinary conflation of fish physiology, biology and ecology with social sciences such as economics and anthropology in order to probe new frontiers of applied science. The *Blackwell Science Series* on *Fish and Aquatic Resources* is an initiative aimed at providing key, peer-reviewed texts in this fast-moving field.

While bony fish stem from a great radiation that followed the invention of the swimbladder in the Cretaceous period 100 million years ago, some fish groups, such as the sharks, lungfish and sturgeons are more ancient beasts. Survivors from earlier eras may be more widespread than we think: the deep sea coelacanths, formerly known only from the Indian Ocean, have recently turned up in Indonesia. Also, these fishes may be more effectively adapted to specialized niches than their ancient body plans would suggest. For example, rays and angel sharks have perfected the art of the ambush predator, while most cartilaginous fishes can detect electric discharges in the nerves of their prey.

Bony fish themselves have evolved into an amazing array of habitats and niches. As well as the open sea, there are fish in lakes, ponds, rivers and rock pools; in deserts, forests, mountains, the great deeps of the sea, and the extreme cold of the Antarctic; in warm waters of high alkalinity or of low oxygen; and in habitats like estuaries or mudflats, where their physiology is challenged by continuous change. Air-breathing climbing perch (regularly found up trees), walking catfish and mangrove mudskippers are currently repeating the land invasion of their Carboniferous ancestors. We can marvel at high-speed swimming adaptations in the fins, tails, gills and muscles of marlins, sailfish and warm-blooded tunas; gliding and flapping flight in several groups of fish; swinging, protrusible jaws providing suction-assisted feeding that have evolved in parallel in groupers, carps and cods; parental care in mouth-brooding cichlids; the birth of live young in many sharks, tooth carps, rockfish and blennies; immense migrations in salmon, shads and tunas; and even the so-called four-eyed fish, with eyes divided into upper air and lower water-adapted sections.

In addition to food, recreation (and inspiration for us fish freaks), it has, moreover, recently been realized that fish are essential components of aquatic ecosystems that provide vital services to human communities. But, sadly, virtually all sectors of the stunning biodiversity of fishes are at risk from human activities. In fresh water, for example, the largest mass extinction event since the end of the dinosaurs has occurred as the introduced Nile perch in Lake Victoria eliminated over 100 species of endemic haplochromine fish. But, at the same time, precious food and income from the Nile perch fishery was created in a miserably poor region. In the oceans, we have barely begun to understand the profound changes that have accompanied a vast expansion of human fishing over the past 100 years.

This book is the second in the *Blackwell Science Fish and Aquatic Resources Series*. It summarizes more than 30 years of research on salmon family fish in the Russian far east and north, and puts this work in context with publications arising in North America and Europe. The three eminent authors, Yuri Altukhov, Elena Salmenkova and Vladimir Omelchneko, are from the Russian Academy of Sciences, and the book has been refereed and edited by two leading British salmon researchers, biologist John Thorpe and molecular population geneticist, Gary Carvalho.

The book makes accessible a huge body of research formerly only available in Russian; of some 1100 references cited in the book, about 40% are from the Russian literature. But, as well as being useful to salmon biologists in reviewing research literature that encompasses allozymes, nucleic acids and mathematical population genetics in nine salmonid species, the book has a powerfully expressed theme. This central message has a dreadful simplicity: the preservation of salmonid diversity is essential for future sustainable exploitation. If we do not manage salmonid populations to do this, a stark future of massive extinctions likely awaits us.

In North America, the bleak status of salmonids in the lower 48 states of the USA is an unpleasant reality – in one river in the Columbia system that teemed with salmon 50 years ago, the sperm of a lone male that made it back to the spawning grounds in 1998 is cryopreserved, awaiting a future mate it will never see. Such disasters are often represented as largely the consequence of freshwater habitat destruction such as dam building and logging, but these Russian authors suggest that the real horror is the destruction of genetic diversity. If they are right, recovery of salmon populations, even when habitats are restored, will be next to impossible.

Salmonids, migrating between streams, lakes, rivers and the sea, have evolved a semi-isolated sub-population structure, both within and among watersheds, that balances the genetic load of mutations and poorly adapted genotypes. Each unit has its own morphology, physiology, resistance to disease and parasites, and unique migratory traits in both the marine and freshwater phases of its life history. Anthropogenic influences are invariably maladaptive, and reduce this intrapopulation genetic diversity. The authors suggest that such dismal influences have brought salmonid populations to the limits of their integrity.

To be sustainable, fisheries on salmonids should aim not to alter the balance among sub-population units, but, unsurprisingly, they generally they do. Hatcheries, often vaunted by their progenitors as the salvation of wild salmon fisheries through 'stock enhancement', actually make this problem much worse by increasing fishing pressure on weak and endangered wild stocks. Moreover, salmon hatcheries typically provoke inbreeding or homozygosity, leading to an irreversible degradation of population structure. In addition, salmonid populations are challenged not only by direct human influences such as fishing and habitat loss, but also by indirect anthropogenic effects, such as global warming. Warmer water reduces their survival when swimming up rivers to spawn, a feat performed entirely with stored energy in Pacific salmon, and by altering the latitude of profitable feeding grounds in the ocean and hence increasing the fishes travel costs to get there.

Only genetic diversity stands a chance of hedging against these perils. So much so, that the authors wish to found a new research field of conservation genetics. This book opens up the discussion of Altukhov, Salmenkova and Omelchneko's agenda for the genetic conservation and sustainable exploitation of salmonids to a much wider readership, and will, I am sure, become essential reading for all salmon researchers and a helpful source for conservation biologists.

Professor Tony J. Pitcher
Editor, Blackwell Science Fish and Aquatic Resources Series Director,
Fisheries Centre, University of British Columbia, Vancouver, Canada
March 2000

Preface

Salmonid fishes inhabiting the Pacific and Atlantic basins are characterized by unique biological features and have provided for many human populations since ancient times. As European civilizations spread to the American and Asian continents, salmonids became the object of comprehensive biological research and, to date, they have been studied in detail from virtually all aspects. This book is devoted to the biology and genetics of salmonids. It is based both on data available from world scientific literature and on our experience gained during 30 years of studies on the genetic processes in Pacific and Atlantic salmon populations in different regions of the Far East and European North of Russia.

Historically, even the earliest steps in our research were made taking into account at least two important circumstances which determined the general course of this work. First, we understood clearly that fish species do not represent structureless, panmictic populations but rather are complex mosaics of semi-isolated subpopulations exchanging genes with one another. An important emergent property that does not follow from the properties of components involved is long-term genetic stability over time intervals commensurate with the length of existence of populations. Second, it was apparent that, to evaluate reliably the specific features of the biological and genetic processes operating in populations exposed to different anthropogenic influences over time, it is necessary to have a reference point, that is, to know the parameters characterizing the state of the populations before the onset of such an influence, or at least at its early stage. In other words, it is necessary to have a concept of the 'normal' state of the 'normal' process. As an example of such a relatively undisturbed natural population system, we chose the sockeye salmon stock of Lake Azabachye (the Kamchatka Peninsula) and, on this basis, interpreted the pattern of genetic processes operating in relatively undisturbed populations and those under anthropogenic impact (primarily, artificial salmon propagation, transplantation, commercial fishing and habitat alterations).

The results showed that the main condition for the long-term existence of populations consists of *preserving their gene diversity and the ratio of its intra- and intergroup components at the optimal level characteristic of each population*. Any disturbance of this ratio has an adverse effect on the normal functioning of a subdivided population, leading to a decrease in its fitness and even to degradation. These findings allowed us to develop recommendations for efficient management on the basis of population genetic data.

It is apparent that this scientific work could be accomplished only through the joint efforts of many researchers, although it originated in 1968 from an expedition to Sakhalin, launched by a few scientists who worked in the Laboratory of Genetics organized in the Institute of Marine Biology (Far East Division, Russian Academy of Sciences, Vladivostok).

In the course of these studies, many specialists gained knowledge and experience in a novel research field. Today, they continue successfully their research works in Moscow, Magadan, Vladivostok, Sakhalin and Kamchatka. The data collected concern the biology and genetics of more than 100 large geographic populations of pink, chum, sockeye and masu salmon. In addition, certain problems of salmonid taxonomy and phylogeny were analysed. Important contributions to progress were made by G.D. Ryabova, V.I. Slynko, R.M. Viktorovsky, L.G. Volokhonskaya, Yu.F. Kartavtsev and N.V. Varnavskaya, A.I. Pudovkin and B.A. Kalabushkin helped us with calculations. Our work would have been impossible without the unwavering support of Glavrybvod (Central Department for Protection and Reproduction of Fish Resources and Management of Fishery: I.V. Nikonorov, L.V. Polikashin, V.A. Izmailov and others), Sakhalin Department for Protection and Reproduction of Fish Resources and Management of Fishery (S.A. Ugryumov, I.M. Zolotareva, Yu.S. Zolotarev, G.T. Kochetkov, L.K. Polozova, T.N. Lyubaeva and others), and the administration of the Sakhalin salmon hatcheries. Directors and personnel of several institutes supervised by the Far East Division of the Russian Academy of Sciences and the former USSR Ministry of Fish Industry – A.V. Zhirmunsky, S.M. Konovalov, Yu.S. Roslyi, F.N. Rukhlov and many others – were of great help to us. We use this opportunity to express our sincere gratitude to them, to researchers of the Laboratory of Genetics (Institute of Marine Biology, Far East Branch, Russian Academy of Sciences) and the Laboratory of Population Genetics (Vavilov Institute of General Genetics, Russian Academy of Sciences), and to all the members of our expeditions to the Far East.

This treatise would not have been possible without the parallel population genetic studies of salmonids conducted by American, West European, Canadian and Japanese scientists. We hope that their important contribution to this field is covered by the book rather adequately.

The long-term investigations of salmonid fishes by scientists from various countries afford striking evidence of the international character of science that can develop successfully only on the basis of cooperation. Furthermore, the additional effort characterized through national scientific schools provides Western readers here with new useful information obtained on the basis of Russian experience. However, not all the issues raised in the book are fully elaborated. Some of them are only outlined and some of the authors' ideas should be considered as open to discussion. The authors take responsibility for the concept and approaches proposed here and welcome useful criticism.

We express our particular gratitude to Dr John Thorpe and Dr Gary Carvalho for their comprehensive scientific editing of the translation. They also helped us with some most recent publications. John Thorpe arranged our contact with Blackwell Science Ltd. We would like to thank the staff of Blackwell Science and in particular Nigel Balmforth for his assistance during the project.

The views expressed in this book are those of the authors and not necessarily those of the editors.

For technical assistance with illustrations and manuscript preparation we thank T.A. Frish and A.V. Rychkov.

Yu.P. Altukhov
E.A. Salmenkova
V.T. Omelchenko

Vavilov Institute of General Genetics
Russian Academy of Sciences
Moscow

Introduction

Population genetics, which originated at the beginning of the twentieth century, focused attention on the dynamics of the evolutionary process. Deviations from the Hardy–Weinberg equilibrium had been studied theoretically from every possible aspect, and hence both individual and combined effects of the main factors of microevolution (random genetic drift, migration, mutation and selection) were characterized in detail. However, the practical application of these models to natural populations was limited because the necessary data were lacking. This concerns information on:

(1) The systems of genetic polymorphism, as population genetics deals primarily with populations of genes.
(2) Animal populations as discrete intraspecific self-reproducing groups characterized historically and geographically.
(3) Demographic parameters of these groups, i.e. their effective size, migrations, and sex and age structure.

Evidently some species cannot always be characterized in this way; however, when such data are available, it is necessary to include them in population genetic analysis. This approach and corresponding logic are the basis of this book. In Chapter 1, the main theoretical principles of population genetics are summarized. Chapter 2 is a review of salmonid taxonomy, population biology and demography. In Chapter 3, the data on systems of hereditary protein and nucleic acid polymorphism are discussed. Chapters 4 to 6 are distinctive because they deal with intra- and interpopulation genetic differentiation and the mechanisms responsible for the maintenance of biochemical polymorphism in populations. These topics fall outside the scope of population genetics of salmonids proper and, in our opinion, are of more general significance, especially in the context of discussion between selectionists and neutralists. Chapter 7 deals with the genetic consequences of anthropogenic impact on populations and the role of population genetics in improving salmonid fishery management. The final summary brings together the main points covered in the book.

Chapter 1
The Theoretical Principles of Population Genetics

Before we turn to an analysis of genetic processes in salmon populations, it will be useful to introduce some general theoretical principles of population genetics. Population genetics, the most theoretically advanced field of genetic research, occupies a special place among the fields of modern genetics and biology as a whole.

In population genetics, quantitative theory has existed for a long time and continues to improve; it is a theory of how natural or anthropogenic factors influence the stability of population changes in important biological traits over successive generations. In other words population genetics is expressed in mathematical models. The models vary in their relevance to biological systems, but they are important, since they permit research to be planned in a particular way. If a model conforms with the natural situation, the changes in populations can be quantified and their possible consequences predicted. Because a large number of diverse papers and books have been published on this theme in recent years, we shall examine only the main population genetic terms, models and approaches. We base our further discussion on works by Sewall Wright (1931, 1951, 1969, 1977, 1978), Neel & Schull (1958), Ehrlich & Holm (1963), Dobzhansky (1970), Kimura (1983), Kimura & Ohta (1971), Cavalli-Sforza & Bodmer (1971), Nei (1975, 1987), Li (1976) and several others mentioned in the text.

1.1 Estimation of gene frequencies

As a first approximation, a population may be defined as *an aggregate of freely interbreeding individuals that share a common gene pool* (Dobzhansky). Because the number of segregating loci in the genome is large, one can understand the difficulties that confront the researcher in attempting to give an adequate description of this pool of inherited information. At the same time though, it is evident that, however great these difficulties may be, there is only one way of obtaining such a description, which is by defining the frequencies of allelic genes at each single locus. Knowledge of the spatial and temporal distribution of gene frequencies enables a quantitative assessment of how genetic processes in populations are influenced by given external and internal factors.

There are several carefully formulated methods of evaluating gene frequencies in a population. We shall examine two of them, the first applied to a situation without dominance and the second to inheritance with dominance. The situation relates to a pair of alleles at a single autosomal locus.

1.1.1 Absence of dominance

Let us assume that of N diploid individuals N_1 carry only the allele A, that is, are AA, N_2 the heterozygotes AB, and N_3 the homozygotes BB, so that $N_1+N_2+N_3=N$, and the total number of genes is $2N$. Each AA homozygote has two A genes, and each AB heterozygote has only one gene of this kind. Consequently, the total number of A genes in the group under study equals $2N_1+N_2$ and the fraction (frequency) of this gene is:

$$pA = \frac{2N_1 + N_2}{2N} = \frac{N_1 + \left(\frac{1}{2}\right)N_2}{N}$$

The frequency of gene B is defined in exactly the same way:

$$qB = \frac{2N_3 + N_2}{2N} = \frac{N_3 + \left(\frac{1}{2}\right)N_2}{N}$$

so that $p+q=1$. We can also apply the same method to loci with more than two alleles.

Of course, a complete examination of all individuals in a natural population is generally impossible, which in practical terms means that sampling is necessary. Hence the reliability of estimates of gene frequency depends very much on the numbers sampled. Such estimates should be characterized by the least possible error or dispersion factor, that is, they should satisfy the so-called *criterion of effectiveness*. Thus the requisite sample size depends on the genetic population structure, which can be established by preliminary research. The most reliable results come from the method of 'directly calculating' genes formulated by Fisher and used in the above example.

1.1.2 Dominance

If there is *dominance* of one allele (A) over the other (a), then only two distinguishable phenotypes are present in a population, and of them only one phenotype – the homozygote (aa) for the recessive allele – corresponds to only one genotype.

The method of directly determining the allelic frequency is inapplicable to a genetic situation of this kind, and one must allow the hypothesis that the Hardy–Weinberg equation (see the next section) holds in the population; namely, the distribution of genotypes in random mating conforms to the coefficients of the binomial expansion $p^2 + 2pq + q^2 = 1$.

It follows that in order to obtain an effective estimate of the frequency (q) of a recessive gene (a) we should extract the square root of the fraction of *aa* individuals in our sampling:

$$qa = \sqrt{N_3/N}$$

Accordingly, $pA = 1 - q$. In principle this method could also be used to estimate allelic frequency in the absence of dominance, but it only gives less-biased estimates at high-frequency values (Fig. 1.1).

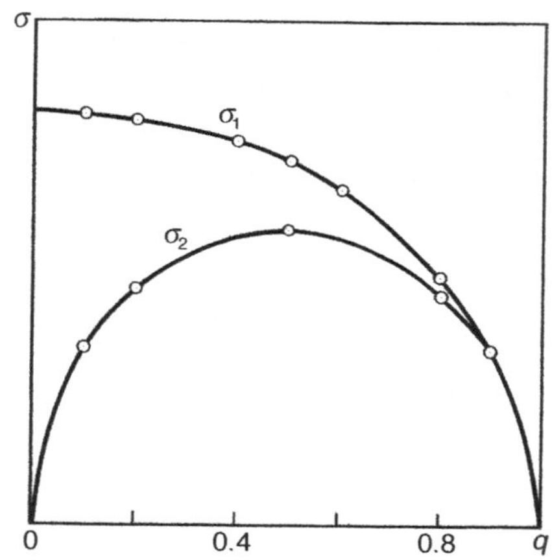

Fig. 1.1 Distributions of the standard errors of the two values of a gene frequency for a single pair of alleles without dominance (based on Neel & Schull, 1958). $\sigma_1 = \sqrt{(1 \quad q^2)/4N}$ (the gene frequency found by extracting the square root of the fraction of homozygous genotype); $\sigma_2 = \sqrt{q(1 \quad q)/2N}$ (the same allele frequency determined by the direct calculation method).

The reader can find more general examples of estimating gene frequencies (e.g. for multiple alleles or linked loci) in C. Li's book (1978) and other publications on theoretical population genetics.

1.2 The Hardy–Weinberg rule

The field of population genetics examines the principles regulating the maintenance and dynamics of population genotypic structure in time and space. The Hardy–Weinberg principle provides a theoretical basis for this view. It reflects the invariability of the genetic composition of a randomly mating (panmictic) population, unlimited in number and existing extra-environmentally, i.e. in the absence of environmental pressure. In this structureless community the genotype frequencies, and hence also the gene frequencies, at an autosomal locus having a pair of alleles A and a, reach equilibrium in the generation that follows random mating.

Inasmuch as random matings signify merely a random association of gametes, it can easily be verified that this combination of $p(A)$ and $q(a)$ male gametes and $p(A)$ and $q(A)$ female gametes, when $p+q=1$, produces an invariable distribution of $p^2\mathrm{AA}+2pqAa+q^2\mathrm{aa}=1$.

		male gametes	
		$p(A)$	$q(a)$
female gametes	$p(A)$	p^2	pq
	$q(a)$	pq	q^2

The algebraic calculations supporting the Hardy–Weinberg equilibrium are given in Table 1.1. As we are concerned here with autosomal genes, the reciprocal crosses (that is of the type *male* $Aa \times$ *female* AA or *male* $AA \times$ *female* Aa etc.) may be combined, and consequently the nine possible crosses reduce to six, given that $p+q=1$.

Table 1.1 Types of matings and proportions of genotypes of population progeny at genetic equilibrium.

Type of mating		Frequency of mating	Proportions of genotypes among progeny		
			AA	Aa	aa
$AA \times AA$	$(p^2 \times p^2)$	p^4	p^4	0	0
$AA \times Aa$	$2(p^2 \times 2pq)$	$4p^3q$	$2p^3q$	$2p^3q$	0
$Aa \times Aa$	$2pq \times 2pq)$	$4p^2q^2$	p^2q^2	$2p^2q^2$	p^2p^2
$AA \times aa$	$2(p^2 \times q^2)$	$2p^2q^2$	0	$2p^2q^2$	0
$Aa \times aa$	$2(2pq \times q^2)$	$4pq^3$	0	$2pq^3$	$2pq^3$
$aa \times aa$	$(q^2 \times q^2)$	q^4	0	0	q^4
Totals for population		1,00	p^2	$2pq$	q^2

It is clear that this equilibrium ratio of genotypes is provided by the symmetry of distribution of allelic genes into male and female gametes and by the free combination of these into the zygotes formed during reproduction. From this it follows that when there are no disturbances affecting a population of unlimited numbers, the frequencies of the genotypes and genes that characterize them remain unchanged in an infinitely long series of generations – the so-called 'absolute zero' of genetic dynamics. Ideal populations of this kind are, however, virtually never encountered in nature, which is why there are always natural factors that shift them from the point of equilibrium, disturbing their stability – random genetic drift, mutations, migration and natural selection. These are the 'factors of evolution' or microevolutionary forces that we shall now examine.

1.3 Random genetic drift

Random genetic drift is a mathematical fact that emerges from the phenomenon of finiteness of numbers of any real population. Of special importance is the difference that exists between the total population number and that part of it which transmits the gene pool to the next generation. A reproductive population size (N_r), and even more so its genetically effective population size (N_e) (and of the species as a whole), are virtually always less, and most often considerably less, than its total or census population size (N_t). Concepts of the substantial differences between genetically effective numbers and total population number were developed both theoretically and experimentally during studies of the stochastic processes of gene frequency dynamics (Serebrovsky, 1927, 1930; Fisher, 1930; Dubinin, 1931; Romashov, 1931; Wright, 1931; Dubinin & Romashov, 1932; Kolmogorov, 1935).

Extreme age groups are excluded from reproduction and the N_e value is influenced by population parameters such as the balance in numbers of the sexes during the reproductive period, individual fertility differences and periodic fluctuations of numbers.

1.3.1 Sex ratio and fluctuations in population size

If N_m represents the male and N_f the female reproductive part of the population, then the effective population size is

$$N_e = \frac{4N_m N_f}{(N_m + N_f)} \qquad (1.1)$$

When the male and female fractions deviate sharply from equilibrium, the N_e value depends more on the less numerous sex. The same effect occurs whether the mean value of N_e is valid for associated populations scattered in space and differing in numbers, or for one and the same population during fluctuations in time (generations) of the number of mating individuals (Wright, 1938; 1939). In the case of cyclical variations at an interval of n generations, $N_e = \tilde{N}$, where

$$\tilde{N} = \frac{n}{\sum_1^n 1/N_i} \qquad (1.2)$$

that is, N_e is the harmonic mean. For instance, if the effective number in each of five generations of one population is 10, 10^2, 10^3, 10^4 and 10^5, the average harmonic N_e

value represents only 45 individuals. The longer the time interval of observations, or the greater the number of populations distributed in space, the more stable the N_e estimate.

1.3.2 Variability of individual fertility

When population numbers are stable (the average number of progeny reaching reproductive age per one pair of parents, $\bar{k}=2$) and there is individual variation in the number of gametes (k) produced by the parent population (N), then:

$$N_e = \frac{(4N \quad 2)}{(V_k + 2)} \tag{1.3}$$

where V_k is the variance k.

When the number of progeny has a Poisson distribution ($\bar{k}=V_k=2$), the genetically effective population size is approximately equal to its reproductive number, $N_e=N_r$. However, in the majority of natural populations $V_k > \bar{k}$; hence N_e is always less than N_r. According to the estimates of Crow and Morton (1955), the N_e/N_r ratio is approximately 0.75 for many species. There are grounds, however, for considering this estimate to be a maximum, since in those cases where this kind of definition has been applied to actual populations, the difference has been much greater (Cavalli-Sforza *et al.*, 1964; Kerster, 1964; Tinkle, 1965; Rychkov, 1968; Kerster & Levin, 1968; Frankham, 1995) and the N_e/N_r ratio could even be as low as 0.30 (Rychkov & Sheremetyeva, 1979). In salmon populations the N_e/N_t ratio may be 0.10 and still less (Bartley *et al.*, 1992a; see Chapter 2).

Crow (see Crow, 1954; Crow & Morton, 1955; Kimura & Crow, 1963) introduced the concept of *inbreeding effective number*, $N_e(f)$ and *variance effective number*, $N_e(v)$. The *inbreeding effective number* is defined as

$$N_e(f) = \frac{(N_{t \ 2} \ V\bar{k} \quad 2)}{(\bar{k} \quad 1 + V_k/\bar{k})} \tag{1.4}$$

in which N_{t-2} is the number of individuals two generations back, and k is the number of gametes introduced by them.

When the population number is constant, $N_{t-2}=N$, $k=2$ and consequently

$$N_e(f) = \frac{(4N \quad 4)}{(V_k + 2)} \tag{1.5}$$

It is clear that when N is sufficiently large, this equation differs little from Equation 1.3

The *variance effective number* is defiend as

$$N_e(v) = \frac{p(1 \quad p)}{2V_{\sigma p}} \tag{1.6}$$

where p and $1 - p$ are the frequencies of any pair of alleles in a population and $V_{\delta p}$ is the change in allele frequency due to random sampling of gametes in a given generation:

$$V_{\sigma p} = \frac{p(1 \quad p)}{2N} \tag{1.7}$$

It is obvious that if the N_e values designated by Equations 1.1, 1.2, 1.3 and 1.4 are derived from field observations of numbers, the sex ratio and other biological parameters of a real population, then Equation 1.6 represents a method of defining an N_e value by analysing the genetic parameters on the hypothesis of the selective neutrality of the allelic genes being studied.

All the preceding equations for N_e were derived for non-overlapping generations. According to Nei & Imaizumi (1966a,b), when generations overlap in time

$$N_e = \tau N_a$$

where N_a is the number of individuals who reach mean reproductive age, and τ is the generation time or mean reproductive age. For example, as these authors have noticed, the N_e value assigned in this way to cover the general Japanese human population represented approximately 40% of the total number, including non-reproductive individuals.

If a population is stable, then $N_a = Nbp$, where N is the total population size, b the annual birthrate, and p the probability of a newborn individual reaching reproductive age.

In principle, the *variance effective number* under conditions of stability is nothing other than the correlation between the random variance of gene frequencies and the effective population size; this theory was developed by S. Wright in the 1930s. The relationship of these parameters was used as a basis for estimating effective population size N_e from temporal changes in allele frequencies, i.e. from their standardized variance (Krimbas & Tsakas, 1971; Nei & Tajima, 1981; Pollak, 1983; Waples, 1990 a,b). The approaches were elaborated for applying this method to populations with discrete (Nei & Tajima, 1981) and overlapping generations (Jorde & Ryman, 1995).

Changes in gene frequency caused by sampling error during the formation of gametes in the subsequent generation has a random, stochastic character. For this reason only a probability approach is feasible when assessing genetic drift based on the genetic structure of a population, making it possible to determine only the range of variance in gene frequencies. According to Wright, these non-directional fluctuations depend exclusively on the N_e value and are described by Equation 1.7.

The distribution of probabilities of all possible q values in a series of generations of a population limited in numbers approaches normal, and an allele's ultimate fate is fixation ($q = 1$) or loss ($q = 0$); this process is essentially irreversible.

1.3.3 *Modelling of random genetic drift*

Dubinin and Romashov (1932) were the first to model random genetic drift by providing 'drawings' of varicoloured balls which imitated different alleles in a population of randomly crossing organisms with no selection. They used a statistical population model to demonstrate that the process of loss (or fixation) during genetic drift is dependent entirely on effective population size.

In the absence of selection, the loss of genetic diversity (that is, reduction in heterozygosity) may be described by an equation enabling one to reconstruct evolutionary time:

$$H_T = H_0 e^{\;T/2N_e} \tag{1.8}$$

where H_0 and H_T are the frequency of heterozygotes at the zero point in the process and at the instant of time T, and N_e is the effective size of populations.

Although fluctuations in gene frequencies during drift are completely random, they acquire 'direction' (or vector) after a certain number of generations: the frequencies of rarer alleles in each successive generation diminish with greater probability (the loss process), whereas frequencies of other alleles increase (the fixation process).

The range of possible changes in the random variance of gene frequencies (Equation 1.7) during the life of one population may be regarded as exactly the same as that for a community of many populations roughly equal in numbers and having identical allele frequencies q_0 in the t_0 generation. In the following generation the gene frequencies in each population change randomly to become q_i. In the process, the average frequency $\bar{q}$ for the community remains equal to q_0, but the interpopulation variance will increase and in the t-th generation will be:

$$V_q = p_0 q_0 \left[1 \quad (1 \quad \frac{1}{2N})^t\right] \tag{1.9}$$

In the final stage of the process all populations become homozygous, and the intergroup variance reaches a maximum of $V_q = p_0 q_0$; that is, in some populations allele A reaches a state of fixation ($p = 1$) and in others it is lost ($p = 0$); at the same time the average allelic frequency remains unchanged as before. Thus, during random genetic drift the process of changing the probability distribution of the frequencies depends on only two factors – the population's effective size and the duration of the process, measured by the number of generations. When the effective population size is small, the distribution rapidly becomes U-shaped.

Buri (1956) modelled random genetic drift in experimental populations of *Drosophila melanogaster*. He studied the distribution frequencies of *bw* and bw^{75} alleles at the 'brown' locus in 19 consecutive generations of 107 lines with 16 individuals in each. The picture he obtained was very similar to what might have been expected (Fig. 1.2).

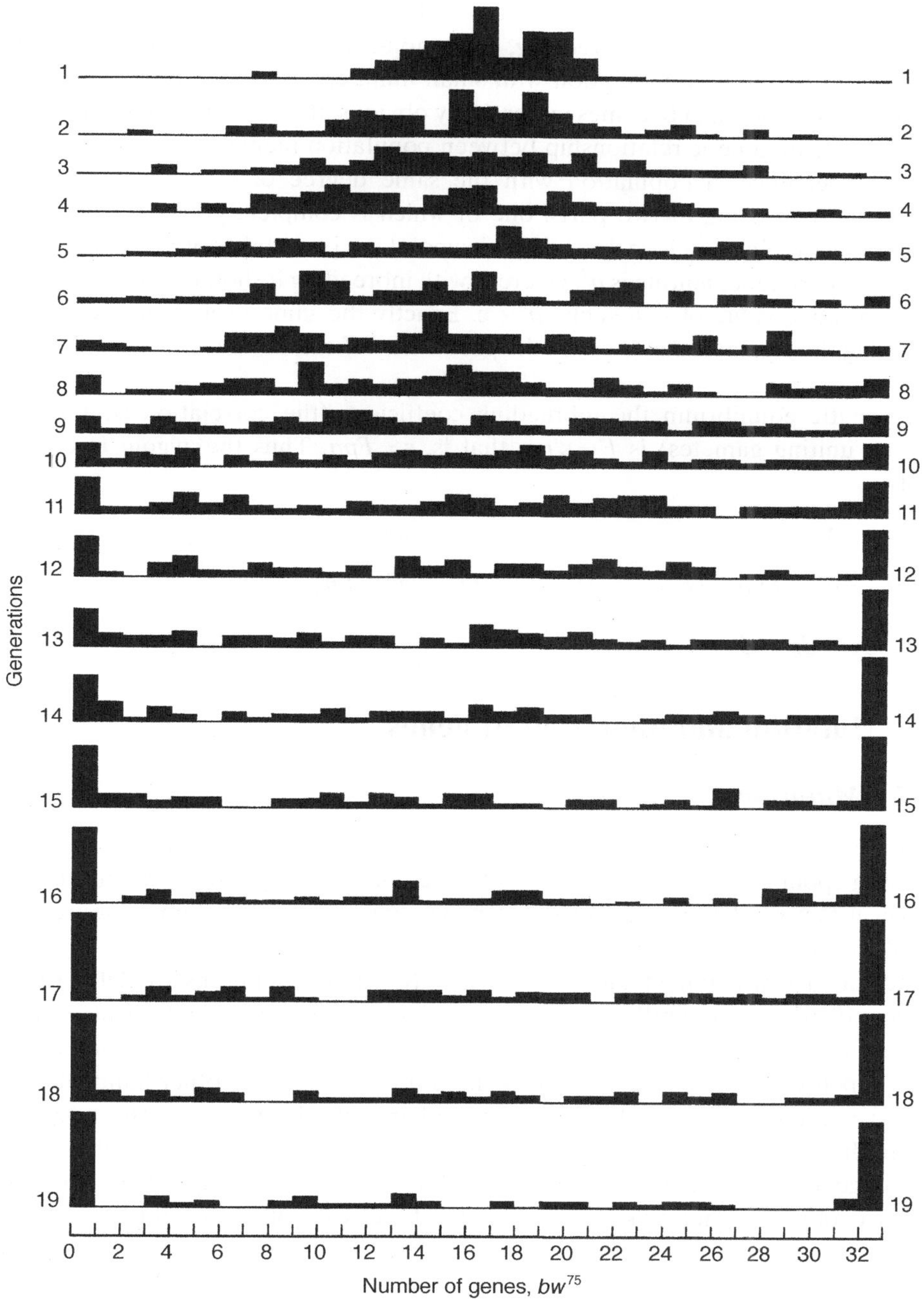

Fig. 1.2 The distribution of gene bw^{75} frequencies in 19 generations of 107 *Drosophila melanogaster* experimental populations (Buri, 1956).

The increase in the homozygosity of a population through random genetic drift indicates a direct link with *inbreeding*, since non-random association of gametes increases with time in a population with small numbers. By virtue of this fact the deviation from panmixia becomes increasingly obvious; this also implies an increase in the degree of genetic relationship between population members.

For convenience, a population with the same degree of inbreeding may be regarded as consisting of two parts, one of which is completely inbred while the other is completely panmictic. From this viewpoint the probability of combining two A gametes in random mating is p^2, whereas with inbreeding it should be more than a certain positive value of ε, namely, $p^2 + \varepsilon$. Exactly the same probability holds for combining the a gametes: $q^2 + \varepsilon$, and the probability of combining A with a is $2pq - 2\varepsilon$.

At genetic equilibrium the inbreeding coefficient (the correlation coefficient between uniting gametes) is $F = \varepsilon/pq$, that is, $\varepsilon = Fpq$. Thus the zygote fractions equal

$$\begin{array}{ccc} AA & Aa & aa \\ Q_1 = p^2 + Fpq & Q_2 = 2pq(1 \quad F) & Q_3 = q^2 + Fpq \end{array} \tag{1.10}$$

When $Q_1 + Q_2 + Q_3 = 1$.

1.4 Mutation and migration of genes

1.4.1 Mutations

Mutations are heritable, non-directional, random changes of genetic material occurring spontaneously or under the influence of specific physical, chemical and biological factors. The process of spontaneous mutation is the ultimate source of new alleles, resulting in increased genetic diversity.

Since the overwhelming majority of *de novo* (or ‘new’) mutations are deleterious (Muller, 1950), one naturally asks why a considerable number of alleles at a wide range of loci are to be found in populations? Such diversity is obviously maintained at the expense of part of the selectively neutral alleles. On the other hand, alleles that have recently arisen and that display no harmful effects in the heterozygotes may gradually be incorporated, step by step, in a species’ gene pool, which, as the Russian geneticist S. Chetverikov wrote, absorbs them like a sponge throughout its lengthy evolution (Chetverikov, 1926). In the latter case of mutation it is the raw material for natural selection which adjusts to coexist with ancient wild-type allelic genes that have already been ground and polished by evolution, though their effects on fitness depends critically on the environment. These questions will form the subject of discussion in Chapters 5 and 6. Here, however, it is only the dynamics of selectively neutral mutations, in generating deviations from Hardy–Weinberg equilibrium, that are discussed.

The fate of a single mutation was first analysed by Fisher (1930), who showed that when a population remains constant in size (the average number of progeny per family $m = k = 2$), the probability of total loss of a mutant gene in the first generation is approximately 0.37. If this mutant gene is not lost in the first generation, it again risks being lost in the second generation, and so on; thus the loss probability limit (l_n) is equal to 1.

Because the loss of *de novo* mutations is an essentially irreversible process, the bulk of them have no chance of becoming fixed in a population. Table 1.2 represents the overall dynamics of this process.

Table 1.2 Probability of loss of a neutral mutant gene (from Fisher, 1930).

Generation	Probability of loss, l_n	Probability of persistence, $1-l_n$
1	0.3679	0.6321
2	0.5315	0.4685
3	0.6259	0.3741
4	0.6879	0.3121
5	0.7319	0.2691
6	0.7649	0.2351
7	0.7905	0.2095
...	...	...
31	0.9411	0.0589
63	0.9698	0.0302
127	0.9847	0.0153
...	...	...
∞	1.0000	0.0000

It should be remembered, however, that mutations constantly recur in every generation, and it is quite likely that so-called 'new' mutations have re-occurred repeatedly. When an allele has a low frequency in a population it is practically impossible to identify its mutation into another allele. It is only when a certain frequency has been reached that one can take the effects of the mutation process on this allele into account. Thus, if μ is the mutation rate from allele A to allele a, and v is the reverse mutation rate, a to A, then the value of the change in gene frequency per generation is

$$\Delta q = \mu p \text{ (increase)} \quad vq \text{ (decrease)} \tag{1.11}$$

Hence the increase or decrease in a gene's frequency is determined by the relative values of such an increase or decrease. For instance, if the increase is more than the decrease, then at any given moment q will grow. But as, due to an increase in q, vq also increases, it can only continue to the point of equilibrium, after which no more changes of gene frequencies will be observed in the generations, that is, at the equilibrium point ($\hat{p}$, $\hat{q}$) $\Delta q = 0$.

Consequently,

$$\Delta q = 0 = \mu(1 \quad \hat{q}) \quad \nu\hat{q}, \text{ and } \hat{q} = \frac{\mu}{\mu + \nu} \tag{1.12}$$

Solving this equation in exactly the same way for p, we obtain

$$\hat{p} = \frac{\nu}{\mu + \nu}$$

This equilibrium is stable; moreover the allele equilibrium frequency $\hat{q}$ is independent of its initial frequency, being determined solely by the relationship between forward and backward mutation rates: if $q > \hat{q}$ in a certain generation, the gene frequency will be reduced in the following generations; whereas if $q < \hat{q}$ it will, on the contrary, increase.

When the q value deviates from $\hat{q}$, then the Δq per generation assigned by Equation 1.11 may be expressed in terms of the deviation $(q - \hat{q})$.

If in accordance with Equation 1.12 we write $\mu = (\mu + \nu)\hat{q}$, then:

$$\Delta q = \mu(1 \quad q) \quad \nu q = \quad (\mu + \nu)(q \quad \hat{q}) \tag{1.13}$$

Thus, the rate at which equilibrium is reached is proportional to the deviation of the actual value of q from its value $\hat{q}$ at equilibrium.

Two important facts should be taken into account when discussing the effect of the mutation process on genetic equilibrium in a population. First, the rate of forward gene mutation is an order of magnitude higher than the rate of back mutation. Second, although the mutation rate may vary substantially for different loci, the value is extremely low, in the order of 1×10^{-5} to 10^{-6} per locus per generation. Thus when describing genetic population structure of any species, the effects of newly occurring mutations can be largely disregarded, compared with the effects of such factors as natural selection, random drift and migration.

1.4.2 Migration

Migration, or gene flow, involves the introduction of new genes into a population through the transfer of gametes and/or propagules via interbreeding with the recipient population. It is important to distinguish migration in the population genetic sense from that employed in the ecological literature; gene flow does not occur simply as a result of the movement of individuals, it is crucial that the migrants interbreed successfully with the recipient population for gene frequencies to be effected.

Migration is an important factor influencing population dynamics, since each population in nature interacts with other groupings of the same kind through exchanging genes (except in cases of extreme isolation). Thus if immigrants differ genetically from the recipient population, they cause a corresponding change in gene frequency in each generation.

$$\Delta q = \quad m(q \quad q_m) = \quad mq + mq_m \tag{1.14}$$

where m is the number of immigrants divided by the size of the population receiving them, q is the gene frequency of the population, and q_m is the gene frequency of the immigrants.

The formula obtained is identical to Equation 1.13, reflecting the similarity with the results of direct and reverse mutation. To be convinced of this, one need only substitute the constant μ for mq, v for $mp = (1 - q)$, and $\mu + v$ for m. An equilibrium is established in precisely the same way during equal rates of gene flow to and from the population.

We shall return to the effects of gene migration later when we examine the effects of natural selection and the combined action of well-known evolutionary factors on the genetic population structure of salmon. Problems of the mutation process will not be considered.

1.5 Natural selection

In the theory of population genetics, natural selection is regarded as an extremely important factor of evolution that causes adaptive changes in genetic structure. These changes result from differences in the relative contributions of genotypes from reproductive individuals in a population, through differential reproduction or survival.

1.5.1 Basic equations and types of selection

If populations inhabited space of unlimited area and resources, and if at any instant of time t the birth rate a exceeded the death rate b by a constant value r, then the population number would grow continuously in an exponential manner:

$$N_t = N_0 \, e^{rt} \tag{1.15}$$

where the parameter $r = a - b$, denoting the coefficient of the population growth, is called the Malthusian parameter; Fisher (1930) introduced it in *The Genetical Theory of Natural Selection.* In nature however, space and resources are always limited and the coefficient r does not remain constant, so the exponential dependence may only be observed over limited sections of time (and space), ultimately giving way to an S-shaped logistic curve

$$\Delta N_t = \frac{k}{1 + C_0 \, e^{\quad rt}} \tag{1.16}$$

where k is the maximum number of individuals capable of living in a given specific environment, and the constant $C_0 = (k - N_0)/N_0$ is a correction factor – the 'environmental resistance' to population growth (Fig. 1.3).

The curves plotted in Fig. 1.3 relate to models which are *continuous in time*, that is, across generations, and their most frequent application is in ecology. Models with *discrete time* (non-overlapping generations) are chiefly used in population genetics for quantitative descriptions of natural selection, and usually operate with a similar quantity – Wright's fitness coefficient W. Thus, if N_t is the number of adult individuals in a generation t, k and v are their fertility and viability, and $W = kv$, then the growth of numbers in a population is:

$$\Delta N_t = N_{t+1} \quad N_t = (W \quad 1)N_t, \text{ and } N_t = W^t N_0 \tag{1.17}$$

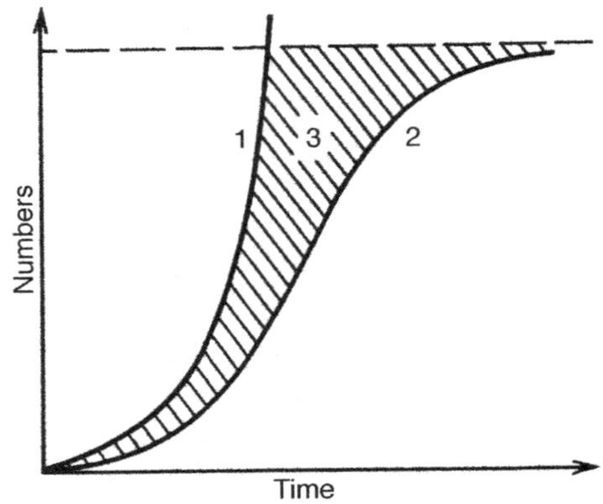

Fig. 1.3 Theoretical curves of population growth. 1 – exponential; 2 – logistic, 3 – environmental pressure.

Consequently, $W^t = N_t/N_0$, that is, the population's fitness at a certain instant of time is equal to the ratio of its numbers in the subsequent and previous generations (when the coefficient of fitness and the state of the environment are stable).

Thus, the parameter W is also important in a biological sense: when $W > 1$ the population size grows, when $W < 1$ it falls, and when $W = 1$ it remains the same.

In the theory of natural selection, developed by Fisher, Wright and Haldane, the fitness of genotypes is taken most frequently as a constant throughout the selection cycle, and its relative value is more important than its absolute value. Here we are only concerned with a change in the ratio of genotypes, which lends itself very conveniently to the study of the dynamics of gene frequencies. In this sense the symbol W_t signifies the i-th genotype's relative fitness; in other words, a quantity that reflects its reproductive contribution to the genetic pool of the following generation through differential fertility or survival, compared with other genotypes when population numbers are stable.

If the consequences of selecting out genotypes in a population with two alleles at one locus are considered from these viewpoints, then the gene frequencies after one selection cycle are changed in the following way (Table 1.3)

Clearly, under selection only certain genotypes can have relative fitness equal to unity, and hence the average fitness of a segregating population is always less than unity (that is, less than the 'optimal' genotype's fitness):

$$\bar{W} = fW_i = p^2W_1 + 2pqW_2 + q^2W_3 \tag{1.18}$$

Table 1.3 Change in gene frequencies in a randomly mating population after one generation of selection.

Genotypes	Frequency before selection	Relative fitness	Frequency after selection	New gene frequencies	Frequency of genotypes before selection
A_1A_1	p^2	W_1	p^2W_1		p'^2
A_1A_2	$2pq$	W_2	$2pqW_2$	$p' = \dfrac{p^2W_1 + pqW_2}{\bar{W}}$	$2p'q'$
A_2A_2	q^2	W_3	q^2W_3	$q' = \dfrac{pqW_2 + q^2W_3}{\bar{W}}$	q'^2
Sum	1		$\bar{W}$		1

A population's fate subsequently is as follows: its evolution towards the loss ($p=0, q=1$) or fixation ($p=1, q=0$) of allele A, or else its transition to a condition in which both alleles remain, depends on the relative W_i value of the genotypes. If $W_1 > W_2 \geq W_3$ or, on the other hand, $W_3 > W_2 \geq W_1$, the population will inevitably reach a trivial equilibrium (a stationary state when either $p=1, q=0$ or $q=1, p=0$), corresponding to what is known as *directional selection*. A non-trivial equilibrium point ($0<p<1$) is reached if $W_1 < W_2 > W_3$ or $W_1 > W_2 < W_3$; that is, when the heterozygote fitness is more than or less than that of both homozygotes. In both cases, despite the continuing effects of selection, no genetic changes occur in the population. However, only in the first instance ('overdominance') will there be a stable ratio of genotype frequencies at equilibrium

$$\hat{p} = \frac{W_3 \quad W_2}{(W_1 \quad W_2) + (W_3 \quad W_2)} \tag{1.19}$$

This type of selection, one of the forms of balancing selection, is called *stabilizing*.

When, however, the heterozygote is less adaptive than both homozygotes the state of equilibrium in the population is unstable and a gene with a higher frequency moves in the direction of fixation, while a gene with lower frequency tends towards a reduction or loss. This type of selection is called *disruptive* or *diversifying*. Figure 1.4 shows how a population's average fitness changes in these cases (Li, 1976).

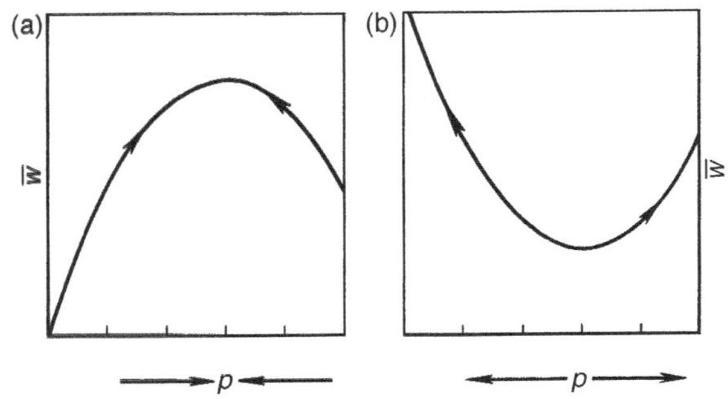

Fig. 1.4 The average fitness of a population ($\bar{W}$) as a function of gene frequencies (Li, 1967). (a) $w_1 = 2$, $w_2 = 4$, $w_3 = 1$, $\bar{w} = 2p^2 + 8pq + q^2$. Point 0.6 on the parabola corresponds to maximum fitness – stable equilibrium; the trivial equilibrium points ($P = 1.0$) are unstable. (b) $w_1 = 3$, $w_2 = 1$, $w_3 = 4$, $\bar{w} = 3p^2 + 2pq + 4q^2$. Point 0.6 corresponds to minimum fitness. As unstable equilibrium is affected by disruptive selection, a population moves towards gene frequencies of $p = 0.1$; these trivial equilibrium states are stable.

The rate at which a population approaches the equilibrium point depends on the intensity of selection ($s = 1 - W$), and is determined by the magnitude of the change in gene frequency per generation (Δp), based on the expressions for p' and q' (see Table 1.3):

$$\Delta p' = p' \quad p = (pq/\bar{W})[p(W_1 \quad W_2) + q(W_2 \quad W_3)] \tag{1.20}$$

Here is a numerical example: population 0.25 *AA*: 0.50 *Aa*: 0.25 *aa*, moving towards the equilibrium frequencies of $\hat{p} = 0.60$, $\hat{q} = 0.40$ and influenced by the effects of selection of the relative values of *W* for the three genotypes *AA*, *Aa* and *aa* in the four following situations: (1, 3, 0), (3, 7, 1), (8, 10, 7), (101, 103, 100). It is clear that the types of selection in these four cases differ greatly from each other. Whereas in the first case (1, 3, 0), genotype *aa* is lethal and the reproductive contribution of the heterozygote *Aa* is far in excess of that of the *AA* homozygote, the fourth type of selection (101, 103, 100) displays only very insignificant differences in the fitness of genotypes. The second (3, 7, 1) and third (8, 10, 7) models occupy an intermediate position between the two extremes. The character of the gene frequency changes in these four situations is shown in Table 1.4.

It can be seen that in the first case the rate at which the population reaches equilibrium is very great, while it is negligibly small in the last case. When $p = 1$ or $p = 0$, $\Delta p = p' - p = 0$ (see Equation 1.20), the population reaches trivial equilibrium state. The factor pq in Equation 1.20 also shows that the rate at which equilibrium is approached is fairly high at intermediate allele frequencies and becomes smaller the smaller the value of p and q.

Let us turn again to Fig. 1.4 to convince ourselves that the mean fitness of the population $\bar{W}$ under the pressure of selection always increases to its maximum value in the equilibrium state. This rule, known as the 'fundamental theorem of natural

Table 1.4 Gene frequency in a population after one generation of selection as a function of fitness coefficients of genotypes.

Selection type	Fitness coefficient values			Initial gene frequency, p	Gene frequency after selection, p'	Change value, $\Delta p = p'\ \ p$
	W_1	W_2	W_3			
1	1	3	0	0.5000	0.57143	0.07143
2	3	7	1	0.5000	0.55556	0.05556
3	8	10	7	0.5000	0.51429	0.01429
4	101	103	100	0.5000	0.50123	0.00123

selection', was formulated by Fisher in 1930 and was originally defined as: 'The rate increase in fitness of any organism at any time is equal to its genetic variance in fitness at that time.' In the new edition of 1941, it was rephrased: 'The rate of increase in the average fitness of a population is equal to the genetic variance [that is, its additive component] of fitness of that population.'

Fisher based this premise on models having continuous time and logarithmic fitness. Li (1967) extended it to include models of populations with non-overlapping generations. This principle, which confirms the monotonic increase of a population's fitness at any instant of time under selection pressure, plays a major role in Wright's 'adaptive topography' – a component of his shifting balance theory of evolution.

Fisher's fundamental theorem of natural selection has several limitations (Levins, 1968; Zhivotovsky, 1981). A population's fitness may be reduced as a result of inbreeding, frequency-dependent selection, mutations and gene recombinations; hence selection is capable of making a population extinct. (Frequency-dependent selection is a form of balancing selection (see Li, 1978), in which a genotype's selective advantage changes as its frequency in a population changes. Of particular interest is the situation in which fitness and frequency are inversely correlated.)

Nevertheless, Fisher's theorem is justified for constant values of genotypic fitness (Zhivotovsky, 1984). Obviously a natural population existing today has reached maximum adaptation in previous stages of evolution that are unknown to us, and now maintains a dynamic balance with its surroundings. The totality of these adaptations to the specific environment that populations encountered previously is recorded in patterns of contemporary genotypic structure. It represents their reserve of genetic stability despite the changing conditions of the environment.

The evolution described above of the genetic structure of a single-locus, diallelic population is certainly a simplification. Each natural population of a bisexual species evolves simultaneously at a large number of loci, among which there may also be multiallelic genetic systems. The corresponding genotypes can interact in various ways; genes may be linked, or their combinations non-random, under the effects of selection (where there is very close linkage, the situation may be reduced effectively to a single locus and the complex of several or many separate genes regarded as a supergene in inheritance). All this creates many additional difficulties, and it follows that no mathematical description is adequate to provide a complete quantitative picture of genetic processes influenced by selection. None the less, existing models are proving exceptionally useful even today: they enable one to plan research along definite lines and to quantify the results obtained. For all this information we refer the reader to several sources that discuss comprehensively these problems and the tasks ahead (see, for instance, Wright, 1969, 1970; Lewontin, 1978a; Li, 1978; Zhivotovsky, 1984; and also Chapter 5).

1.5.2 Genetic load of populations

The constant pressure of mutations and the migration of genes, as well as the segregation of biologically less adapted genotypes at balanced polymorphic loci, give rise to the problem of so-called 'genetic loads' which is of paramount scientific and practical importance.

The concept of genetic load was introduced by Muller (1950), but the first researches – revealing a saturation of natural populations with non-adaptive mutant phenotypes that are produced in each generation from apparently normal individuals – were carried out in the 1920s and 1930s (Chetverikov, 1926; Timofeev-Ressovsky & Timofeev-Ressovsky, 1927; Dubinin *et al.*, 1937; and see also Dobzhansky, 1970).

In his paper 'Our load of mutations', Muller (1950) showed that weakly deleterious mutant genes can inflict greater damage on a population than individual mutant genes having a strong negative effect. Moreover, he suggested that each of us is on average the bearer of at least eight harmful genes concealed in the heterozygous state.

Morton *et al.* (1956) determined the amount of genetic load in humans by investigating inbreeding depression. They proposed the special term 'lethal equivalent' and indicated an approach to evaluating the relative contribution to the total amount of genetic burden of the mutation processes (mutational load) and segregation (segregational load) of genes; the latter relates to loci whose polymorphism is maintained at the expense of the selective advantage of heterozygotes.

Although many unresolved points remain in the problem of genetic load (Dobzhansky, 1970; Cavalli-Sforza and Bodmer, 1971), the concept of genetic load is of fundamental interest in quantifying selection intensity and as a parameter linked with a population's fitness. From this point of view the following definition by Crow (1958) deserves attention. According to Crow (1958), the genetic load is the proportion by which the fitness of the average genotype in a population is reduced in comparison to the best (optimal) genotype.

In this instance, the genetic load for a model population with discrete time is:

$$L = \frac{W_{max} - \bar{W}}{W_{max}} \qquad (1.21)$$

where W_{max} is the fitness of the best (optimal) genotype, and $\bar{W}$ is the average fitness of a population.

It is obvious that this approach can easily be implemented by taking into account the segregational load arising on the basis of the heterozygotes' increased fitness compared with both homozygotes ('overdominance').

Let us assume two alleles at one locus with constant and positive selection coefficients:

Genotype	A^1A^1	A^1A^2	A^2A^2
Fitness	$W_1\,(1-s_1)$	W_2	$W_3\,(1-s_2)$
Frequency	p^2	$2pq$	q^2

In random mating, equilibrium is reached when $s_1p = s_2q$ and hence the equilibrium frequencies are:

$$\hat{p} = \frac{s_2}{s_1 + s_2}, \quad \hat{q} = \frac{s_1}{s_1 + s_2}$$

The total amount of load equals a decrease of population fitness resulting from the selective elimination of both types of homozygote, that is, $s_1\hat{p}^2 = s_2\hat{q}^2$, and consequently:

$$L_{OD} = s_1 \frac{s_2^2}{(s_1 + s_2)^2} + s_2 \frac{s_1^2}{(s_1 + s_2)^2} = \frac{s_1(s_2)^2 + s_2(s_1)^2}{(s_1 + s_2)^2} = \frac{s_1 s_2}{s_1 + s_2} \qquad (1.22)$$

The population's total load, segregating at n loci, is

$$L_T = 1 \quad e^{\Sigma l_i}$$

where l_i is the load at the i-locus.

The concept of genetic load in connection with the problem of inbreeding is discussed in the book of Lynch & Walsh (1998), which we recommend for interested readers. It is necessary only to emphasize in our context that in evaluating segregational (balanced) load, great importance is attached to the question of the number of loci whose polymorphism may be maintained in a population by overdominance. It is quite obvious that a population's maximum genetic load will come to bear when both homozygotes are lethal; in that event 50% of the descendants will perish in each generation.

If the homozygotes are not lethal, the load is reduced; but even in this case the number of overdominant loci is limited by selection coefficients and population numbers.

For example, if $s_1 = s_2 = 0.02$, the equilibrium frequency $\hat{p} = 0.5$ and the number of independent overdominant loci $n = 1000$, the probability of a specific individual reaching reproductive age is 4.3×10^{-6}. In a selection type of this nature the view of a maximally heterozygous individual as the best-adapted, optimal genotype has no meaning, since such an individual is simply not encountered in a population of finite numbers. If, for instance, the number of polymorphic loci $n = 40$, then the expected frequency of a genotype, heterozygous at all loci, is less than 2^{-40}, that is, one in a trillion.

Taking these calculations into account, Kimura and Ohta (1971) estimated, on the basis of the population size N and the number n of overdominant loci with multiplicative fitness effects, by how many times ($e^{\tilde{L}}$) the fertility of the most adapted heterozygous genotype would be greater than the population mean:

$$\tilde{L} = \sqrt{\sum_{1}^{n}\left(\frac{s_1 s_2}{s_1 + s_2}\right)^2 \times 2\log_e(0.4N)} \qquad (1.23)$$

If $s_1 = s_2 = 0.01$, $n = 1000$ and $N = 25\,000$, then $\tilde{L} = \sqrt{0.46} = 0.68$ and $e^{\tilde{L}} = 1.97$. This estimate is completely realistic and corresponds to many natural situations (for example, many populations of mammals). However, when $s_1 = s_2 = 0.1$, other conditions being equal, the segregational load is already exceptionally large ($\tilde{L} = \sqrt{46} = 6.8$) because a maximally adapted individual should produce $e^{6.8} \approx 898$ times more progeny compared with the population average. Clearly, this scale of fertility does not exist in populations of most species. We shall return to the problem in Chapter 6 when we examine the data on the role of selection in maintaining biochemical polymorphisms of salmon populations.

A population's segregational load is a constant 'cost', in the form of the less adapted genotypes appearing in each generation, that a population is forced to pay for its stable existence at the maximum point of fitness.

Under directional selection, usually regarded as the most important adaptive evolutionary factor linked with the replacement of 'less adapted' alleles, the total volume of the genetic load increases still more at the expense of the so-called 'substitutional load' (Kimura's term). This problem was first posed and then mathematically investigated by Haldane (1957, 1960). His logic in solving this problem is closely allied to what forms the basis of calculating the number of overdominant loci of a population in equilibrium. By his calculations, the number of deaths resulting from a changed vector of selection are not connected with its intensity, but are determined exclusively by the initial frequency p of an unfavourable allele. Nevertheless, the number of generations to which elimination distributions apply depends on the intensity of selection.

If the effects of genes are additive, the substitutional load in one generation for a single locus is $L_i = -2 \log p_0$.

For $p = 0.5$, the approximate equation for the relationship between selection intensity and the number of generations necessary for an allele's substitution is $s = 30/n$, where 30 is the 'cost' of a single gene replacement. This quantity shows by how many times in the selection process the total number of deaths in all generations exceeds the number of individuals in a given generation. Assuming that $s = 0.1$, corresponding to 10% selective mortality per generation, then $n = 300$; in other words, about 300 generations are required to replace a single allelic gene in the adaptive evolutionary process. Thus, according to Haldane, the rate of evolution must be very slow and the numbers of simultaneously evolving genes fairly small for there not to be a sharp reduction in fitness threatening the life of a population under the conditions described above. Haldane suggested that allelic substitutions at 1000 loci are sufficient for the emergence of a new species, necessitating not less than 300 000 generations, the number of concomitantly evolving genes not exceeding 12 (see also Kimura, 1960a,b).

The speciation rate along the lines of this model generally accords with palaeontological findings about evolutionary tempos, at least insofar as they apply to mammals. However, numerous examples are known of exceedingly rapid speciation, which contradicts Haldane's calculations (see, for instance, Mayr, 1968, 1974; Dobzhansky *et al.*, 1977; Altukhov, 1990; Coyne, 1992; Fontdevila, 1992; Hurst, 1992; Orr, 1995; Gavrilets & Hastings, 1996; Slatkin, 1996).

In the preceding sections, we have concentrated on the microevolutionary factors which influence a population's genetic structure but which are essentially external to it. Such a view, however, is incomplete without taking into consideration a population's organizational structure. In many instances natural populations are not panmictic groups but represent historically evolved communities of semi-isolated subpopulations. Subpopulations are constantly exchanging genetic material with each other, being subjected as much to genetic random drift as to different forms of selection.

Several original models have been suggested to describe these more realistic situations in population genetics; Sewall Wright (1931, 1938, 1943a,b, 1951), Masatoshi Nei (1975, 1987) and other authors (Kimura, 1953; Kimura & Weiss, 1964; Malecot, 1955, 1967; Bodmer & Cavalli-Sforza, 1968; Maruyama, 1970, 1971, 1972; Nagilaki, 1985; Rannala & Hartigan, 1995) have distinguished themselves most in this field.

A population's subdivision into subpopulations of a limited size has important consequences on genetic structure. Let us examine this factor.

1.6 The influence of population subdivision on genetic structure

1.6.1 Subdivision and inbreeding – the Wahlund effect

Wahlund (1928; quoted by Li, 1955) was the first to show that if a large population is subdivided into K panmictic groups, an effect is observed in this community similar to the consequences of inbreeding in a randomly mating population: the fraction of homozygotes increases to the value of the interpopulation variance of gene frequencies through a reduction in the heterozygote fraction.

Indeed, if we denote by q_i the frequency of the gene in the i-th group ($p_i + q_i = 1$) and the frequency of this same gene in the subdivided population as a whole by q, then the average gene frequency characteristic of it and its variance will be:

$$q = \frac{\sum q_i}{K}, \; V_q = \frac{\sum(q_i \quad q)^2}{K} = \frac{\sum q_i^2}{K} \quad q^2 \tag{1.24}$$

The corresponding zygote (or the genotype) frequencies equal:

$$(AA) \quad \frac{\sum p_i^2}{K} = p^2 + V_q$$

$$(Aa) \quad \frac{2\sum p_i q_i}{K} = 2pq \quad 2V_q$$

$$(aa) \quad \frac{\sum q_i^2}{K} = q^2 + V_q \tag{1.25}$$

By comparing the frequencies of the genotypes in Equation 1.25 with their frequencies in a population, with an inbreeding coefficient F (Equation 1.10), we obtain the equation:

$$V_q = Fq(1 \quad q), \text{ or } F = \frac{V_q}{\bar{q}(1 \quad \bar{q})} \tag{1.26}$$

Since the F value characterizes the subdivided population as a whole, the corresponding frequencies of the genotypes in the population are equal to those frequencies that would characterize a *separate* inbreeding colony. In other words, 'a population's subdivision into separate crossing groups is equivalent in formal terms to inbreeding of the total population' (Li, 1978, p. 467).

The degree of this differentiation is directly connected with the scale of the interpopulation differences of the gene frequencies – the greater the genetic differences between subpopulations, the higher the variance of q.

Sewall Wright (1943a,b, 1951) played a leading part in describing the local differentiation of the gene frequencies of a subdivided population in terms of F-statistics, by establishing several F-coefficients as indicators for measuring genetic differentiation:

(1) F_{IT} – inbreeding coefficient of an individual relative to the total (T) population
(2) F_{IS} – inbreeding coefficient of an individual relative to the subpopulation (S) to which it belongs
(3) F_{ST} – inbreeding coefficient of a subpopulation (S) relative to the total (T) subdivided population.

The relationship between these values is given by the equation

$$F_{IT} = F_{ST} + (1 \quad F_{ST})F_{IS} \tag{1.27}$$

Nei's (1975, 1987) G_{ST} statistics are equivalent to F_{ST} statistics, relating the total and intrapopulation gene diversity (H_T and H_S) by the following formula:

$$G_{ST} = \frac{(H_T \quad \bar{H}_S)}{H_T} \tag{1.28}$$

Here:

$$H_T = 1 \quad \sum \bar{p}_i^2$$
$$\bar{H}_S = 1/n \sum H_S$$
$$H_S = 1 \quad \sum p_{is}^2$$

where P_{is} is the frequency of the i-th allele in subpopulation S and $\bar{p}_i$ is the mean allele frequency in the total subdivided population consisting of n subpopulations. Hence, $\bar{H}_S$ is the mean heterozygosity of the subpopulation, and H_T is the heterozygosity of the total subdivided population if it were converted to a single randomly mating unit.

The F_{ST} coefficient described in Equation 1.26 was suggested by Wright as long ago as 1943, since when it has been repeatedly used to analyse the distributions of gene frequencies in natural subdivided populations (Cavalli-Sforza *et al.*, 1964; Nei & Imaizumi, 1966a,b; Rychkov, 1969; Altukhov, 1974; Altukhov *et al.*, 1975a,b). This coefficient is of great interest as it enables one to analyse certain important effects of population subdivision on genetic structure.

Wright proposed two original population models for this purpose: an 'island model' and 'isolation by distance'.

1.6.2 The island model of population structure

There are two known versions of the island model:

(1) A species' subdivision into a large (infinite) number of randomly mating subpopulations and with the same effective size N, each of them having equal probability and identical intensity (m) of exchanging genes with the common gene pool.

(2) A large panmictic population ('mainland') surrounded by a host of isolated, genetically differentiated small colonies ('islands') each of which receives genes from the mainland with an intensity of m per generation (Fig. 1.5). The effects of back migration may be disregarded.

The intergroup variance of gene frequencies serves as a measure of the random differentiation of subpopulations in the system.

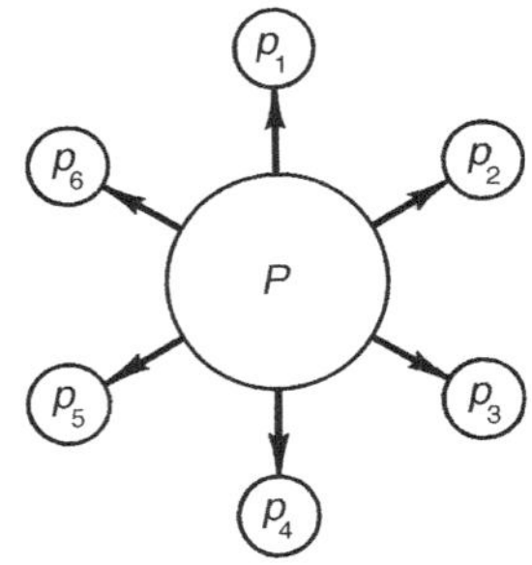

Fig. 1.5 A version of the island population model. p = the gene frequency in the system averaged out for the peripheral sub-populations ('islands', 1–6) and equal to the value of P in the population 'core' ('mainland').

$$V_q = \frac{\bar{q}(1 \quad \bar{q})}{4N_e m + 1} \tag{1.29}$$

and hence, the state of equilibrium between genetic drift and migration in terms of F-statistics (see Equation 1.26) may be written as:

$$F_{ST} = \frac{1}{4N_e m + 1} \tag{1.30}$$

Wright's (1943a) more exact treatment gives:

$$V_q = \frac{\bar{q}(1 \quad \bar{q})}{2N_e \quad (2N_e \quad 1)(1 \quad m)^2} \tag{1.31}$$

At small values of m ($m << 1$) the difference between Equations 1.29 and 1.31 is negligible.

Thus the local differentiation of gene frequencies is conditioned by the parameter Nm. In other words, the deciding factor is not the migration coefficient or the size of the population itself, but their product, equalling the number of individuals entering a population in a generation.

As a result of interactions between drift and migration we have a probability distribution of gene frequencies. At any instant of time T it is a function of $\Delta q = -m(q \;-\; qT)$ as a measure of the effect of systematic migration and $V_{\delta q} = q(1-q)/2N_e$ is the sampling variance of a gene frequency in one generation through isolation, that is, random drift (Wright, 1938, 1939):

$$\Phi(q) = (C/V_{\delta q}) \exp[2 \int (\Delta q / V_{\delta q}) dq]$$

The general formula for the stationary distribution of gene frequencies in the island model represents the β-function of the probability density as follows:

$$\Phi(q) = \frac{C}{q(1 \quad q)} \exp\,[4N \int \frac{\Delta q}{q(1 \quad q)} dq] \tag{1.32}$$

where C is a normalizing constant chosen so that $\int_0^1 \Phi(q)dq = 1$:

$$C = \frac{\Gamma(4Nm)}{\Gamma(4Nmq)\,\Gamma[4Nm(1-q)]}$$

Equation 1.30 takes the following forms, depending on what combination of random and systematic factors is assigned it:

- For equilibrium of random drift by gene migration:

$$\Phi(q) = Cq^{4Nm\bar{q}-1}(1-q)^{4Nm(1-\bar{q})-1} \quad (1.33)$$

  where p and q are the allelic frequencies in the subpopulation, $\bar{p}$ and $\bar{q}$ are the mean allelic frequencies for a subdivided population as a whole, N is the effective population size, and m is the migration coefficient.
- With the combined effects of isolation, migration and selection:

$$\phi(q) = C\bar{W}^{2N}q^{Nm\bar{q}-1}(1-q)^{4Nm(1-\bar{q})-1} \quad (1.34)$$

  where the notation is the same as for the previous formula and $\bar{W}$ is the population's average intralocus fitness determined by summation of the genotypes' fitness according to their frequencies (see Equation 1.18).

The relationship between the effects of random and systematic evolutionary factors upon population genetic structure can vary widely by virtue of the fact that the corresponding stationary distributions of gene frequencies may take a different form (Fig. 1.6).

Stationary distributions can describe:

(1) The distribution of allelic frequencies of many loci in the same population in the case of neutrality or when subjected to nearly identical pressure of selection
(2) The distribution of gene frequencies at any locus in successive generations of the same stationary population
(3) The distribution of allelic frequencies of one or several loci in the community of isolated populations.

All three types of distribution are mathematically equivalent.

In the island model the gene migration coefficient is independent of the degree of population isolation. Wright (1943a) and Malecot (1955, 1967) also employed mathematical methods to investigate a population in which the intensity of exchange between the subpopulation depends on distance.

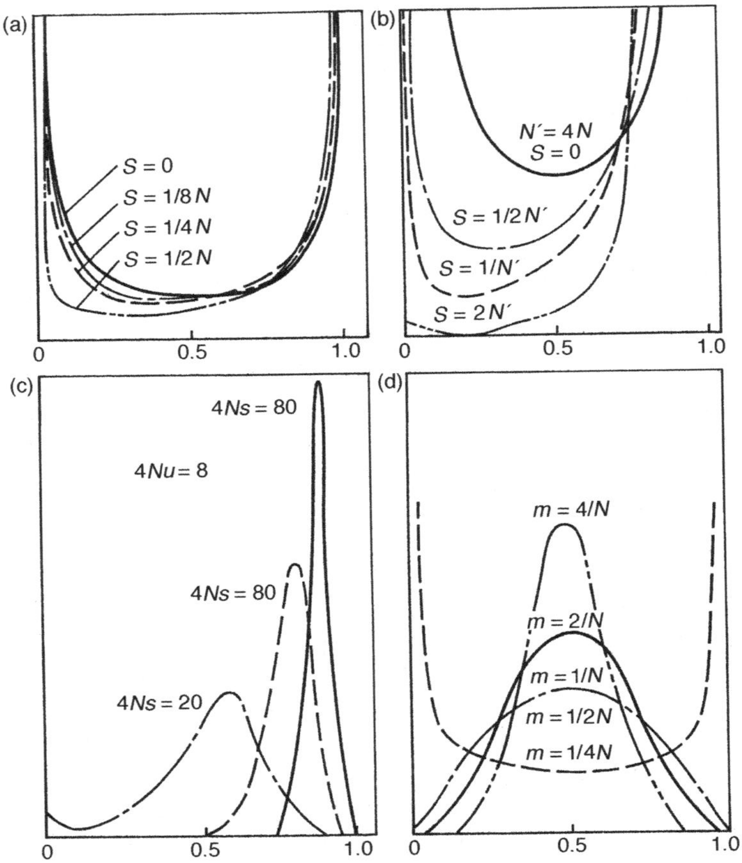

Fig. 1.6 Equilibrium (stationary) distributions of gene frequencies for different relationships between population size and the effects of migration, mutation and selection. The x-axis represents gene frequency, the y-axis represents number of subpopulations. (a) Distribution for small and completely isolated populations affected by selection. (b) Distributions for the same selection s coefficients as in (a), but population size is fourfold ($N' = 4N$). (c) Equilibrium frequency distributions of mutant alleles when $u = 2/N$; s as in (a); $N' = 40N$. (d) Interaction between genetic drift and migration at a mean frequency of 0.5; a U-shaped distribution reflecting allelic fixation in many populations is observed only when $m = 1/4N$, i.e. when a population of no more than one migrant is received for every four generations.

1.6.3 *Isolation by distance*

This model postulates a population spread evenly over a large territory substantially exceeding the radius of an individual's reproductive activity. The special features of local differentiation in such a system depend upon the effective size of the 'neighbourhood' N_N from which parents arise at random, as well as upon the dimensions of the area. In particular, local differentiation arises sooner or later in a one-dimensional area, if it is sufficiently large in extent. However, on a plane (two-dimensional area) the possibility of this differentiation is much less.

According to S. Wright (1951), the size of a neighbourhood corresponds approximately to the number of genetically effective individuals within a circle whose radius equals twice the standard deviation (σ) of the extent of migration in one direction in a given generation (or, in other words, the distance between the birthplaces of parents and progeny).

Differentiation is very great when $N_N \approx 20$, somewhat less but still quite pronounced when $N_N \approx 200$, and verges on panmixia when $N_N \approx 2000$.

Kimura (1953; see also Kimura & Weiss, 1964) has proposed another model, called the 'stepping-stone model', representing a situation intermediate between Wright's island model and that devised by Wright and Malecot for evenly distributed populations.

1.6.4 The stepping-stone structure of gene migration

In this, as in the island model, colonies are considered together, but the exchange of individuals proceeds only between neighbouring colonies, as shown in Fig. 1.7; thus it is directly dependent on the colonies' distance from each other.

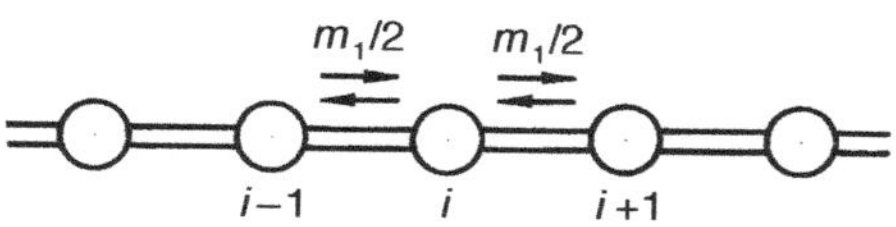

Fig. 1.7 One-dimensional stepping-stone model of gene migration.

At equilibrium the interpopulation variance of gene frequencies is:

$$V_q = \frac{\bar{p}(1 \quad \bar{p})}{2N_e \quad (2N_e \quad 1)\{1 \quad 2R_1R_2/(R_1+R_2)\}} \tag{1.35}$$

where $R_1 = [(1+\alpha)^2 - (2\beta)^2]^{\frac{1}{2}}$; $R_2 = [(1-\alpha)^2 - (2\beta)^2]^{\frac{1}{2}}$, in which $\alpha = (1-m)(1-\bar{m}_\infty)$ and $\beta = m_1\ (1-\bar{m}_\infty)/2$; in these equations m_1 stands for the intensity of migrations between adjacent colonies (*short-range migration*) and $\bar{m}_\infty$ for the external effects of gene migration between all the colonies together (*long-range migration* corresponding to the coefficient m in Wright's island model). If $m = 0$, then $\alpha = 1 - \bar{m}_\infty$, $\beta = 0$ and Equation 1.35 reduces to Wright's Equation 1.26. Thus the island model represents a special case of a stepping-stone model in which there is no gene exchange between neighbouring colonies.

In cases where m_1 is much higher than $\bar{m}_\infty$, Equation 1.35 reduces approximately to:

$$V_p = \frac{\bar{p}(1 \quad \bar{p})}{1 + 4N_e\sqrt{2m_1\bar{m}_\infty}} \tag{1.36}$$

assuming that $\bar{m}_\infty << m_1 << 1$.

The standardized genetic variance in this case is:

$$F_{ST} = \frac{V_p}{\bar{p}(1 \quad \bar{p})} = \frac{1}{1 + 4N_e\sigma\sqrt{2\bar{m}_\infty}} \tag{1.37}$$

where $\sigma = \sqrt{m_1}$, since $m_1 = V_m$ is the migration dispersion for a distance of more than one step apart.

According to the formula obtained by Kimura & Weiss (1964), under stationary conditions where there is no selection, the variance of gene frequencies may also be found from the expression:

$$V_p = \frac{\bar{p}(1 \quad \bar{p})}{1 + 4N_e m[1 \quad \mathrm{r}(1)]} \tag{1.38}$$

where r(1) is the correlation coefficient of gene frequencies between adjacent subpopulations.

The tendency in the stepping-stone model and in isolation by distance towards local differentiation depends greatly on the dimensions of the area. Following Kimura & Weiss (1964), the correlation coefficient of gene frequencies between colonies in a one-dimensional model diminishes with distance from the exponent and is described by the formula:

$$\mathrm{r}(d) \approx \mathrm{e}^{\left(\sqrt{\frac{m_\infty}{m_1}}\right)d}$$

where d is the distance between subpopulations (in 'steps') and r(d) is the correlation coefficient between them.

The mathematical apparatus becomes more complicated when it involves analysis of two- or three-dimensional models, each colony exchanging genes with four or six adjacent ones (Kimura & Ohta, 1971). For our purposes it is enough to stress that local differentiation of gene frequencies, all other conditions being equal, is at a maximum in one dimension and quickly diminishes as the number of dimensions increases. This dependence manifests itself strikingly in research on the genetic correlation r(d) between colonies based on the distance between them. When the distance increases, the correlation quickly declines, a feature which is particularly characteristic of the three-dimensional model (Fig. 1.8). In accordance with Kimura & Maruyama's (1971) calculations, local differentiation is especially marked in a two-dimensional stepping-stone model when $Nm < 1$, while when $Nm > 4$ a population acts as a single panmictic unit.

For the one-dimensional case, the condition for local differentiation is $Nm < k/\pi^2$, in which k is the number of subpopulations (Maruyama, 1970).

Thus, apart from the dimensions of an area, local genetic differentiation in populations having a stepping-stone structure depends both on the intensity of

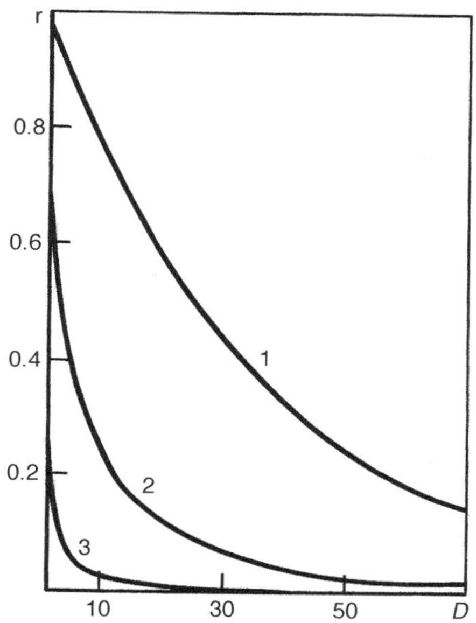

Fig. 1.8 Decrease of correlation r(d) between gene frequencies in colonies with increased distance (d) between them (Kimura & Weiss, 1964). $m = 0.1$; $m_{\infty} = 4 \times 10^{-5}$; 1, 2, 3 = one, two and three dimensions, respectively.

short-range (m) and long-range (m_{∞}) migration (the parameter m_{∞} can combine all possible stabilizing factors: the mutation process, migration from a constant external gene pool and selection), and on the size of the colonies and their numbers.

Conclusion

Almost all population genetic models and methods considered above will be employed in our following analysis. Along with this, the main estimates and calculations will be connected with the subdivided population models that are mostly applicable to species structure in salmon, and at least in those sections dealing with the theory of stationary distribution of gene frequencies.

In essence the theory of stationary distributions emphasizes yet again the important fact that, although individual evolutionary factors are also capable of causing directional genetic changes, the interaction of these factors (for example, forward and reverse mutations, genetic drift and migration, etc.) leads finally to reciprocal balance, engendering a stationary type of gene frequency dynamics. This stability may be particularly great when all the known factors of evolution act simultaneously on a population.

Another important feature of subdivision, which has also been examined theoretically, is the ability of subdivided populations to maintain significantly greater genetic diversity compared with panmictic populations of similar total size (Wright, 1951; Kimura, 1968). It is believed that this diversity allows a subdivided population to react more effectively to changes in the environment, through adaptive changes in genetic structure. This thesis, which plays a decisive role in Wright's concept of evolution, is known as the shifting balance theory in which the 'surface' *W* is represented on a topographical map with peaks and valleys for one landscape of gene combinations (Fig. 1.9). The most important conclusion from this model is that the 'evolutionary process depends on a continually shifting but never obliterated state of balance between factors of persistence and change, and that the most favourable condition for this occurs where there is a finely subdivided structure in which isolation and cross-communication keep in proper balance' (Wright, 1951).

Our further task is to compare this concept with what is observed in nature. The models of subdivided populations are especially expedient for studying genetic processes in salmon populations. However, before making this comparison it is necessary to examine the principal features of the biology and demography of salmonid species, that are essential for further population genetic analysis.

Fig. 1.9 Wright's two-dimensional contour maps with adaptive peaks and non-adaptive valleys in the field of gene combinations occupied by different types of populations (Wright, 1932). The isolines relate to different levels of adaptation: the thick broken line is the population's 'starting' position; arrows indicate evolutionary trends. (a) Reduced selection (or increased mutation rate) produces increased genetic variability and reduced average fitness. The population's (= species') evolutionary plasticity is fairly large, and with increased numbers it can occupy the lower slopes of another, more highly adaptive peak and then also conquer it entirely ($4Nu$ and $4Ns$ are very large). (b) The consequences of increased selection (or reduced mutation rate). The amount of the population's genetic variability is narrowed by selection, and the population's average fitness grows. Correspondingly, its evolutionary plasticity is reduced and its chances of conquering a neighbouring peak are diminished ($4Nu$ and $4Ns$ are very large). (c) Sharp changes in the adaptive landscape: peaks are transformed into valleys, and vice versa. The result of the genetic process is determined exclusively by selection intensity and environmental rates of change. A population having a small adaptive peak prior to these shifts and lacking an adequate reserve of genetic plasticity (variability) may remain in a valley and become extinct ($4Nu$ and $4Ns$ are large). (d) The consequences of drastic reduction of numbers and close inbreeding. A population falls from the adaptive peak, and as a result of drift its genetic pool undergoes random fluctuations leading, however, eventually to inbreeding and, as a rule, to degradation ($4Nu$ and $4Ns$ are very small). (e) An average-sized population subject to moderate mutations, leading to increased genetic diversity. The population shifts from its adaptive summit but cannot go far away from it. This means that the occupation of any new adaptive peak would be an exceptionally slow process ($4Nu$ and $4Ns$ have values intermediate between extreme situations). (f) The population of an extensive area subdivided into a multiplicity of interacting subpopulations, evolving rapidly by migration and mostly inadaptive. When one population or another comes into the environs of any adaptive peak it may colonize it. After increasing in numbers, it begins to 'pour' corresponding genes into other subpopulations, and entire species may become 'stretched out' in the zone of this new peak. Such a picture is considered to be evolutionarily optimal in the sense that genetic variability and the capacity for further change are preserved ($4Nm$ has an intermediate value).

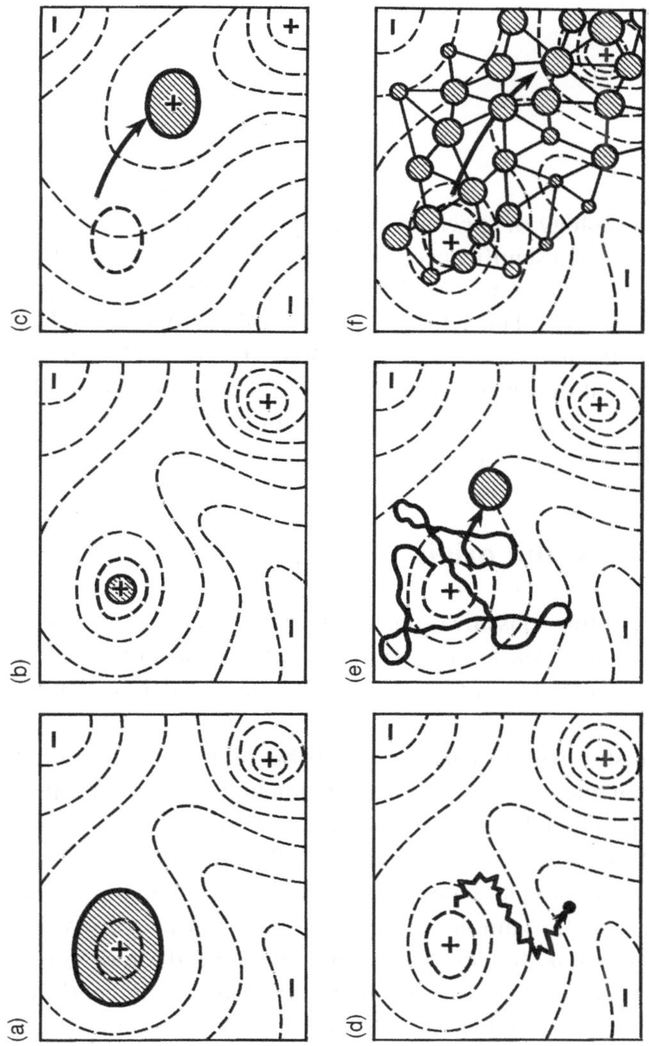
(a)
(b)
(c)
(d)
(e)
(f)

Chapter 2
Natural History, Biology and Demography of Salmonids

The population biology of species belonging to the family Salmonidae – owing to their great commercial significance – has been studied more thoroughly than that of many other species (Dvinin, 1952; Neave, 1952, 1958; Aro & Shepard, 1967; Foerster, 1968; Ricker, 1972; Hart, 1973; Smirnov, 1975; Hoar, 1976; Konovalov, 1980; Rukhlov, 1982; Birman, 1985; Zelinsky, 1985; Groot & Margolis, 1991; Glubokovsky, 1995; Kazakov 1998). On the basis of these and other publications, we will discuss only those features of salmonid life history, biology and demography that are directly related to the topic of this book.

2.1 Taxonomy and origin

According to existing classification, salmon comprise the subfamily Salmoninae and, along with whitefishes (Coregoninae) and graylings (Thymallinae), are included in the family Salmonidae (suborder Salmonoidei, order Salmoniformes) (Nelson, 1984).

Many specialists attribute salmon, whitefishes and graylings to individual families on the basis that these fishes, although belonging to the same polyploid branch, represent independent lines of development (Norden, 1961; Simon, 1963; Dorofeeva, 1982, 1985; Reshetnikov, 1980; Soin, 1980; Ivankov, 1985).

The family Salmonidae comprises seven genera: *Salmo* (Atlantic salmon), *Oncorhynchus* (Pacific salmon), *Salvelinus* (charr), *Brachymystax* (lenok), *Hucho* (taimen), *Salmothymus* (endemic Balkan genus) and the new genus (and species) *Salvethymus* (*S. svetovidovi*) Chereshnev and Skopetz, recently described as an endemic of the central Chukchi Peninsula (Chereshnev & Skopetz, 1990), although the taxonomic status of the latter is discussed (Savvaitova, 1995).

There is conclusive palaeontological evidence that the existing salmonid species appeared in the late Tertiary period, in the Pliocene, and became widespread in the Pleistocene, i.e. several million years ago (Berg, 1940; Vladimirov, 1946; Yakovlev, 1961, 1964; Kimmel, 1975; Smith, 1975: Smith *et al.*, 1982). Their ancestors probably belonged to the clupeiform fishes (Nikol'sky, 1971).

Most specialists agree that the salmonids originated in fresh waters (Chernavin, 1921; Berg, 1948a; Vladimirov, 1948; Neave, 1958; Vladykov, 1963), although some authors have proposed a marine origin (Smitt, 1895; Zenkevich, 1933; Schmidt, 1947). Studies on the evolution and origin of salmonids are numerous, and many of them provide schemes of phylogenetic relationships between the genera which are based on comparative morphological, karyological and molecular genetic data (Norden, 1961; Simon, 1963; Viktorovsky, 1978; Glubokovsky & Glubokovskaya, 1981; Dorofeeva, 1982; Vasiliev, 1985; Oleinik, 1994; Glubokovsky, 1995; Phillips & Oakley, 1997). Figure 2.1 shows one of the latest phylogenetic schemes. Virtually all authors agree that *Salmo* and highly specialized *Oncorhynchus* are the most evolutionarily advanced genera, whereas the position of other genera is subject to discussion.

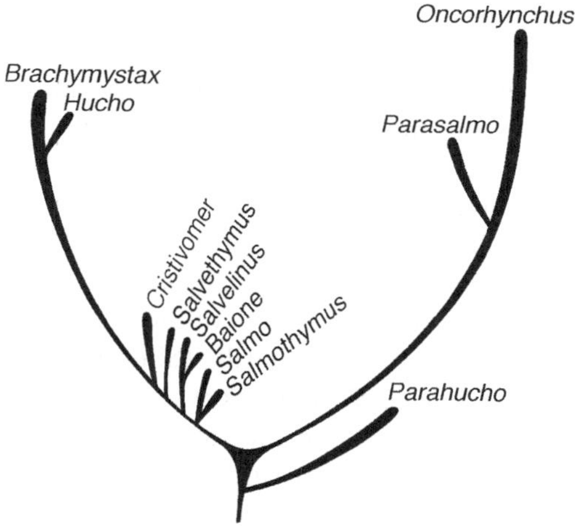

Fig. 2.1 Phylogram of salmonid fish genera (Glubokovsky, 1995).

Our attention will be focused on two genera, *Salmo* and *Oncorhynchus*, which have been studied in the greatest detail from all aspects. Pacific salmons of the genus *Oncorhynchus* include the chum salmon *O. keta* (Walbaum), the pink salmon *O. gorbuscha* (Walbaum), the sockeye salmon *O. nerka* (Walbaum), the chinook salmon *O. tshawytscha* (Walbaum), the coho salmon *O. kisutch* (Walbaum), and the masu salmon *O. masou* (Brevoort). The genus *Salmo* is represented by the Atlantic salmon *S. salar* L. and the brown trout *S. trutta* L. Until recently, it also included the Pacific trout: the steelhead (rainbow) trout *S. gairdneri* (Richardson) (with its Kamchatkan trout forms *S. penshinensis* and *S. mykiss*) and the cutthroat trout *S. clarki* (Richardson). The Pacific branch of *Salmo* is classified as the subgenus (or genus) *Parasalmo* (Vladykov, 1963; Behnke, 1972; Glubokovsky & Glubokovskaya, 1981; Dorofeeva, 1985). Smith & Stearley (1989) proposed that Pacific trout be included in the genus *Oncorhynchus* and the name *O. mykiss* given to all forms of steelhead trout. However, many ichthyologists objected to such a classification (e.g. Dorofeeva, 1994; Glubokovsky, 1995).

It is considered that the genus *Oncorhynchus* originated from an ancestor of Pacific *Salmo*: Neave (1958) derived it from the ancestor of the existing *S. gairdneri*, whereas Dorofeeva (1982) considered that the ancestral form was probably endemic to the Sea of Okhotsk and close to *S. mykiss*. Analysing the problem of Pacific salmon evolution, Neave (1958) and subsequently Glubokovsky & Glubokovskaya (1981) concluded that the conditions favourable for divergence in this genus were created during the periods of Pleistocene oceanic regressions. During these periods, the effects of geographic isolation of the Far Eastern seas (Lindberg, 1972) and also climate cooling and glaciation most probably led to disjunct distributions. According to Glubokovsky & Glubokovskaya (1981), the ranges of salmonids shifted to the south during these periods, promoting segregation of the American and Asian stocks. After the separation of species, subsequent intraspecific divergence was also closely associated with similar processes occurring in the Upper Pleistocene, which provided for the long-term isolation of some parts of the species in glacial refugia.

The analysis of various traits, including the composition of nuclear and mitochondrial DNA, showed that masu salmon is the most ancient species similar to *Salmo* in many respects, coho and chinook are close to masu, whereas chum, pink and sockeye salmon stand apart from the previous species but are similar to one another (Tsuyuki & Roberts, 1966; Yamanaka *et al.*, 1967; Utter *et al.*, 1973; Gorshkov *et al.*, 1979; Glubokovsky & Glubokovskaya, 1981; Dorofeeva, 1982; Oleinik, 1994; Domanico & Phillips, 1995). Moreover, some traits of the latter two species indicate that they are the most advanced evolutionarily (Simon, 1963; Viktorovsky, 1978; Glubokovsky & Glubokovskaya, 1981; Glubokovsky, 1995).

In the genesis of salmonids, one of the key features is that they are of polyploid (tetraploid) origin, as first proposed by Svärdson (1945). Convincing evidence was obtained in studies on the quantity of DNA, gene markers and cytological features in different representatives of the family Salmonidae. The main role in this work belongs to Susumu Ohno and his colleagues (Ohno *et al.*, 1968, 1969; Ohno, 1970a, 1970b, 1974) who formulated the hypothesis of polyploidization that occurred in an ancestor of salmonids. This hypothesis was based on the following facts:

(1) DNA content per cell in salmonids reaches about 80% of that in mammals and is approximately twice as high as in phylogenetically related fish groups.
(2) The usual number of chromosome arms in salmonids (about 100) is twice as great as in related fish groups.
(3) Preparations of meiotic germ cells of different salmonid species usually contain multivalents.
(4) Salmonids carry numerous duplicated enzyme loci.

Allendorf & Thorgaard (1984) reviewed cytogenetic and biochemical studies confirming this hypothesis. Analysing and generalizing the corresponding experimental data, they concluded that salmonids are probably autopolyploids in which the process of polyploidization is not yet accomplished. Other specialists do not share this view and have produced evidence that this group had an allotetraploid ancestor, i.e. is of hybrid origin (Massaro & Markert, 1968; Bailey *et al.*, 1969, 1970; Wilkins, 1970; Altukhov *et al.*, 1970, 1972; Altukhov, 1974; Bushuyev *et al.*, 1975; Reinitz, 1977; Vasiliev, 1977, 1985; Omelchenko & Gerasimenko, 1981).

The opinion expressed by Vasiliev (1977), and corroborated by more recent cytological data (Lozano *et al.*, 1992), is apparently compromising. He proposed that polyploids of this type should be *segmental tetraploids*: as genomes of the initial species involved in hybridization are similar, this makes it possible for chromosome pairing with the formation of tetravalents by chromosomes belonging to different genomes.

Some manifestations of polyploidy in the genetic variation of salmonids will be discussed in detail in Chapter 3.

2.2 Geographic distribution: spawning ranges and sea migration

2.2.1 Pacific salmon

Natural populations of Pacific salmon are confined to the Pacific basin. Their spawning ranges differ in width and are located on the Asian and North American continents.

The spawning range of chum salmon is the most extensive (Fig. 2.2). This species spawns in virtually all rivers of the Asian coast from the middle part of Honshu and Korea in the south to the Lena River in the north. In North America, its range extends from Monterey Bay to the Mackenzie River (Sano, 1966; Smirnov, 1975; Salo, 1991). Chum salmon populations reach the greatest size in the Amur River basin; rivers discharging on the continental coast of the Sea of Okhotsk, northern Sakhalin and Kamchatka; the Anadyr and Yukon river basins; and rivers of Alaska and British Columbia.

The range of pink salmon is only slightly smaller than that of chum salmon, and their areas of mass spawning overlap (Ishida, 1966; Smirnov, 1975; Neave *et al.*, 1976; Takagi *et al.*, 1981). Spawning stocks of pink salmon are especially numerous in the rivers of Sakhalin, the continental coast of the Sea of Okhotsk, Kamchatka (especially in the west), and the Amur and Anadyr river basins; on the American coast, from the rivers flowing into Puget Sound to northern Alaska.

The spawning range of sockeye salmon is narrower than those of chum and pink salmon and is mainly confined to the regions where lakes are a typical component of river basins (Ricker, 1966, 1972; Foerster, 1968; Smirnov, 1975; French *et al.*, 1976). This is why sockeye populations are especially numerous on the North American coast, from the Bering Strait to the Klamath River (southern California). On the Asian coast, this species expands to the Anadyr River but is more abundant on

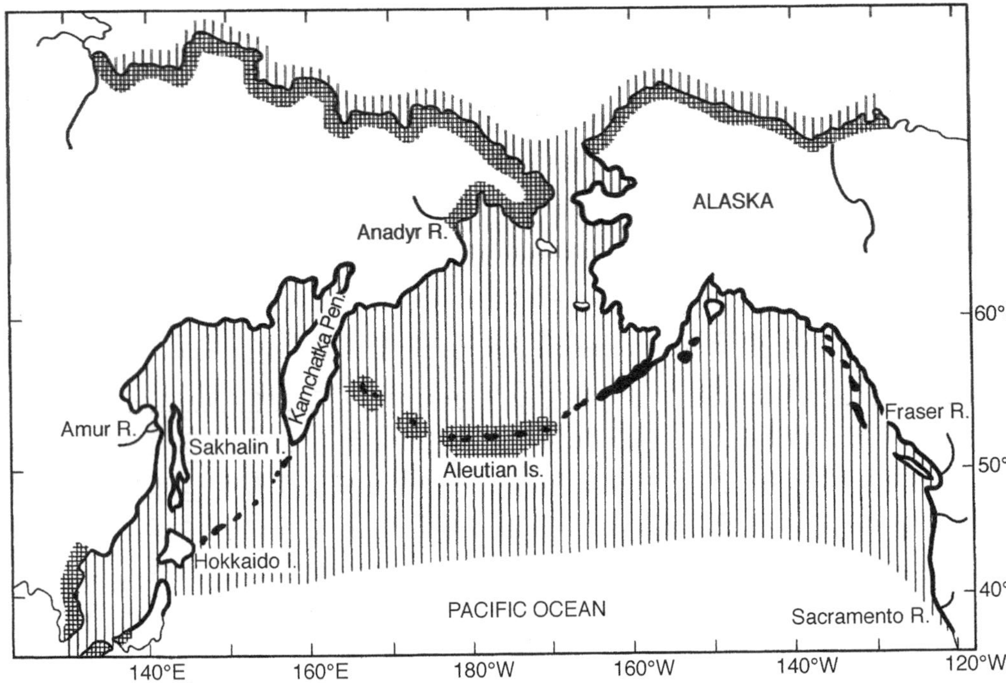

Fig. 2.2 Generalized freshwater and oceanic distribution of chum salmon. Oceanic distribution is shaded by vertical lines; bold line indicates relatively continuous spawning area; crosshatching indicates extreme locations where chum salmon have been reported. (Adapted from Salo, 1991. Reproduced with permission of University of British Columbia Press).

western and eastern Kamchatka. It inhabits the northern coast of the Sea of Okhotsk (to the Okhota River) and enters the rivers of the Commander and Kuril islands but does not occur in Sakhalin, Primorye (Maritime territory), or the Amur River.

The next extensive range is that of coho salmon. In Asia, this fish occurs in the area from the Anadyr to rivers flowing into the Amur estuary. It enters the rivers of the Commander, Aleutian and Kuril islands, eastern Sakhalin, and northern Hokkaido, being especially abundant in the Kamchatka River. On the American coast, coho salmon populations are significantly more numerous and inhabit the area extending from Norton Bay in Alaska to Monterey Bay in California (Smirnov, 1975; Godfrey *et al.*, 1975; Sandercock, 1991).

The spawning range of chinook salmon has almost the same size, though this species is less abundant and spawns mainly in large rivers and main tributaries. In Asia, chinook salmon is found mostly in the Kamchatka River. The main chinook stocks spawn on the American coast, in the Sacramento, Columbia and Fraser rivers. In small numbers, chinook salmon occurs in more northern regions, up to the Arctic Ocean coast (Smirnov, 1975; Major *et al.*, 1978; Healey, 1991).

Masu salmon is confined to the Asian Pacific coast, and its range is generally more southern than the ranges of other *Oncorhynchus* species. Masu salmon spawns in the rivers of the Japanese archipelago, Primorye, and Sakhalin, i.e. it is abundant in the basin of the Sea of Japan. This species extends to 38°N in the south and rarely occurs north of the Amur River (Smirnov, 1975; Kato, 1991).

Chum and pink salmon are the most numerous representatives of the genus.

After seaward migration, chum, pink and sockeye salmon fry remain in the coastal zone for several weeks. Migrations of these species in the first summer are very similar, but chum and pink salmon come to the sea slightly later than sockeye salmon (Neave, 1966a,b). To date, subsequent dispersal and migrations of these salmon species in the sea have been studied in sufficient detail with the aid of tagging and the analysis of catch composition by the structure of scales, marker parasites, and by other methods (for reviews, see *International North Pacific Fisheries Commission Bulletin* **31** (1975), **34** (1976), **35** (1976), **38** (1978), **40** (1981), **43** (1984); Konovalov, 1971, 1980; Birman, 1981, 1985; Groot & Margolis, 1991). It was found that each stock is usually characterized by a specific pattern of dispersal and oceanic migration, which is repeated between years.

Thus, pink salmon of American origin migrate from the southern regions northward and westward, counterclockwise, occupying by the next summer the entire Gulf of Alaska, neighbouring water areas on the east (near the Aleutian Islands) and on the south, and the south-eastern Bering Sea. Pink salmon from the Amur River and rivers of Primorye and western Sakhalin spend winter and forage in the northern Sea of Japan and the waters of the Pacific and the Sea of Okhotsk off Hokkaido and the southern Kuril Islands. Pink salmon from other regions of the Asian coast migrate to the Pacific Ocean and occupy the water area south-east of the Kurils, reaching 42°N and 175°E (fishes from more northern regions migrate even farther eastward, to 155°E), and their migrations are mostly directed clockwise. In some places, prespawning migrations of pink salmon extend northward of their reproductive areas; thereafter, they change their course, and then migrate to spawning grounds from the north southward, e.g. along the coasts of western Sakhalin, western Kamchatka, south-eastern Alaska, and British Columbia.

During the initial marine period, chum salmon stocks of Asian and American origin remain relatively close to the coast and do not overlap in the ocean. In the next spring and summer, the Asian stocks migrate to more northern regions, reaching far into the Bering Sea and using its food resources. At the same time, American chum salmon migrate westward along the eastern Aleutians, and only a small proportion of these fishes come to the Bering Sea. Chum salmon of different age groups migrate for wintering to more southern regions and occupy the Gulf of Alaska and the waters south of the Aleutians, to 140°W. The wintering ranges of Asian and American stocks overlap in this zone. Northward migration of chum salmon in summer and southward migration in winter are repeated until maturation. In the final summer, maturing spawners leave the northern regions and move to spawning grounds.

The spawning range of sockeye covers a vast area of the North Pacific south of the Commander Islands and the Aleutian Islands, which extends from 51°N to 46°N and from approximately 165°W to 174°W. In summer, this species migrates for feeding to northern regions of the Gulf of Alaska, the Commander and Aleutian islands, and the Bering Sea (up to 60°N), where food is more abundant. Konovalov (1980) noted that both feeding and wintering ranges of sockeye salmon belonging to the Kamchatka, Bristol Bay and Pacific–Alaskan complexes largely overlap, although the boundaries of corresponding regions are specific for each stock.

Chinook salmon smolts coming to the sea remain near the coast during the first summer–autumn season and migrate to the open ocean waters in winter. Subsequently, they spread over the Bering Sea, the Sea of Okhotsk, and the entire North Pacific down to approximately 40°N (Major *et al.*, 1978; Healey, 1991). According to Healey (1991), Asian chinook salmon concentrate west of 180°E, whereas the more numerous American stock penetrates westward to approximately 160°E; moreover, the ocean-type chinooks, which spawn in the American rivers to the south of the Alaska Peninsula, remain significantly closer to the coast during their feeding migration.

Sea migration routes of coho salmon lie in approximately the same regions of the Pacific as those of chinook salmon. The Asian stocks of coho salmon spread eastward to 177°E and southward to 45°N, and the American stocks spread westward to 177°E and southward to 43°N (Manzer *et al.*, 1965; Godfrey *et al.*, 1975; Sandercock, 1991).

Immature masu salmon migrate in the Sea of Okhotsk, the northern Sea of Japan, and Pacific waters off the Kurils in summer and the first half of autumn. When the weather becomes colder, they enter for wintering to the Sea of Japan, and a part of the stock apparently remains near the Pacific coast of Honshu (Kato, 1991).

The most widespread species of Pacific trout, *Oncorhynchus mykiss*, currently comprises American and Asian forms, i.e. the former *Salmo gairdneri* (steelhead, or rainbow trout), *S. penshinensis* and *S. mykiss* (Kamchatkan trouts) (see above). Anadromous steelhead trout during the reproductive period live in the rivers of North America from the Alaska Peninsula to central California. Resident river and lake forms of this species inhabit the waters of the west coast and the neighbouring inland regions of the United States (MacCrimmon, 1971; Sutherland, 1973, cited from Okazaki, 1986a). These forms have been introduced and are cultivated on all the continents. The Asian forms, both anadromous and resident, inhabit northern regions of East Asia (mainly Kamchatka), the Commander Islands, and some rivers flowing into the Sea of Okhotsk (Berg, 1948b; Savvaitova, 1975a,b). During sea migrations, anadromous forms occur both in the West Pacific, mainly south of the Aleutians and east of the Kurils, and in the East Pacific, south of the Gulf of Alaska (Okazaki, 1986a).

The cutthroat trout, *Oncorhynchus clarki*, is a polytypic species comprising the coastal subspecies *O. c. clarki* and several subspecies (or forms) inhabiting inland water bodies of western North America (Cope, 1957). The coastal subspecies includes anadromous and resident forms, and its spawning range extends along the west coast from California to Alaska (De Witt, 1954; Scott & Crossman, 1973; both works cited from Campton & Utter, 1987; Allendorf & Leary, 1988). The resident form inhabits lakes and river sections upstream from obstacles preventing migration. The anadromous form does not migrate to the open ocean but remains near the coast, within a 100 km range from the parent river estuary, for 3–6 months (Wydosky & Whitney, 1979, cited from Campton & Utter, 1987). These fishes return to fresh waters both for spawning (mature) and for wintering (immature individuals).

All specialists note that migrations of different Pacific salmon species are connected with sea and ocean currents: the Oyashio, the Alaska Current, the Subarctic Current, etc. It is remarkable that the areas of sea and ocean migrations of many Asian salmon populations are generally much greater than those of American populations and extend toward the American coast, especially in the case of chum salmon. Such differences in distribution are apparently accounted for by a milder winter climate in the latter region.

2.2.2 Atlantic salmon

The range of Atlantic salmon (*Salmo salar*) is located on both sides of the Atlantic Ocean (Fig. 2.3). In Europe, this fish in the reproductive period occurs in the area from the Kara River (the Kara Sea) to the Douro River (northern Portugal) and in the Baltic Sea basin; in North America, from the Connecticut River in the south to the Hudson Strait and Ungava Bay. In the north, Atlantic salmon spreads to higher than 70°N, inhabiting the rivers of northern Norway, Iceland and southern Greenland (Berg, 1948b; Vladykov, 1963; Zelinsky, 1985). In addition to the

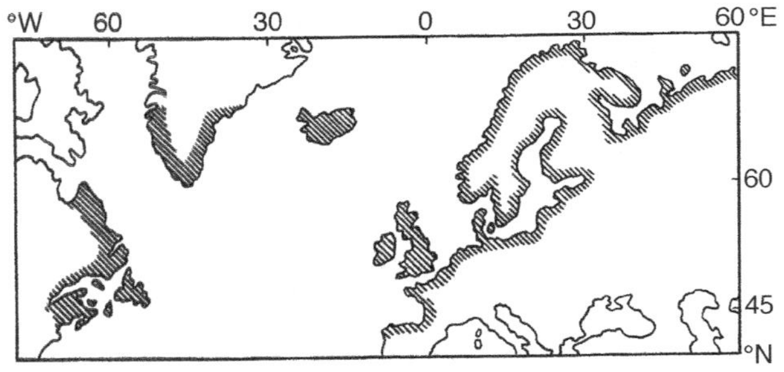

Fig. 2.3 Spawning range of Atlantic salmon (from Vladykov, 1963).

anadromous form, there is a resident freshwater form of Atlantic salmon, which lives in lakes and spawns in the rivers flowing into these lakes (Berg, 1948b; Nikol'sky, 1971).

North American populations of Atlantic salmon migrate to the the waters off western Greenland, in the shelf area off Newfoundland, and in the Labrador Sea, where they stay in winter. After wintering, some fish move to spawning grounds, whereas others return to the northern sea areas for feeding. European populations feed along the polar front from western Greenland (most fish from the south-western part of the species range) to the Norwegian Sea (salmon from north-eastern Europe). Atlantic salmon from northern Russia migrate mainly to the latter region (Bakshtansky *et al.*, 1991).

As follows from publications of the late nineteenth and early twentieth centuries, the natural range of brown trout (*Salmo trutta*) covered the area from Iceland in the north to Algeria in the south, and from the Atlantic coast in the west to the northern slopes of the Himalayas in the east (Fig. 2.4). Today, this formerly continuous range

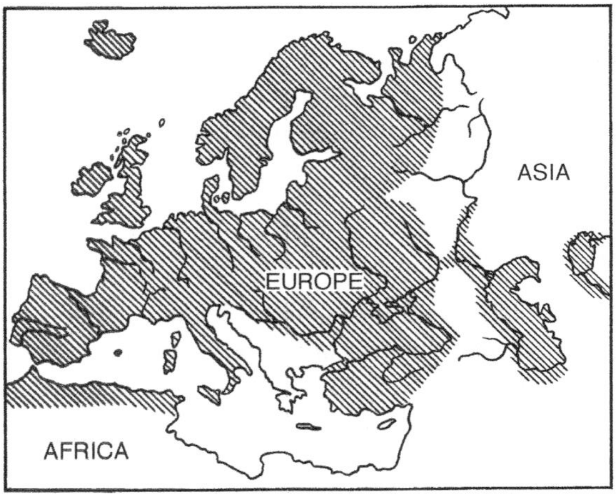

Fig. 2.4 Natural range of *Salmo trutta* (from MacCrimmon & Marshall, 1968).

is reduced to localities inhabited by anadromous and freshwater populations (MacCrimmon & Marshall, 1968; MacCrimmon *et al.*, 1970; Nikol'sky, 1971) confined to different lake and river basins, including those of the Barents, White, Baltic, Black and Caspian seas in the territory of Russia.

Most of the existing natural and cultivated populations of brown trout originated from either the European anadromous form *S. trutta trutta*, or the resident *S. trutta fario* (MacCrimmon *et al.*, 1970).

Long-range migrations in the open sea are not characteristic of brown trout: smolts coming to the sea can spend the first year in estuaries or move upstream for wintering in rivers (Nikol'sky, 1971). Fishes spawning in the rivers flowing into the Baltic Sea can migrate virtually throughout the sea (Skrochowska, 1969).

2.3 Reproductive biology

2.3.1 Homing

After long-term migrations in the ocean, both Atlantic and Pacific salmon return for spawning to the water bodies in which they were born, i.e. demonstrating the *homing* instinct. Considerable research has been performed on this phenomenon (Hartman & Raleigh, 1964; Hasler, 1966; Ricker, 1972; Nordeng, 1977; Hasler & Scholz, 1983; Stabell, 1984; Quinn, 1984, 1990, 1993; deGroot & Roberts, 1994). It should be noted that homing is one of the basic elements of animal behaviour (e.g. see Petras, 1967; Madison, 1969; Baker, 1978; Meylan, 1982; Carr, 1986; Pezzack & Duggan, 1986; Sokolov, 1991), and the point at issue is its stronger or weaker expression in different species. Homing aberrations, i.e. the return for spawning to places other than the natal (parent) water body, are referred to as *straying*.

Salmonids have always impressed people with their precise homing: spawning stocks confine themselves strictly to the same water body or even the same spawning site. Homing made possible the artificial propagation of salmon with the release of fry into natural water bodies. Such behaviour creates conditions favouring strong intraspecific differentiation with respect to the history and geography of the species range. Homing-dependent isolation during spawning, promoted by a complex reproductive behaviour, facilitates the formation of numerous reproductive populations, or stocks, dispersed over a vast area (see Ricker, 1972; Konovalov, 1980). Table 2.1 shows estimates of straying intensity (or homing accuracy) in six *Oncorhynchus* and two *Salmo* species, according to the published data available. In compiling this table, we tried to use the data on populations least exposed to anthropogenic pressures. As shown in Table 2.1, straying intensity was estimated by different methods. In most cases, researchers tagged (marked) fry, and only in sockeye salmon, rainbow trout and brown trout were the spawners tagged. Variations in average values of straying intensity ranged from 1.25% for coho salmon to 8.1% for rainbow trout.

Table 2.1 Estimates of salmon straying intensity.

Species	Straying intensity (%)	Method of estimation (region)	References
Oncorhynchus gorbuscha	3–4	Release of marked fry from a hatchery and mark recovery in returning adults (British Columbia, Canada)	Bams, 1976
	0.4	Analogous method (Canada)	Boyd, 1964
	0.1–11.5	Release of marked fry from the hatchery, reproducing only 'own' fish, and monitoring of mark return (Kuril hatchery, Russia)	Omelchenko & Vyalova, 1990 (review)
Mean	3.2		
O. keta	2–10	Release of marked fry from hatcheries and mark recovery in adult return (Japan)	Sakano, 1960 (cited by Okazaki, 1982b)
	5.3	Release of marked fry from incubators and mark recovery in adult return (Canada)	McQuarrie & Bailey, 1980 (cited by Quinn, 1993)
Mean	5.8		
O. nerka	3	Marking and transfers of spawners in basins of two lakes (Alaska)	Hartman & Raleigh, 1964
	0.59	Experiments with marking and transfer of spawners (Canada)	Foerster, 1968
	3	Observation of kokanee spawning (Canada)	Vernon, 1957
	0–5.3	Marking and transfer of spawners in lake basin (Kamchatka, Russia)	Il'in *et al.*, 1983
Mean	2.3		
O. kisutch	0.5	Release of marked fry and mark recovery in adult return (USA)	Donaldson & Allen, 1957
	2	Marking of juveniles from hatchery and natural populations and mark recovery (Canada)	Labelle, 1992
Mean	1.25		
O. tshawytscha	8.4	Release of marked fry from hatcheries and mark recovery in adult return (USA)	Major *et al.*, 1978 (review)
	10	Analogous method (USA)	Sholes & Hallock, 1979
	1.4	Analogous method (USA)	Quinn & Fresh, 1984

Cont.

Table 2.1 Cont.

Species	Straying intensity (%)	Method of estimation (region)	References
O. tshawytscha (cont.)	3.2–10.3	Release of marked fry from natural and hatchery populations and mark recovery in adult return (USA)	McIsaak, 1990 (cited by Quinn, 1993)
Mean	6.6		
Salmo salar	0–2	Release of marked juveniles and mark recovery in adult return (west and Baltic coast of Sweden)	Carlin, 1969
	2.7–3	Release of marked juveniles and mark recovery in adult return (Britain, Ireland)	Thorpe & Mitchell, 1981
	4.1	Experiments on release of marked wild juveniles into native rivers and mark recovery (Iceland, Scotland, Canada)	Stabell, 1984 (review)
	11.5	Experiments on release of marked hatchery juveniles and mark recovery (Norway, Iceland, Canada)	Stabell, 1984 (review)
Mean	3.9		
Oncorhynchus mykiss	6 (year 1) 0.4 (years 2, 3)	Marking of young fish and mark recovery in spawners for 3 years (Canada)	Lindsey *et al.*, 1959
	23	Marking of adult fish and transfer within lake basin (Canada)	Kwain, 1981
	3	Analogous method (Canada)	Swanson, 1985 (cited by Krueger & May, 1987b)
Mean	8.1		
Salmo trutta	0.1	Recapture of marked spawners in subsequent spawning runs (Scotland)	Stuart, 1957 (cited by Hartman & Raleigh, 1964)
	6.0	Release of marked smolts and monitoring of mark return (Sweden)	Carlin, 1965 (cited by Svärdson & Fagerstrom, 1982)
Mean	3.1		

The methods used in these studies are imperfect, and the corresponding parameters of straying are apparently overestimated. In particular, this follows from the results of a careful experiment performed by Hartman & Raleigh (1964). Analysing straying intensity in male sockeye salmon spawning in Brooks and Karluk lakes (Alaska), they tagged fishes, transferred them to the 'foreign' spawning sites, and followed their return to the initial sites. These authors made an important observation that *the fish not yet reaching a certain critical level of sexual maturity demonstrated a remarkable accuracy in returning to their 'native' spawning sites.* Ample experimental material allowed the conclusion that no less than 97% of fishes returned to their native sites. We evaluated these data and, excluding from analysis some contradictory results, obtained a straying coefficient of 2%, i.e. 98% homing accuracy. Approximately the same results were obtained in similar experiments performed in 1978 and 1979 with sockeye salmon from Lake Azabachye (Il'in *et al.*, 1983).

Similar estimations were also made for other salmonid species. In rainbow trout, for example, about 94% of mature fishes spawn for the first time in natal waters, whereas the proportion of fishes returning to these spawning sites in the second and third years reaches 99.6% (Lindsey *et al.*, 1959). In cutthroat trout, about 97% of spawners spend their second reproductive season at the sites where they spawned for the first time (Cope, 1957). Of 1131 kokanee (the resident form of sockeye salmon) studied by Vernon (1957), less than 3% spawned at foreign sites.

How can we estimate the biological significance of these values? Apparently the only possible way, based on the concepts of population genetics, is to transform the parameters of straying intensity into coefficients of gene migration.

Taking into account doubt expressed by some specialists (Mina, 1978), it should be emphasized that we never assumed quantitative estimations of homing in sockeye or other salmonid species with a similar structural pattern to be exactly equivalent to the value of the migration coefficient in its strict population genetic sense. However, observations of this kind are important in principle and provide the natural basis for evaluating the degree to which subpopulations are isolated. Moreover, it is apparent that if there was an error in using these values, it would lead to overestimation of migration coefficients, as the intensity of 'physical' fish migration usually proves to be higher than that of their effective gene migration (Wallace, 1966, 1979; Ehrlich & Raven, 1969; Raven, 1979).

We regard the parameters of straying in Table 2.1 as maximal estimates of potential gene migrations. Experimental data on salmonids confirm that the level of these migrations is indeed lower than that of straying. Thus, gene exchange between two sockeye populations, estimated using allozyme markers, proved to be less than 1%, which corresponded to the lowest values of straying intensity as revealed by tagging (Quinn *et al.*, 1987). In studies on three chum salmon populations, straying intensity among marked fishes was approximately one order of magnitude higher than the level of gene exchange (Tallman & Healey, 1994).

Several hypotheses were proposed to explain the phenomenon of homing in salmonids. In recent years, two of them have become more popular:

(1) The hypothesis of imprinting, i.e. remembering the parent river odour during the downstream migration, combined with orientation by the position of the sun during navigation in the open sea (Hasler & Wisby, 1951; Hasler & Scholz, 1983).
(2) The pheromone hypothesis (Nordeng, 1977), based on the ability of fishes to

> distinguish the odour of individuals belonging to the same species or even the same genetic line, which has a special attraction for them. This hypothesis takes into account the existence of specific genetic features in different populations inhabiting the same river system and asynchrony of their migrations.

Subsequently, Hansen *et al.* (1989) tested the pheromone hypothesis of homing, together with hypotheses about single and sequential imprinting, using transplantation experiments with Atlantic salmon. They showed that the pheromone hypothesis was untenable, and that the fishes required sequential imprinting of local information throughout early life if they were to home successfully.

In experiments with pink salmon, Bams (1976) obtained direct evidence for the genetic basis of homing: in interpopulation hybrids (transplanted × native fish), homing was less perfect than in the native population but superior to that in the transplanted group.

In artificially reproduced *S. salar*, the percentage of fishes returning to the parent river was lower and the level of straying was higher than in native fishes (Vivert & Laurent, 1972, cited from Stabell, 1984). This was attributed to a high probability of interpopulation hybridization in the course of artificial breeding and the presence of several spawning subpopulations in the river.

Quinn (1984) interpreted homing and straying as evolutionary alternatives: homing provides for the maximal adaptation of the species to relatively stable environmental conditions, whereas straying promotes species dispersal and helps to avoid the effect of catastrophic changes in the environment. Both these types of life strategy – homing and straying – are in a state of dynamic equilibrium with the prevalence of homing (see also Bakshtansky, 1980; Thorpe, 1994).

As will be shown in Chapter 6, homing and straying in the population genetic sense are major mechanisms that provide for the maintenance of balance between the processes of differentiation and integration of populations and regulate the ratio of intra-and interpopulation gene diversity. Artificially induced changes in this ratio may lead to adverse genetic consequences (Altukhov & Salmenkova, 1994).

2.3.2 Iteroparity and semelparity

In Atlantic salmon, some fish die after the first spawning, whereas others survive and can spawn repeatedly. The proportion of the latter, referred to as the *residue*, differs depending on the population. It is usually higher in lake populations (Zelinsky, 1985), while in Atlantic salmon from the Pechora River, for example, it amounts to only 5–6% and is represented mostly by females (Antonova, 1976). In some populations, the proportion of repeatedly spawning individuals has been decreasing recently (Antonova & Chuksina, 1985; Kulida, 1985). Male parr of Atlantic salmon maturing in fresh waters can spawn repeatedly, either remaining in the same water body after the first spawning or migrating to the sea and then returning (Leyzerovich & Mel'nikova, 1979; Zelinsky, 1985).

In Pacific trout, most fish die after spawning, although some individuals survive and can spawn three or even four times (Smith & Stearley, 1989; Pavlov, 1989).

Finally, all the Pacific salmon species are typically semelparous (monocyclic), i.e. they spawn only once and then die. Exceptions are very rare and occur only in species with a sufficiently long freshwater stage of the life cycle. Thus, sockeye, coho, chinook and masu salmon have small freshwater (residual) forms of males that spawn repeatedly (Krogius & Krokhin, 1956; Ricker, 1972; Smirnov, 1975) (see Table 2.2).

In Pacific salmon, as in other biological species, the generation length depends on the average age of all individuals involved in reproduction. The generation length and degree of generation overlap in time and space are the basic demographic parameters which are assessed by analysing the age structure of the reproductive part of a population.

2.3.3 Age structure

The age of an individual in salmonids is determined as the sum of the freshwater period (from hatching to the migration to the sea) and the marine period of life (sea migration until the spawning migration to fresh water). There are several systems of recording the age in which either the total life span or the duration of each period are indicated. The authors of recent publications have followed the European notation system and designate the age by two numerals, showing the number of winters spent in fresh water and in the sea, e.g. 1.1, 1.2 or 0.3, where 1.1 refers to 3-year-old fishes, 1.2 and 0.3 to 4-year old fishes, etc. We use the same system in this book.

The duration of both periods in salmonids is subject to strong inter and intra-specific variation (except for pink salmon).

Pink salmon has the simplest age structure among species of the genus *Oncorhynchus*: young fish migrate to the sea soon after completing their embryonic development and reach maturity in the second year of life, so that spawning populations consist only of fishes at the age 0.1 (Heard, 1991). Generations are represented by lines of odd and even years, which do not overlap in time, and gene exchange between them is practically excluded. The generation length of pink salmon is 2 years. Single catches of small mature individuals under 1 year of age, reported by some researchers, are obviously a rare exception (Ivankov *et al.*, 1975, 1987; Efanov & Kochneva, 1980; Foster *et al.*, 1980). The opinion that pink salmon from natural populations may spawn at the age of 3 years (Anas, 1959; Lapin, 1964, 1971; Turner & Bilton, 1968) has not been confirmed conclusively. Three-year-old individuals appeared only in pink salmon populations transplanted to the Kola Peninsula and the Great Lakes and spawning in fresh water (Bakshtansky, 1962; Kwain & Chappel, 1978). Thus, species-specific features of pink salmon include

Table 2.2 Dwarf and freshwater (non-anadromous) forms in salmonid fishes.

Species	Early maturation in fresh water (at parr stage), sex, age (years)	Spawning of parr in fresh water		Sea migration after spawning at parr stage	Presence of freshwater (non-anadromous) forms
		Single	Repeat		
Pink salmon	No	No	No	No	Appears after introduction in the Great Lakes
Chum salmon	No	No	No	No	No
Sockeye salmon	Females and males, 1.0, 2.0, 3.0, 4.0	Yes	?	No	(1) Kokanee (reproductively isolated from the anadromous form), females and males (2) Residual form, exceptionally or mainly males, spawns with anadromous form
Chinook salmon	Males (rare), 0.0, 1.0	Yes	Yes (seldom)	Yes	Yes
Coho salmon	Females, males (rare), 1.0, 2.0, 3.0	?	No	No	Yes
Masu salmon	Males, 1.0, 2.0, 3.0	Yes	Yes	?	River and lake forms
Steelhead trout	Males	Yes	?	?	River and lake forms
Atlantic salmon	Males, 0.0, 1.0, 2.0, 3.0	Yes	Yes	Yes	River and lake forms
Brown trout	Males, 1.0–7.0	Yes	Yes	?	River and lake forms

Data from Vladykov, 1963; Foerster, 1968; Osterdahl, 1969; Ricker, 1972; Thorpe, 1975; Savvaitova, 1975a,b; Smirnov, 1975; Krogius, 1979, 1981; Healey, 1986; Pavlov, 1989; Groot & Margolis, 1991.

extremely reduced periods of life in fresh water and in the sea, a 2-year life cycle and uniform age structure.

Chum salmon, like pink salmon, totally migrate seaward in the first year of life but remain in the sea for a longer period. The maximal age of maturity, revealed in single individuals, is 0.6–0.7. Three- to five-year-old fishes (age groups 0.2–0.4) comprise more than 95% of the spawning stock, and the group 0.3 is dominant in most regions (60–90%). Thus, the generation length is 4 years. As in other species, however, the age structure of chum salmon populations depends on latitude: 4- to 5-year-old spawners prevail in northern regions, whereas the proportion of 3-year-olds is greater in the south (Salo, 1991).

In *masu* and *coho salmon*, compared to chum salmon, the average marine period of life is shorter and the freshwater period is longer: 1, 2 or (rarely) even 3 years. Coho salmon populations proved to consist of up to ten age groups of which only two – 1.1 and 1.2 – are dominant, and the major part of the stock remains in the sea for 1 year. The maximal recorded individual life span is 5 years. In masu salmon, populations include eight age groups, and the bulk of the anadromous stock consists of 3- to 4-year-olds (age groups 1.1, 1.2 and 2.1). The proportion of fish remaining in the sea for 2 years is greater than in coho salmon, although the period of sea migration in the majority of masu salmon is 1 year (Smirnov, 1975; Konovalov, 1980; Sandercock, 1991; Salo, 1991).

Chinook and sockeye populations are characterized by a more complex age structure. In the former species, 14 age groups are known. Young fish may migrate to the ocean in the year of birth (ocean type) or remain in fresh water for up to 2 years (stream type), and the marine period in the life of chinook salmon lasts from 1 to 5 years. The maximal life span is 7 years; dominant age groups are 0.3, 0.4, 1.2, 1.3 and 1.4 (Vronsky, 1972; Healey, 1991).

In sockeye salmon, the number of age groups reaches 21 because the freshwater and marine periods last from several months to 4 years and from 1 to 3 years, respectively. The maximal individual life span is 8 years, and spawning populations consist mainly of fishes that spent 1–2 years in fresh water and 1–3 years in the sea, i.e. of age classes 1.2, 2.2, 1.3 and 2.3 (Foerster, 1968; Smirnov, 1975; Konovalov, 1980; Burgner, 1991). The generation length in sockeye salmon is usually 4–5 years, with variations in different parts of the species range.

Anadromous *steelhead, Oncorhynchus mykiss*, live 1–4 years in fresh water and up to 3 years in the sea, and can form 15 age groups. Dominant age groups are 2.1, 3.1 and 3.2, and the maximal life span is 7 years (age group 3.3) (Savvaitova & Maksimov, 1969; Sheppard, 1972; Savvaitova, 1975a,b; Okazaki, 1986a).

Atlantic salmon live in fresh water for 1–5 years and in the sea for several months to 5 years (Zelinsky, 1985). The maximal individual age is 12 years. Because of repeated spawning (up to four or even five times; Berg, 1935) and intermediate sea migration for 1 or 2 years, the number of age groups is high. For example, one of the largest Atlantic salmon stocks of the Pechora River includes 24 age categories belonging to eight age classes: from 4- to 11-year-old fish (Antonova, 1976). Most of the anadromous migrants belong to two or three age classes (Jarvi, 1948; Azbelev, 1960; Antonova, 1976; Gardner, 1976; Kulida & Derets, 1985; Martynov & Kuznetsov, 1985; Nekludov, 1985; Zelinsky, 1985). Lake salmon (foraging in lakes) has principally the same age structure as the anadromous form, and its spawning populations include many fish that spawn repeatedly.

Brown trout live in fresh water for 3–7 years and in the sea for 1–4 years, and in lakes for up to 7 years. Long-range sea migrations are not characteristic of this species, and sea-run fish in the first year may live in estuaries (Nikol'sky, 1971; Dirin, 1980). The stream form, which does not migrate seaward, usually has a life span of 5–7 years in the European north (Ershov, 1985) and up to 8 years in the rivers of Armenia (Vladimirov, 1948).

The analysis of age structure of salmon populations makes it possible to formulate some general rules that partially apply to other fish species as well.

(1) In all salmonids, the life span of fish in northern populations is longer than in southern populations.

(2) Males and females differ in the rate of sexual maturation. The most rapidly growing and early maturing age classes are mostly comprised by males, up to 100%. This is the case in coho salmon at the age of 1.0 (jacks), which return to the rivers for spawning after feeding in the sea for 4–6 months (Sandercock, 1991); in sockeye salmon of age classes 1.1 and 2.1 (Foerster, 1954a; Konovalov, 1980); in chum salmon of age classes 0.1 and 0.2 (Sano, 1966); in chinook salmon of groups 0.1 and 0.2 (Healey, 1991); and in Atlantic salmon grilse, spending only one winter in the sea (Gardner, 1976; Khalturin, 1978; Zelinsky, 1985).

(3) There exist freshwater forms reaching sexual maturity at the parr stage. These forms are typical for Atlantic salmon, brown trout and masu salmon, less common in sockeye salmon and steelhead trout, and extremely rare in chinook and coho salmon (Table 2.2).

(4) The age structure of different populations of the same species is not identical. In the same population, age structure varies slightly from one generation to another, but one or two age groups usually remain dominant. An illustrative example is the large sockeye salmon population of the Fraser River, for which age structure, size and other demographic parameters were recorded annually from 1915 to 1960 (Killick & Clemens, 1963). In this population, age group 1.2 (4-year-olds) remained dominant: its proportion varied from 65 to 99%, averaging 89.3%. Each of the three remaining age classes accounted for an average of $(100 - 89.3) : 3 = 3.6\%$ of the population. The latter value may be regarded as the parameter characterizing the average level of generation overlap in the sockeye salmon population of the Fraser River.

Salo (1991) surveyed the data on the age structure of numerous spawning chum salmon populations throughout the species range. These data show that the dominant group was at the age 0.3 (on average, 60% of all the fish studied), except for the most southern Asian and American populations with relatively large proportions of age group 0.2. As averaged over the species range, each of the four other age classes accounted for 9.8% (our calculations), which characterizes the average degree of generation overlap in chum salmon. The corresponding proportion for chinook salmon, calculated from the data on populations of the Yukon and Fraser rivers (Healey, 1991, Fig. 37), also equals 9–10%.

According to Pritchard (1940), coho salmon populations of British Columbia included seven age classes, but among a total of 6312 fish studied, 97.9% were of the age 1.1. In the Columbia River, four age classes were found, and 84% of fish were also of the age 1.1 (Marr, 1944, cited from Sandercock, 1991). As follows from these data, the average proportions of non-dominant classes varied from 0.35 to 5.3%. Thus, the degree of generation overlap in Pacific salmon, averaged over the species range or in a large population system, is about 6%.

At the same time, the lower the hierarchical level of intraspecific subdivision, the greater the range of variation in age structure. Consider an example of sockeye salmon, in which the age structure was studied in the greatest detail. In populations inhabiting the eight main lake systems of Bristol Bay, the dominant ages are 2.2, 1.2 or 1.3, and the average proportion of a non-dominant age class varies from 4.5 to 22%, depending on location. In five stream subpopulations of the Lake Aleknagik system, this proportion varies from 9 to 25% (calculated from the data by Rogers, 1987, cited by Burgner, 1991), and in subpopulations from Lake Azabachye, from 10 to 22.5% (Konovalov, 1980). Correspondingly, the degree of generation overlap at lower hierarchical levels of intraspecific subdivision is also higher and more variable.

Note that *the degree of generation overlap is comparable to straying intensity*. This fact is very important because both straying and generation overlap are mechanisms providing for gene exchange between the structural components of population systems separated temporally and spatially. The data discussed above show that the intensity of gene exchange decreases with an increase in the level of population hierarchy (see Altukhov, 1989a). It is noteworthy that the same conclusion can be drawn from the results of estimating the intensity of gene migration at different levels of population hierarchy in pink salmon by means of allozyme analysis (Zhivotovsky *et al.*, 1994).

2.3.4 Sex ratio and size structure

Considering the reproductive season in general, salmon species are characterized by a balanced sex ratio. At the same time, the sex ratio, as with the age structure of the population, considerably varies in time and space at the lower levels of population hierarchy (e.g. for chum salmon, see Altukhov, 1974, 1981a; for sockeye salmon, Foerster, 1954b; Altukhov *et al.*, 1975a; Konovalov, 1980; for pink salmon, Hunter, 1959; Ivankov, 1968; Enyutina, 1972). As mentioned above, the sex ratio varies depending on age class because of differences between males and females in the rate of their development. In spawning stocks, for example, males radically prevail in the youngest age classes, accounting for up to 100% of fish, especially among small forms maturing in fresh water. Hence, in masu, sockeye and Atlantic salmon, brown trout and rainbow trout – the species in which the proportion of such small males can be considerable – females strongly prevail in the anadromous part of populations (Skrochowska, 1969; Krogius, 1979; Ershov, 1985; Zelinsky, 1985; Okazaki, 1986a; Kato, 1991).

In Pacific salmon, males usually dominate at the onset of the spawning run to fresh waters; in the middle of this period, the sex ratio becomes equal; and by the end of the run, females become prevalent (Fig. 2.5, our data; see also Shapovalov & Taft, 1954; Groot & Margolis, 1991). In Atlantic salmon spawning in the Pechora River, females prevail at the beginning of the migration, and the proportion of males increases in its course (Antonova, 1976); fish spawning for the second time include 80% of females, and those rare individuals spawning for the third time are female only.

The more diverse age structure of the male population is associated with a significantly greater variation in male body size. In many species, body length distribution in males is bimodal or even trimodal, whereas in females it is unimodal. This trend is especially apparent in sockeye salmon (Fig. 2.6) (Altukhov, 1983a,

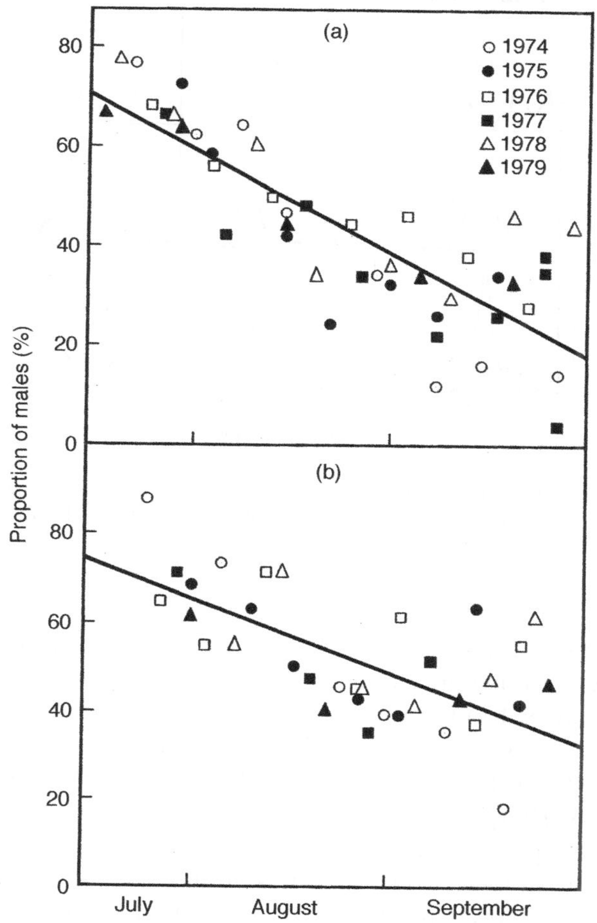

Fig. 2.5 Sex ratio dynamics in samples of pink salmon from (a) Firsovka and (b) Lesnaya rivers (south Sakhalin) during spawning runs in 1974–79. Mean sample size is 50 fish. (a) $y = 70.24 - 1.59x$, r = −0.83, d.f. = 41, $P < 0.001$; (b) $y = 71.21 - 1.28x$, r = −0.68, d.f. = 31, $P < 0.001$.

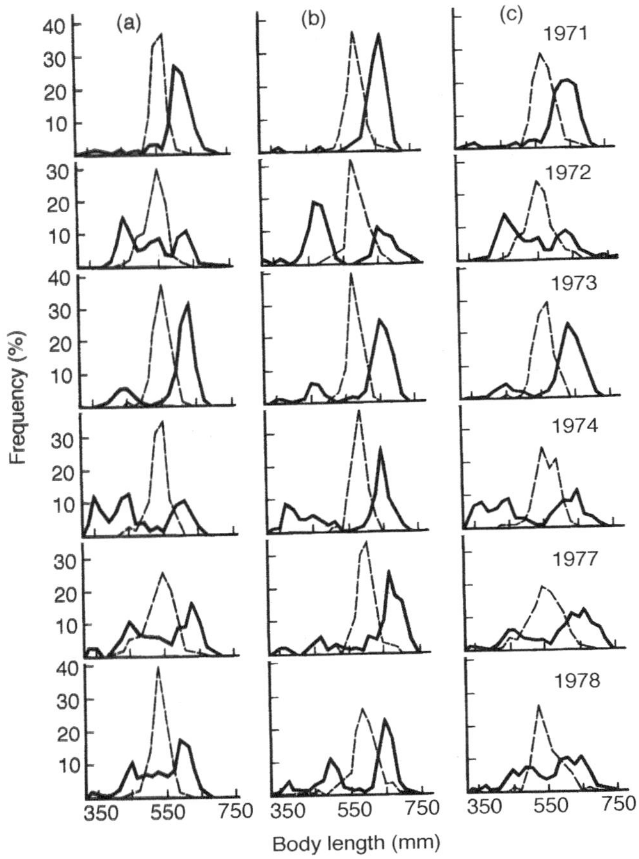

Fig. 2.6 Distribution of body length of males (continuous line) and females (dotted line) in the reproductive part of a population system of sockeye salmon in Lake Azabachye. (a) Spring race; (b) summer race; (c) population system as a whole. Sample sizes: 1971 – 542; 1972 – 654; 1973 – 793; 1974 – 676; 1977 – 800; 1978 – 1090.

1989a; Altukhov & Varnavskaya, 1983). The growth of semelparous salmonids ceases when they reach maturity (Alm, 1959) and hence rapidly maturing fish (as noted above, mainly males) form a cohort of small-size mature individuals. These fish are characterized by the rapid growth at early stages of development in fresh water (Foerster, 1954b; Krogius, 1972, 1978; Saunders *et al.*, 1982; Altukhov & Varnavskaya, 1983; Thorpe, 1986).

Body size as an adaptive trait is related to the type of spawning rivers in the course of natural selection: in large rivers, spawners are usually larger and fatter, which allows them to cover energy expenditures on the way to spawning sites, as they cease to feed in the river (Ricker, 1972; Smirnov, 1975; Beacham, 1984). Body size also depends on the total abundance of salmon stocks feeding in the ocean, including the stocks of closely related species as competitors for food, and on temperature and other climatic factors in the period of sea migration (Birman, 1985; Healey, 1986; Groot & Margolis, 1991), and commercial fishing, especially selective (Ricker, 1981; Altukhov, 1983a, 1989a, 1994a; Altukhov & Varnavskaya, 1983; Birman, 1985; Healey, 1986; Nelson & Soule, 1987; Thorpe, 1993).

2.3.5 *Reproductive ecology and behaviour*

By the time of spawning, salmon, especially males, acquire a specific body colour and shape (breeding dress). The latter is species specific, and this provides for ethological reproductive isolation of salmonids whose spawning ranges usually overlap.

Salmon spawn in rivers and lakes, sometimes in estuaries, burying their fertilized eggs in the sandy-pebble-gravel ground at the sites of upwelling groundwater or underflow (Aro & Shepard, 1967; Smirnov, 1975). Different species and intraspecific groups tend to use spawning sites with specific properties. Thus, chum salmon prefer river sections with pronounced groundwater discharge, which retain temperatures favourable for fish development even in winter. The summer run of Amur chum salmon spawns in rivers and tributaries characterized by underflow, i.e. under conditions also preferred by pink salmon. However, spawning grounds of the latter species are usually located further upstream, although they may overlap occasionally with those of summer chum salmon. According to Smirnov (1975), these typical spawning grounds of pink salmon cover extensive areas with similar hydrological conditions, and populations of large size reproduce in them. In autumn, pink salmon spawn near the sites of groundwater discharge or, sometimes, in springs.

Masu salmon move upstream to spawn at riverheads and in small tributaries. Sockeye salmon may spawn in the upper reaches of large rivers but mainly spawn in the littoral of lakes and in the springs, brooks and streams flowing into these lakes. As a rule, spawning grounds of this species are also characterized by groundwater discharge. Coho salmon spawn in rivers, their tributaries and springs, often at shallow sites with a slow current. Chinook salmon prefer relatively deep sites with a rapid current and large pebbles and stones on the bottom (Smirnov, 1975; Konovalov, 1980).

In Atlantic salmon, spawning is principally similar to that in Pacific salmon but occurs at sites where the water is deeper.

Most salmonids reproduce in late summer and especially in autumn, but late migrating groups of chum, coho and autumn chinook salmon spawn in winter as well. In steelhead and cutthroat trout, spawning usually occurs in late winter and spring (Pavlov, 1989).

Polygamy is typical for salmonids. Females usually lay eggs in batches for several days, and a 'nest group' of one or two leading and several satellite males gathers near the female during each laying. Males also inseminate eggs of several females, moving from one nesting group to another (Hanson & Smith, 1967; McCart, 1969; Simonova, 1975; Chebanov, 1979, 1982; Konovalov, 1980; Helle, 1981; Foote, 1988).

According to Helle (1981), eggs laid by one pink or chum salmon female are sometimes inseminated by more than a dozen males.

Although matings in many salmonids are usually selective, i.e. females prefer larger males, small males also become involved in reproduction under conditions such as spawning at sites with a very low water level, in low-water years, or at certain ratios between size groups of males and females in populations (Chebanov, 1979, 1982; Konovalov, 1980; Altukhov & Varnavskaya, 1983; Parensky & Kovalev, 1983; Gross, 1985; Parensky, 1992).

In masu salmon, Pacific trout, and especially Atlantic salmon, mature male parr actively participate in reproduction (Smirnov, 1975; Zelinsky, 1985; Okazaki, 1986a; Kato, 1991). With respect to the type of sexual maturation, Zelinsky (1985) distinguished in Atlantic salmon three categories of males:

(1) Male parr that die after single or repeated maturation in fresh water
(2) Male parr that mature and spawn in fresh water, migrate to the sea, and return to the river for repeated spawning
(3) Males that reach maturity only after migration to the sea.

Table 2.2 provides information about the existence and certain features of mature parr forms in salmonids.

2.4 Population structure

Information about intraspecific heterogeneity in fish had been accumulating even before the approaches and methods of population genetics were first applied to fish biology. This led to the formation of concepts that species are characteristically subdivided into groups confined to different geographic localities and diverging from one another in a variety of morphological and ecological parameters (Heincke, 1898; Gilbert, 1914–25, cited from Konovalov, 1971; Schmidt, 1917, 1930; for more detail, see Altukhov, 1974). Thereafter, theories concerning the adaptive nature of such differences and their hereditary basis began to develop (see Kirpichnikov, 1933, 1935, 1943; Altukhov, 1969a, 1974; Ricker, 1972). Among different terms referring to intraspecific groups in fish, the most widespread are *local population*, or *stock*; the latter was introduced as early as 1938 at the conference on salmonids in Canada (Anon, 1939, cited from Ricker, 1972). However, at times the term 'stock' is used with different meaning and the situation is complicated further by interactions with political, social and economic factors (for more details, see Carvalho & Hauser, 1995).

It is necessary to discuss here only the most genetically important problem of term usage. The point is that the English term *stock* and its Russian translation equivalent *stado* differ in their meaning. English-speaking authors often apply this term to the smallest self-reproducing population unit, which Russian specialists usually designate as a *subpopulation*, or *elementary population*; in Russian scientific literature, 'stado' refers to the established system of such population units that remain more or less isolated from one another temporally or spatially during the reproductive period (for more detail, see Lebedev, 1967; Altukhov, 1974, 1981a). Probably, for such a community, 'population system' (Altukhov & Rychkov, 1970) or the English term 'metapopulation' (Levins, 1970), corresponding to a subdivided population in the strict genetic sense, would be more adequate. Such a community in spacial coordinates corresponds to a local or geographical population, consisting of many interrelating semi-isolated subpopulations.

The definition of such a fish community was formulated for the first time by K.P. Yanulov in 1962, and we still regard it as appropriate:

> 'Geographical population or *stado* is a group of fish sufficiently segregated with respect to the inhabited area and morphological, anatomical, and biological features. Individuals comprising the population should be united by more or less obvious kinship, which is maintained owing to the exchange of hereditary traits among small constitutive biological units (subpopulations) of the population, on the one hand, and the absence of any significant flow of hereditary traits from other genetic groups, on the other. An important feature of the population is that its abundance (number and biomass of fish) is determined by biogenic and abiogenic environmental factors operating within the boundaries of the population range and, hence, affecting the replenishment of reserves, growth, and mortality in the population' (Yanulov, 1962, pp. 285–6).

This fully applies to isolated subpopulations of salmon which, owing to well-developed homing, spawn repeatedly in the same river or lake and are represented by many reproducing subpopulations.

According to a conservative estimate, the total number of salmon stocks on both coasts of the Pacific Ocean may exceed 10 000 (Ricker, 1972).

Major intraspecific subdivisions are usually related to the geographic regions of spawning or feeding ranges. In Asian pink salmon, for example, stocks feeding in the Sea of Japan, the Pacific Ocean and the Sea of Okhotsk are distinguished (Gritsenko, 1981; Efanov, 1989); Atlantic salmon are subdivided into the West Atlantic, East Atlantic and Baltic groups (Ståhl, 1987). Conspecific salmonids from different river or lake basins differ from one another in many biological parameters (Ricker, 1972; Smirnov, 1975; Dirin, 1980; Birman, 1981; Thorpe, 1986; Beacham *et al*, 1987, 1988; Groot & Margolis, 1991). Many species are represented in the same large basin by different seasonal races, such as the summer and autumn races of Amur chum salmon; the early- and late-spawning runs of pink salmon in some regions; the summer and autumn (winter) races of Atlantic salmon; the spring and summer races of sockeye salmon in many lakes; and the spring, summer and autumn races of chinook salmon (Berg, 1948b; Ivankov, 1967, 1971; Foerster, 1968; Ricker, 1972; Smirnov, 1975; Gritsenko, 1981; Birman, 1981; Konovalov, 1980; Groot & Margolis, 1991).

To date, it has been proven by many researchers that natural local salmon populations are far from being panmictic but represent population systems organized hierarchically (Altukhov, 1971, 1973, 1974, 1981a, 1982 1990; Altukhov & Salmenkova, 1981, 1991; Salmenkova, 1989), and their subpopulation structure is characterized by temporal and spatial orderliness. Using examples of different salmonid species, it is possible to demonstrate that spawning migrations of subpopulation groups from the ocean to rivers and lakes are characterized by the biological dynamics reproduced from generation to generation and from year to year. In chum and pink salmon, for example, males usually prevail in spawning stocks at the onset of migration to fresh waters; in the middle of this period, the sex ratio becomes equal and by the end of migration, females become prevalent (see Fig. 2.5) (Bakkala, 1970; Enyutina, 1972; Altukhov, 1974). In south-eastern Alaska, pink salmon demonstrate three major spawning runs, based on temperature adaptations. They also differ microgeographically: the early runs occur in mainland streams which freeze in winter; the late runs are mostly found in warmer streams on the outer islands; and the middle runs occur in streams mainly in the inner islands with intermediated temperature regimes (Sheridan, 1962). In many sockeye salmon populations, spring and summer races are distinctly separated in time; the onset, middle and end of the run are easily distinguished in each race (Foerster, 1968; Konovalov, 1980; Killick, 1955, cited from Burgner, 1991), and each is represented generally by separate spawning groups. Such a temporal or spatial isolation of populations strongly correlates with their unique biological features. These are unequal sex ratios in subpopulations (with an equal sex ratio in the native population system as a whole; Altukhov, 1974); differences in the physiological condition of fish from different subpopulations; and most importantly, differences in age and size structure. Significant characteristics of respective population systems (abundance, sex–age structure, the time of spawning migration, growth and development rates, the tempo of generation change) remain rather stable in normally fluctuating environments.

The hierarchy of population structure of salmon is specified by the natural history of species ranges and involves many levels, though the detection of them is rather complex, and is not always taken into account. In reality only the sample (i.e. collection) level appears in most studies, while necessary historical and geographical information is either missing or incomplete. As a consequence, complex population structure is not described as a whole or, on the contrary, some samples may consist of fish belonging to different reproducing units. We will illustrate this thesis later (Chapter 4), when showing how important historical, geographical and demographic data are in the study of genetic processes on the microgeographical scale of spatial differentiation of a species.

In terms of macrogeographical scale or population hierarchy of a species as a whole, an example of such a multiple-level organization may be shown for sockeye salmon. If we take a species as a population system, the natural subdivision into spawning population groups, inhabiting American and Asian continents, is obvious.

Within the latter, some regions with river and lake systems may be distinguished, and so on up to discrete subpopulations, coinciding with spawning grounds of a certain type within rivers and lakes and represented by actual samples (Fig. 2.7). For the construction of such a hierarchy, corresponding to the historically formed structure of a species, natural proportionality of the components of a level is important: for example, at the level of regions, we attribute both a group of rather small rivers of western Kamchatka, and basins of distinct large river systems, such as the Kamchatka River, the Fraser River and the Skeena River. Thus, as a first approximation, in sockeye salmon we can separate at least four natural levels of intraspecific hierarchical structure (Fig. 2.7).

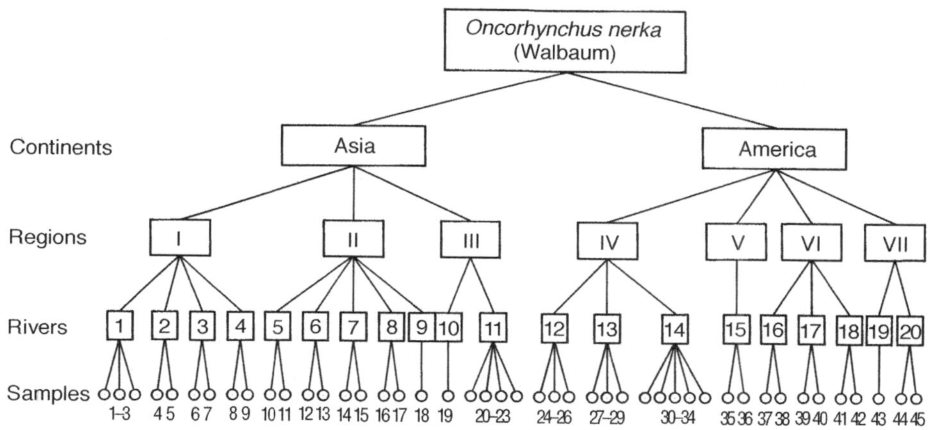

Fig. 2.7 Scheme of hierarchic intraspecies subdivision in sockeye salmon. Regions: I – west Kamchatka, II – Kamchatka River basin, III – east Kamchatka, IV – Cook inlet (Alaska), V – Skeena River basin (Northern British Columbia), VI – Fraser River basin (British Columbia), VII – southern part of American range. Rivers: 1 – Palana (three samples, 1992), 2 – Khairuzova (two samples, 1988, 1992), 3 – Bol'shaya (two samples, 1988, 1990), 4 – Ozernaya (two samples, 1988, 1989), 5 – Kamchatka (two samples, 1988, 1990), 6 – Andrianovka (two samples, 1988, 1990), 7 – Kirganik (two samples, 1988, 1990), 8 – Kimitina (two samples, 1988, 1990), 9 – Kitil'gina (one sample, 1988), 10 - Avacha (one sample, 1987), 11 – Pakhacha (two samples, 1988, and two samples, 1990), 12 – Kenai (three tributaries, 1975), 13 – Susitna (three tributaries, 1975), 14 – Kasilof (five tributaries, 1975, 1976), 15 – Skeena (two samples, 1980, 1981), 16 – Fraser, upper part (two lakes, 1982), 17 – Fraser, middle part (lake and tributary, 1982), 18 – Fraser, lower part (lake and tributary, 1982), 19 – Quinault (one sample, 1982), 20 – Columbia (lake and tributary, 1981). (Data from Grant *et al.*, 1980; Utter *et al.*, 1984; Pustovoit, 1993, 1994).

It should be emphasized that species other than salmon also have a complex population structure, involving several hierarchical levels. Since such levels formed historically, and the subdivided population as a whole is characterized by geographic and ecological unity, such communities were defined by us as population systems (Altukhov & Rychkov, 1970; see details in Altukhov, 1990). At the same time the term 'metapopulation' began to spread in the biological literature, implying also a subdivided population (Levins, 1970). Evidently, any population system with an internal subpopulation structure conforms to some extent to the model of a subdivided population. But not every subdivided population is the system. If we consider an isolated community of elements interrelating non-randomly as a system, then such a community possesses some emergent properties, that do not result from the properties of its structural components. These properties applied to salmon population systems are the stability of their biological and genetic parameters in time, despite significant variability of the composing subpopulations. We will give attention to these effects in the following chapters. In Chapter 4, we will consider the relationships between the concept of population systems and the concept of metapopulations.

2.5 Total, reproductive and genetically effective size of salmon populations

The overall abundance of salmonids can be estimated on the basis of statistical data on commercial catches. According to these data, the total annual catch of Pacific salmon from the mid-1980s to the early 1990s averaged 800 000 metric tons, i.e. was almost twice as great as in the 1950s. In 1995 the catch approached 1 million metric tons and exceeded the previous historical maximum in the late 1930s (Kaeriyama, 1998; Low, 1998). The proportions of five salmon species in the total catch were pink – 34%, chum – 30%, sockeye – 27%, coho – 5% and chinook – 3%. This catch corresponds to more than 400 million fish.

The total number of adult Pacific salmon, including fish escaping capture, should be approximately 1.5–2 times as great as the catches, taking into account that commercial fishing removes 50–80% of fish migrating to spawning sites (Aro & Shepard, 1967; Foerster, 1968; Ricker, 1972). The total number of steelhead trout can be inferred from the data by Sheppard (1972), who estimated the stock of this species on the American Pacific coast to be at 1.5 million individuals. The maximum total catch of Atlantic salmon amounted to 14 000 metric tons (Alexeev, 1988), which corresponded to 2–3 million adult fish.

The size of individual river populations varies from several hundred to several hundred thousand individuals. Aro & Shepard (1967) provided comprehensive data on the size of virtually all spawning populations of Pacific salmonids registered in British Columbia (for example, see Figs 2.8 and 2.9). British Columbia is a vast region located in the centre of the salmon ranges on the American coast; there are more than 1500 rivers, lakes and streams where salmon spawn and, hence, typical features of their populations can be monitored there.

Analysing the detailed maps of the distribution and abundance of the spawning salmon populations over the period from 1951 to 1962 (Aro & Shepard, 1967), we obtained the results shown in Table 2.3. According to them, small populations prevail in all salmonid species, and the most and least abundant populations are those of pink and chinook salmon, respectively.

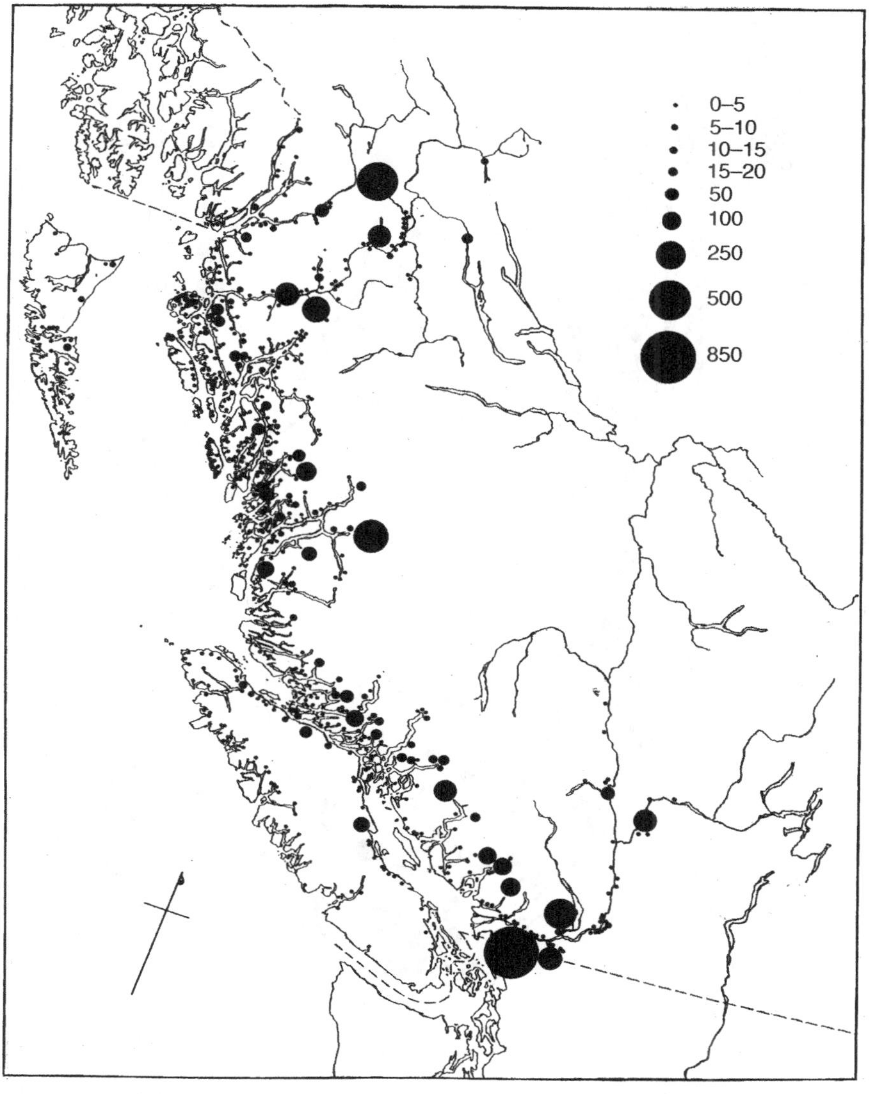

Fig. 2.8 Estimated average annual numbers, in thousands, of pink salmon spawners in British Columbia streams in odd-numbered years from 1951 to 1961 (from Aro & Shepard, 1967)

The above values concern the *reproductive* population size (N_r), i.e. the number of individuals directly involved in reproduction. Of special interest is an estimation of the *genetically effective* population size (N_e), which reflects the efficiency of gene pool transmission to the next generation. This parameter is always smaller than the reproductive size because it is affected by such factors as the sex ratio in the population during the reproductive period, individual variation in fecundity, periodic fluctuations of abundance, etc. As shown in Chapter 1, effective population size

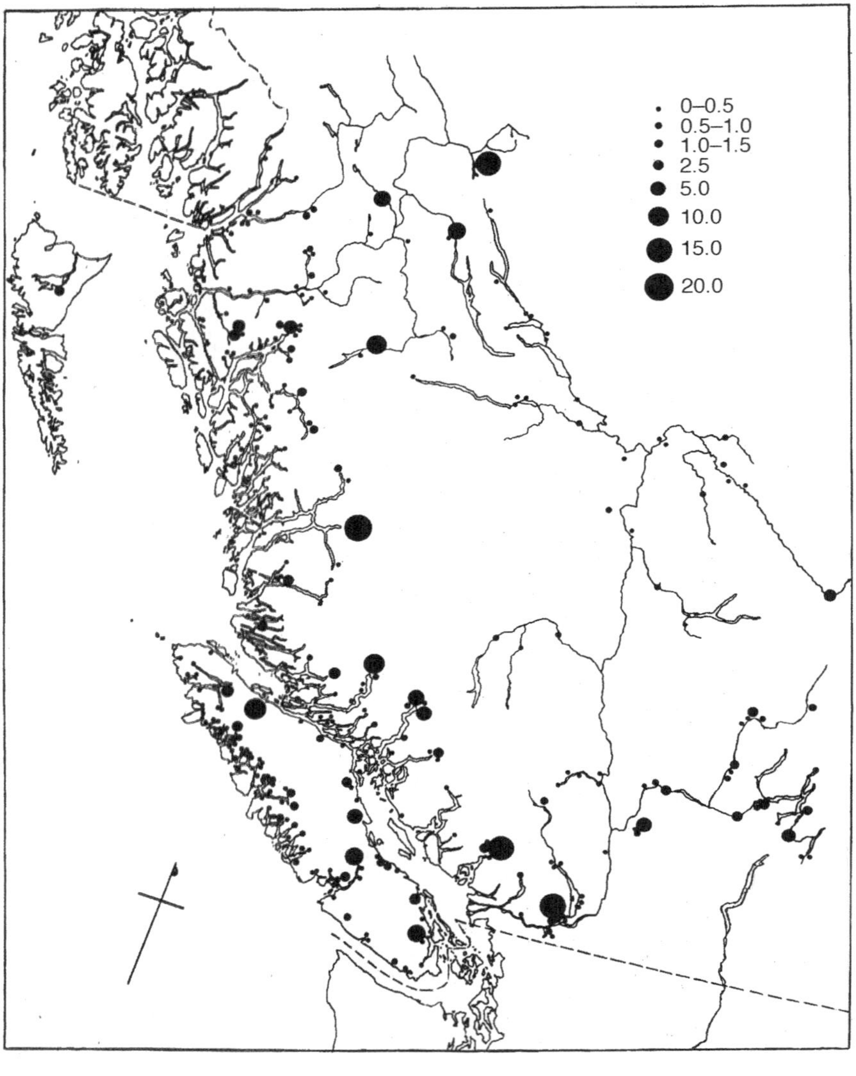

Fig. 2.9 Estimated average annual numbers, in thousands, of chinook salmon spawners in British Columbia streams from 1951 to 1961 (from Aro & Shepard, 1967).

is a significant parameter determining the type and the dynamics of genetic processes in populations.

As to salmonids, the average level of generation overlap in Pacific species is low (see above) and, in some cases, may be ignored while calculating N_e. However, this overlap can be considerable in certain populations containing significant proportions of non-dominant age classes. According to Waples (1990a,b), the effective population size per generation in such cases is equal (in his original notation) to gN_b, where g is a generation length (the average age of maturity), and N_b is the effective

Table 2.3 Population size of Pacific salmon in British Columbia (extracted from data of Aro & Shepard, 1967).

Species	Number of localities observed	Range of population size (thousands)	Proportion of 'small' populations (%)
Pink salmon	1115	<5–850	78.4
Chum salmon	835	<1–75	78.7
Sockeye salmon	298	<1–400	77.2
Coho salmon	857	<1–50	95.8
Chinook salmon	269	<0.5–20	95.2

number of spawners in a given year; this is equivalent to Nei and Imaizumi's formula cited in Chapter 1. Waples (1990a,b), Tajima (1992) and Jorde & Ryman (1995) proposed some statistical approaches for estimating N_b on the basis of temporal variation in allele frequencies in generations of the same population. Jorde & Ryman (1996) used this principle for estimation of N_e in *Salmo trutta.*

To calculate N_e for Atlantic salmon under conditions of generation overlap, L'Abée-Lund (1989) estimated g at 4.4. This author analysed the entire Atlantic salmon population from one of the Norwegian rivers, which totally perished after being treated with rotenone to prevent the spread of the parasite *Gyrodactylus salaris*. N_b and N_e values were estimated taking into account sex ratio and the involvement in reproduction of a large fraction of mature male parr. At the total population size of 2543 individuals, N_b and N_e per generation proved to be 71.7 and 315 individuals, respectively, i.e. $N_e/N_t = 0.124$.

Publications on Pacific salmon provide data on some attempts to estimate the effective population size in hatchery stocks. Simon *et al.* (1986), taking into account sex ratio, mean family size and its variance, estimated N_e for coho salmon at about 150 individuals, which corresponded to approximately 3% of the long-term average number of spawners returning to the Big Creek hatchery (the Columbia River basin). According to Kijima & Fujio (1984), the average N_e value for 37 Japanese hatchery stocks of chum salmon was 949, reaching approximately 35% of the average number of spawners returning to a river. Bartley *et al.* (1992a) used linkage disequilibrium data to estimate N_e (according to Hill (1981)) of some hatchery and natural fish populations: their estimates were equal to 36 for *O. mykiss* in Shasta hatchery, 133 for *O. tshawytscha* in Coleman hatchery, and 85.5 for the threatened winter run of the same species in the upper Sacramento River. These values comprised 89.5%, 1.3% and 4.3% of run sizes, respectively.

We estimated the effective size of wild sockeye salmon subpopulations spawning in Lake Azabachye (Kamchatka) by counting spawners and determining their sex ratio directly at spawning sites. In this case, N_e was approximately 200 individuals, amounting to 10% of the number of spawners per site averaged over several years. The analysis of data on the number of spawners in streams and ponds of the Wood River system (Marriott, 1964) resulted in estimating N_e at 174 individuals, i.e. at about 3.6% of the average size of spawning subpopulations (Altukhov, 1989a; see Chapter 4).

Averaging all these estimates and assuming that the highest values of N_e approach 20% of N_r, we processed the demographic data on five Pacific salmon species from British Columbia, provided in the review by Aro & Shepard (1967). This agrees well with Foerster's (1968) conclusion, that 20–25% of returning sockeye spawners are sufficient to occupy all natural spawning grounds.

Table 2.4 shows the results of this analysis, including the data on the number of populations surveyed and the geometric and harmonic means of their reproductive size (N_e). The values of N_e were calculated as equal to 20% of N_r, i.e. these estimates are conservative. As N_e values for the even and odd generations of pink salmon are virtually the same, we pooled the corresponding data. Moreover, it is seen from Table 2.4 that the highest and the lowest N_e among Pacific salmon species are characteristic of pink and chinook salmon, respectively. Taking into account the great number of spatially separated Pacific salmon populations studied, it is likely that such estimates are representative of the effective size for each of these species.

The ratio of N_e to the arithmetic mean size of a local spawning population is in the interval from 3% for sockeye salmon to 10.9% for pink salmon, with a mean value of 6.5% for five Pacific salmon species. These estimates of effective population size/

Table 2.4 Values of reproductive (N_r) and effective (N_e) sizes of salmon populations (by demographic data, see details in the text).

Species	Number of studied populations	N_r (mean ± SE)		N_e (20% from mean harmonic N_r)	References
		Geometric	Harmonic		
Oncorhynchus gorbuscha (Canada)	1115	3870 ± 22.1	3300 ± 22.2	660	Aro & Shepard, 1967
O. keta (Canada)	835	1392 ± 9.0	850 ± 9.1	170	Aro & Shepard, 1967
O. tshawytscha (Canada)	269	480 ± 5.5	343 ± 5.6	69	Aro & Shepard, 1967
O. nerka (Canada)	298	1809 ± 56.9	1100 ± 57.3	220	Aro & Shepard, 1967
O. kisutch (Canada)	857	800 ± 4.2	630 ± 4.2	126	Aro & Shepard, 1967
Salmo salar (Norway)	17	279	63	52*	Hesthagen & Hansen, 1991

* N_e was calculated from the data on commercial catches in Norwegian rivers, taking into account generation overlap.

adult population size ratio are in keeping with the values obtained by Frankham (1995) for a large sample of species from published data.

For *S. salar*, N_e was calculated from the data on commercial catches in Norwegian rivers, taking the generation overlap into account.

Conclusion

In summarizing the above data, especially on the reproductive biology of salmonids, it is appropriate to emphasize a number of specific features that are especially important in the context of population genetics.

The first is semelparous reproduction in Pacific salmon. Combined with the drastic prevalence of only one age group in the reproductive structure of many stocks, this feature prevents a significant temporal overlap of generations.

The second is the homing instinct, which provides for not only macrogeographic but also microgeographic isolation, down to the level of subpopulations confined to specific types of spawning grounds.

The third is the relatively small effective size of local subpopulations, which does not exclude random genetic drift of moderate intensity in time and space.

The fourth is the interconnection of subpopulations into complex population systems with many hierarchical structural levels.

The following analysis will be conducted in terms of the systemic organization of salmon species. We understand such an organization as a subdivision into many historically formed and relatively independent groups, composed from an assemblage of spawning subpopulations, related by a kinship. The most important property of the system is hierarchical structure, forming naturally (for example, see Fig. 2.7). If we define a system, according to L. von Bertalanfi, as an isolated assemblage of elements interrelating non-randomly, we should expect the manifestation of some emergent properties not resulting from its structural components.

As was shown many years ago, population systems of various biological species, including fish, exhibit significant temporal genetic stability (Altukhov & Rychkov, 1970). This stability is more pronounced the higher the level of population structure hierarchy. Its source is the genetic variability at the level of elementary spawning subpopulations.

As ecological and some theoretical population genetic works show, the higher the level of population hierarchy, the lower the rate of gene flow among populations (see, for example, Konovalov, 1980; Zhivotovsky *et al.*, 1994). It is distinct population systems (local geographic populations) that should be treated as relatively independent objects in artificial propagation, fishing and management. On the contrary, spawning groups, composing the system, are related populations and should be considered as units contributing to the maintenance of ecological and genetic integrity of a population system. Revealing this phenomenon requires the revision of traditional approaches to the exploitation of fish resources and the introduction of a systemic approach to management.

These issues will be considered in detail in the sections dealing with intra- and interpopulation genetic differentiation in salmonids (Chapters 4 to 6). However, the quantitative assessment of this differentiation is possible only using data on genetic polymorphisms in species, which became available through advances in biochemical and molecular genetic research. We review these data in the next chapter.

Chapter 3
Genetic Polymorphism of Proteins and Nucleic Acids

Genetic variation of individuals comprising a population is essentially variation in the structure of individual DNA molecules. This variation partially manifests itself in the primary structure of immediate gene products – proteins – and is detected by biochemical genetic methods. The molecular genetic approach is the basis for the study of heritable variation at the level of DNA itself.

3.1 Proteins

Without focusing attention on the primary role of DNA in the mechanism of heredity (for reviews, see Ayala, 1976; Harris, 1977; Ayala & Kiger, 1984; Lewin, 1997), we should review several basic facts concerning the genetic control of protein synthesis and the nature of heritable variation.

(1) The DNA molecule is the carrier of information about the primary structure of all protein molecules found in an organism.
(2) The amino acid sequence of proteins directly reflects the nucleotide sequence of the corresponding gene (DNA region or gene locus).
(3) One structural gene (cistron) usually codes for the synthesis of one polypeptide chain.
(4) Many proteins with a quaternary structure (e.g. haemoglobin or various enzymes) consist of several polypeptide chains, or subunits, encoded by different gene loci.
(5) Certain mutational changes in the nucleotide sequence of a gene can change the amino acid composition of a corresponding protein, and may be detected by changes in the properties of a protein molecule.

Individual differences at the protein level are analysed both immunologically and by electrophoretic methods. In the 1950s and 1960s, immunological studies on fish revealed individual variation in erythrocyte antigens (blood groups). On this basis, it became possible to reveal discrete molecular genetic differences between individuals and by analysing the ratio of individuals belonging to different genetic types, between populations or their components (see Altukhov *et al.*, 1964, 1969a, 1969b; Altukhov, 1969a, 1974; de Ligny, 1969; Hodgins, 1972; Utter, 1991). During the 1960s, however, immunological tests in population genetic research were superseded by electrophoretic methods, in which genetically determined molecular variants of the same protein were separated on the basis of differences in their electric charge and molecular weight. Electrophoresis was performed in carrier gels (agar, starch, polyacrylamide), which were then stained histochemically to determine enzymatic activity or other specific properties of protein molecules. The resulting gel showing histochemically stained bands at the electrophoretic position of respective enzymes is referred to generally as a zymogram (Markert & Möller, 1959). This method proved to be more simple and convenient because it allowed researchers to separate variants of proteins, including enzymes, directly from biological fluids or tissue extracts of individual organisms and, thus, to reveal genetic polymorphism in any biological species. As a rule, variants of proteins occurring in individuals from the same population are inherited in strict conformity with Mendel's laws, and the inheritance is usually codominant, i.e. both alleles are active in the heterozygous state. Apparently, such variation falls within the definition of genetic polymorphism as formulated by Ford (1940), who interpreted it as

> 'the occurrence together in the same locality of two or more discontinuous forms of a species in such proportions that the rarest of them cannot be maintained merely by recurrent mutations'.

A number of structural genes of higher organisms, primarily the loci responsible for the synthesis of proteins with enzymatic activity, can be defined as multiple cistrons. The multiplicity of structural genes is reflected in the multiplicity of proteins synthesized under their control, and this phenomenon is usually possible to detect electrophoretically.

The genetically determined molecular forms of an enzyme that differ in their primary structure are called *isozymes* (Hunter & Markert, 1957; Markert & Möller, 1959; Korochkin *et al.*, 1977). Isozymes that are encoded by alleles of the same gene and reflect intraspecific polymorphism are called as *allozymes* (Prakash *et al.*, 1969).

Genes encoding isozymes usually demonstrate differential expression in tissues. When several loci function in the same tissue at the same developmental stage, and the protein they encode consists of several subunits (i.e. has a quaternary structure), such isozyme systems produce 'hybrid' molecules readily identifiable electrophoretically.

For example, in many animal species the isozymes of malate dehydrogenase (dimer) and lactate dehydrogenase (tetramer) are encoded in the same tissue by at least two independent loci, and, hence, electrophoretic analysis reveals three and five zones of enzyme activity, which reflect the association of two subunits into dimeric and tetrameric molecules, respectively. If a mutation in any of the loci affects the total electric charge, size or shape of the molecule, the spectrum of isozymes broadens: the genetic interpretation of such variation in codominantly inherited traits is, however, not difficult (Fig. 3.1). Some problems may arise when changes in several loci controlling a given enzyme synthesis in the same tissue occur simultaneously, and some protein zones have similar electrophoretic mobilities. In such cases, it is necessary to analyse other tissues in which these gene loci differ in their expression. The main difficulty is sometimes associated with the presence of so-called null (silent) alleles, i.e. the cases when the protein subunit either is not synthesized at all or has no enzymatic activity. However, it is apparent that the effect of such a gene can be reliably identified in the homozygous state by the absence of the corresponding zone of enzymatic activity in zymograms. In the heterozygous state, this allele can be revealed only in cases of incomplete dominance (codominance) of the normal allele (the dose effect).

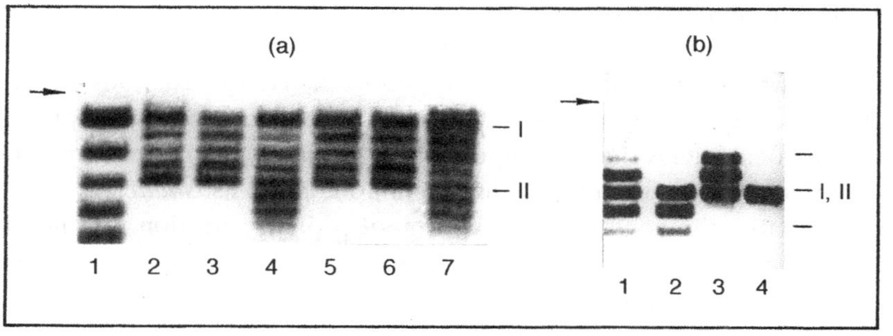

Fig. 3.1 Electrophoretic variants of blood serum lactate dehydrogenase (a) in sockeye salmon and (b) in muscle malate dehydrogenase in pink salmon. (a) I – *LDH-B1** locus; II – *LDH-B2** locus; polymorphism is caused by allele variants of *LDH-B2** locus; genotypes: 1 – 120/120; 2, 3, 5, 6 – 100/100; 4, 7 – 100/120; (b) *MDH-B1,2** isoloci, genotypes: 1 – 75/127; 2 – 100/127; 3 – 100/75; 4 – 100/100. Electrophoresis in polyacrylamide gel; buffer by Peacock *et al.*, 1965. Arrows indicate the sample origin.

The first study demonstrating hereditary biochemical polymorphism of salmonids was published in 1969 (Hodgins *et al.*, 1969). Its authors analysed the activity of lactate dehydrogenase (LDH) in sockeye salmon blood serum using starch gel electrophoresis and corresponding histochemical staining of gel. Each individual fish proved to have one of the three LDH phenotypes revealed in the sample, which the authors attributed to the presence of two alleles at the *LDH-B** locus. Analysis of several samples from different geographic localities showed that they differed in allele frequencies.

However, direct confirmation of inheritance of allelic forms was not undertaken, and genetic information was based on the general principles of biochemical genetics and analogies with other animal species. Since that time, hundreds of species belonging to all the main taxa have been involved in biochemical genetic research, and dozens of polymorphic genetic systems were discovered and characterized to a greater or lesser degree in the most extensively studied species (Altukhov, 1974,

1983a, 1989a; Korochkin *et al.*, 1977; Nevo *et al.*, 1984; Ward *et al.*, 1994), including salmonids (Salmenkova, 1973; May, 1980; Shaklee *et al.*, 1990b; Jorde, 1994).

Because of its tetraploid origin, the genome of salmonids is duplicated, and this is reflected in an increased multiplicity of protein and isozyme systems in these fish. Studies have shown that in the course of evolution from a common ancestor, salmonids have lost approximately 50% of their duplicated genes (Allendorf *et al.*, 1975; Allendorf & Thorgaard, 1984), i.e. the products of these genes are impossible to reveal. Other duplicated genes have diverged to an extent that their protein products, retaining specific functions (e.g. enzymatic activity), acquired structural features allowing their separation by electrophoretic methods. However, some duplicated loci did not demonstrate such a divergence, and their products are impossible to separate by means of electrophoresis.

In rainbow trout, for example, duplication was not revealed at 10 out of 33 protein loci; 15 loci retained expression typical for duplicates and became completely 'diploidized', i.e. with a demonstration of disomic segregation, and 8 loci, retaining duplication, without any divergence (Allendorf & Thorgaard, 1984). Loci of the latter type, i.e. those retained in the course of diploidization and represented by virtually identical alleles that are impossible to definitely attribute to any of the two loci, are referred to as *isoloci*, and their common alleles as *isoalleles*.

For information on genetic interpretation of protein variation revealed by electrophoretic methods we refer the reader to the paper by Utter *et al.* (1987). Here, we present typical electrophoretic patterns (for salmonids) of protein variants encoded by different genes, from our own studies (Figs 3.1 and 3.2).

Let us consider the most recent nomenclature of genes encoding fish proteins, proposed by the Genetic Nomenclature Committee of the American Fisheries Society (Shaklee *et al.*, 1990a). This nomenclature is now generally accepted and will be used in this book.

Fig. 3.2 Examples of enzyme system polymorphism in salmonid fishes. (a) Muscle phosphoglucomutase of sockeye salmon, monomer coded by *PGM-2** locus with two alleles; genotypes: 1, 7 – 100/135; 2–5 – 100/100; 6 – 135/135 (starch gel, buffer by Poulick, 1957). (b) 6-Phosphogluconate dehydrogenase of pink salmon in erythrocytes or muscle; dimer coded by *PGDH** locus with four alleles; genotypes: 1, 3, 4, 6, 8 – 100/100; 2 – 100/94; 5 – 100/89; 7 – 100/106 (polyacrylamide gel, buffer by Peacock *et al.*, 1965). (c) Heart phosphoglucose isomerase of pink salmon, dimer coded by two polymorphic *PGI(GPI)-A** (II) and *PGI-B** (I) loci; genotypes: 1 – 100/100/100/60; 2, 5 – 100/100/100/100; 4, 6 – 100/110/100/100; 3 – 100/80/100/100 (same method as (b)). (d) Muscle aspartate aminotransferase of chum salmon; dimer coded by *AAT-1,2** isoloci; genotypes: 2, 4, 5, 7 – 100/100/100/100; 1, 3 – 100/100/100/118; 6 – 100/118/100/118 or 118/118/100/100 (same method as (b)). (e) Muscle lactate dehydrogenase of chum salmon; tetramer coded by *LDH-A1** (I) locus with two alleles and monomorphic *LDH-A2** (II), *LDH-B1** (III), *LDH-B2** (IV) loci; genotypes: 1, 8, 10 – 100/250; 2 – 4, 6, 7, 9 – 100/100; 5 – 250/250 (same method as (b)). Because of overlapping of some bands in heterozygotes, there are only eight to nine, instead of 15, bands. (f) Liver isocitrate dehydrogenase of chum salmon, dimer coded by two loci, monomorphic *sIDHP-1** (I) and more active polymorphic *sIDHP-2** (II), with four alleles; genotypes: 1, 3, 7, 10, 11 – 100/43; 2 – 43/25; 4, 9, 13 – 100/25; 5, 6, 8, 12 – 100/100; 14 – 43/43; 15 – 100/83; 16 – 43/83 (starch gel, buffer by Cross & Payne, 1971).

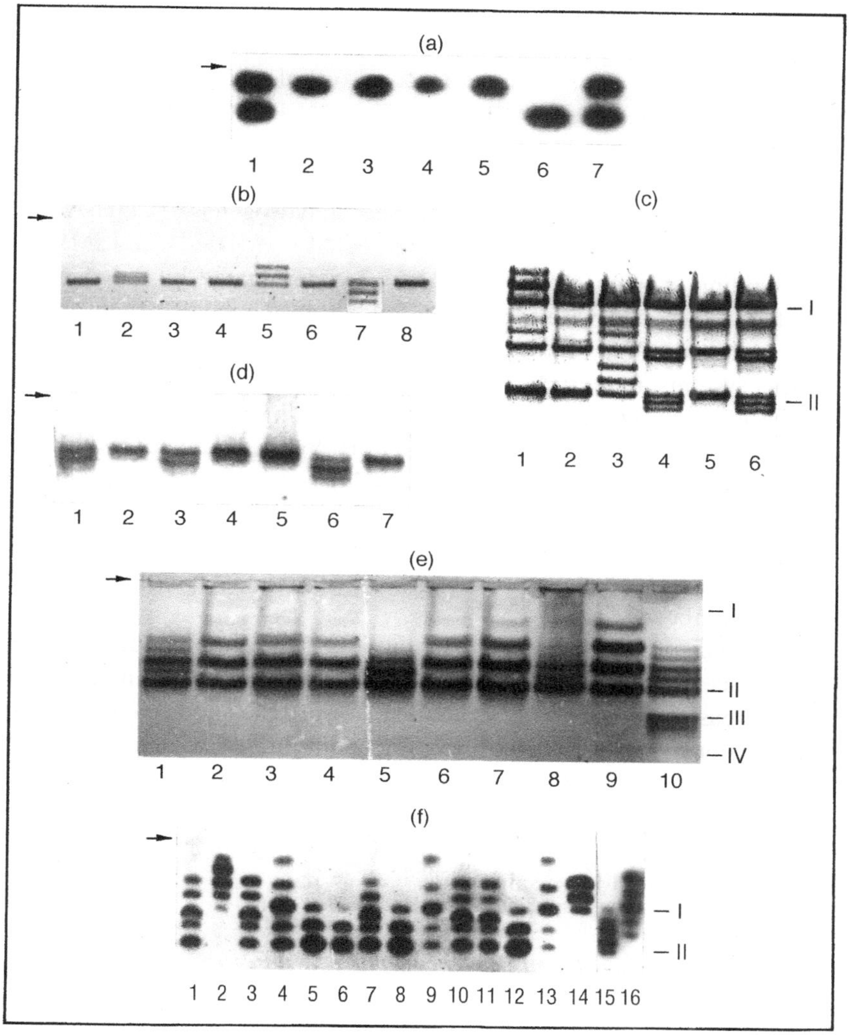

As genes are designated in the same way as the proteins (enzymes) they encode, let us begin with the latter. Enzymes are designated by abbreviations in capitals (LDH, PGM, ADA, etc.) recommended by the Nomenclature Committee of the International Union of Biochemistry (IUBNC; 1984). If the intracellular localization of an enzyme is known, the abbreviation is preceded by a lower-case letter, *s* for cytoplasmic and *m* for mitochondrial enzymes (e.g. *s*AAT, *m*AAT). To designate multilocus enzymes (isozymes), a numeral or a capital letter is added, e.g. *s*IDHP-1 and *s*IDHP-2 or LDH-A and LDH-B. Letters are used only when *orthology* with corresponding enzymes of other taxa is proven. The terms *orthology* and *paralogy* are used to describe molecular homology: the former reflects interspecific homology (e.g. genes encoding the LDH-A subunit in different fish species), whereas the latter reflects common evolutionary origin (e.g. LDH-A, LDH-B and LDH-C subunits in the same species). If no evidence that the loci are orthologous is available, numerals are used in designations (e.g. MEP-1, MEP-2, MEP-3) in order of increasing mobility to the anode in gel. To date, orthology has been established for four isozyme systems: creatine kinase (CK-A, CK-B, CK-C and CK-D); phosphoglucose isomerase (PGI-A, PGI-B and PGI-C); lactate dehydrogenase (LDH-A, LDH-B and LDH-C); and cytoplasmic malate dehydrogenase (*s*MDH-A and *s*MDH-B).

Genes encoding enzymes or their subunits are designated by the same abbreviation, but in italics, and with an asterisk added, to distinguish them from their products: for example, *MPI** is the locus encoding mannose-6-phosphate isomerase. In multilocus systems, numerical designations are used, such as *LDH-A1**, *LDH-A2**; *LDH-B1** and *LDH-B2**; or, for isoloci of salmonids, *sAAT-1,2**, *sMDH-A1.2**, and *sMDH-B1.2**. To indicate regulatory loci, *r* is added as a suffix, e.g. *PGM-1r**. Alleles are designated according to the electrophoretic mobility of their products related to that of the most frequent allele taken as 100, e.g. alleles *LDH-A1* 100* and *LDH-A1*140* or, in the corresponding context, simply alleles **100* and **140*. If the product of the allele migrates to the cathode, its mobility is designated by a negative number: −**100*, −**80*.

For convenience of taxonomic comparison, it is also recommended to designate alleles by the ascending series of literal or numerical indices, beginning with *A* and *1*, in order of increasing electrophoretic mobility (toward the anode) of their products.

Designations proposed for alleles encoding quantitative variants of products are as follows: **Q0*, no product (activity), null allele; **QL*, a decreased product concentration (activity); and **QE*, an increased product concentration (activity).

Below, we describe the most extensively studied polymorphic isozyme systems and their corresponding loci in salmonids.

3.1.1 Enzymes with tetrameric quaternary structure

Lactate dehydrogenase

Lactate dehydrogenase (LDH, EC 1.1.1.27) of salmonids represents a complex system comprising several series of isozymes with well-defined tissue specificity. As noted above, the majority of researchers attribute this multiplicity to the tetraploid origin of salmonids and, hence, to gene duplication. In mammals, birds and many other animals, LDH isozymes are encoded by three loci; in salmonids, some authors initially proposed the existence of more than seven LDH loci (Massaro & Markert, 1968; Williscroft & Tsuyuki, 1970), but subsequent studies showed that this isozyme system is actually encoded by five loci (Sachko, 1973; Allendorf *et al.*, 1975; Bailey *et al.*, 1976; Lim & Bailey, 1977. Each locus controls the synthesis of the corresponding polypeptide subunit, and the assembly of the latter into homo- and heterotetramers of different composition produces a complex array of LDH isozyme molecules.

Figure 3.3 shows the electrophoretic spectrum of LDH isozymes from tissues of three different sockeye salmon individuals.

According to recommendations for the nomenclature of loci encoding fish proteins (Shaklee *et al.*, 1990a), the loci of LDH from muscle tissue were designated as *LDH-A1** and *LDH-A2**; of LDH from the heart, liver and eye tissues and erythrocytes as *LDH-B1** and *LDH-B2**; and of specific LDH from the eye as *LDH-C**.

The results of several studies showed that, in terms of kinetic, biochemical and immunological properties, salmon LDH isozymes of the A series are homologous to the LDH isozyme from mammalian and bird muscle tissue (M_4), and those of the B series are homologous to the isozyme from the heart (H_4). Moreover, homology of isozymes within the same series proved to be significantly higher than that between the series (Bailey & Wilson, 1968; Massaro & Markert, 1968). These facts show that the series of LDH isozymes in salmonids originated via duplication of 'muscle' and 'heart' isozyme loci during polyploidization.

Remarkably, tetraploidy resulted in fixation of apparently constant heterozygosity at the loci *LDH-A1**, *-A2**, *-B1** and *-B2** in most salmonids, whereas the locus *LDH-C** is an exception. The nature of its polymorphism, revealed in brown trout, provides no evidence for its duplication (e.g. see Guyomard & Krieg, 1983; Taggart & Ferguson, 1984b, 1986).

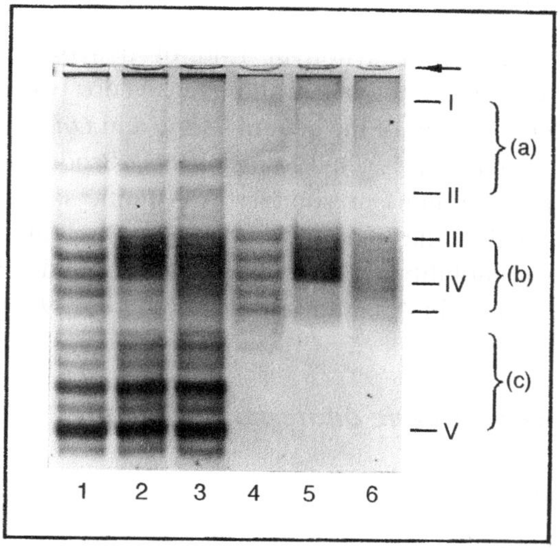

Fig. 3.3 Lactate dehydrogenase isozymes in the eye (1–3) and gut (4–6) of sockeye salmon. Isozymes of the slowest series (a) are coded by *LDH-A1** (I) and *LDH-A2** (II) loci; in the eye only two slow isozymes are visible. Isozymes of series (b) are coded by *LDH-B1** (III) and *LDH-B2** (IV) loci; the latter locus has a fast allelic variant visible in both tissues. The series of iozymes of highest electrophoretic mobility (c) in eye tissue is formed by the interaction of subunits coded by *LDH-C** (V), *LDH-B1** and *LDH-A2** loci (from Sachko, 1973).

It should be noted here that the series of B isozymes in masu salmon, *Oncorhynchus masou*, is represented by only one B_4 isozyme, and the series of A isozymes in white-spotted charr, *Salvelinus leucomaenis*, by only A_4 isozyme. Apparently, either the products of duplicated genes *LDH-B1** and *LDH-B2** in masu salmon and *LDH-A1** and *LDH-A2** in white-spotted charr do not differ in electric charge, or one of the genes in each pair is inactive (Sachko, 1973; Utter *et al.*, 1973).

Malic enzyme

Malic enzyme, or NADP-dependent malate dehydrogenase (MEP, EC 1.1.1.40) has been studied in many salmonid species (Allendorf *et al.*, 1977; Cross *et al.*, 1979; Stoneking *et al.*, 1979; Kijima & Fujio, 1980; May, 1980; Salmenkova & Omelchenko, 1983a; Okazaki, 1986b; Ermolenko *et al.*, 1988; Semenova & Slynko, 1988; Shaklee *et al.*, 1990b). The results showed that the MEP system in salmonids is represented by two forms of the enzyme: the soluble (cytoplasmic) *s* form and the mitochondrial *m* form. In chum and pink salmon, and other species studied more extensively in this respect, the liver, kidney and retina contain only the *s* form of MEP, whereas the heart and skeletal muscles contain both forms. The type of polymorphism in the *m*MEP system shows that it comprises products of two loci, designated *mMEP-1** and *mMEP-2**, which may differ markedly in their activity.

Sorbitol dehydrogenase

Sorbitol dehydrogenase (SDH; synonym L-iditol dehydrogenase, IDDH; EC 1.1.1.14) demonstrates highest activity in the liver, where two loci coding for this enzyme are expressed in species of the genera *Salmo* and *Oncorhynchus* (Allendorf *et al.*, 1975; May, 1980; Okazaki, 1986b; Elo *et al.*, 1994). Products of the two loci may have the same (in pink salmon) or similar (in Atlantic salmon) electrophoretic mobility, forming a single zone of activity in the gel. In chum and masu salmon, these products differ in mobility and form five isozyme bands. Polymorphism may be observed at both loci, as in Atlantic salmon (Elo *et al.*, 1994).

3.1.2 Enzymes with dimeric quaternary structure

Alanine aminotransferase

Alanine aminotransferase (ALAT; synonym glutamic-pyruvic transaminase, GPT; EC 2.6.1.2) polymorphism was studied among salmonids for the first time in sockeye salmon (Grant *et al.*, 1980). One of the loci proved to be polymorphic, with four alleles. To date, polymorphism at this locus, which is expressed in skeletal and heart muscles, has been revealed in pink salmon (Shaklee *et al.*, 1991) and chum salmon (Kondzela *et al.*, 1994; Salmenkova *et al.*, 1994).

Aspartate aminotransferase

The aspartate aminotransferase (AAT; synonym glutamic oxaloacetic transaminase, GOT; EC 2.6.1.1) isozyme system, like the MDH system, includes cytoplasmic (*s*) and mitochondrial (*m*) forms, each represented by tissue-specific products of several loci (May, 1980; May *et al.*, 1980; Shaklee *et al.*, 1990b). The heart and skeletal muscles of salmonid species demonstrate the presence of the isolocus pair *sAAT-1,2** with the pattern of polymorphism characteristic of isoloci. Specific loci are also found in the eye (*sAAT-3**) and the liver (*sAAT-4**). In several species, the same tissues (except for the eye) contain additional 'slower' isozymes, which are products of mitochondrial loci (Shaklee *et al.*, 1990b). In crossings, pseudolinkage of the loci *sAAT-1** and *sAAT-2** has been demonstrated (Allendorf & Utter, 1976; Wright *et al.*, 1980) (see section 3.1.4 below).

In chum salmon, cutthroat trout and brook charr, experiments on inheritance revealed polymorphism at each *sAAT-1,2** isolocus (May *et al.*, 1975).

Creatine phosphokinase

Creatine phosphokinase (synonym creatine kinase, CK; EC 2.7.3.2) accounts for a considerable proportion of the total soluble protein in skeletal muscles. The system of CK isozymes is controlled by several loci. Two of them, *CK-A1** and *CK-A2**, are expressed in muscles, and up to three loci *CK-B**, *CK-C1** and *CK-C2** were revealed in eye tissues (Allendorf *et al.*, 1977; Utter *et al.*, 1979; Cross & Ward, 1980; May, 1980; Taggart *et al.*, 1981; Taggart & Ferguson, 1984b; Shaklee & Varnavskaya, 1994; Winans *et al.*, 1994).

An analysis of the CK isozyme systems in ten different salmonid species and experiments on crossing (Utter *et al.*, 1979) showed that neither the products of different loci nor the allelic products of the same locus form heterodimers. The product of each allele of the *CK-A1** and *CK-A2** loci forms two bands on the zymogram, which are probably accounted for by stable posttranslational modification of their polypeptide chains. In *Oncorhynchus masou* and species of the genus *Salmo*, one of the two bands corresponding to each locus has the same electrophoretic mobility as its counterpart, and the CK phenotype is represented by three bands; other *Oncorhynchus* species have a four-banded phenotype. In *Salmo trutta*, for example, allelic variants of the locus *CK-2** differ only in the ratio between activities of different bands (Allendorf *et al.*, 1977; Taggart & Ferguson, 1984b; Semenova & Slynko, 1988).

Esterase D

Esterase D (ESTD, EC 3.1.1.1) is revealed by its ability to break the ester bond in an artificial substrate (usually 4-methylumbelliferyl acetate), which results in the release of a fluorescent product. This enzyme is active in skeletal muscles, liver and blood serum. It is encoded by one non-duplicated locus, which is polymorphic in many salmonid species, or, possibly, by two loci (Ermolenko & Rudminaitis, 1984; Salmenkova *et al.*, 1986; Semenova & Slynko, 1988; Shaklee *et al.*, 1990b).

Glucose-6-phosphate isomerase

Glucose-6-phosphate isomerase (GPI; synonym phosphoglucose isomerase, PGI; EC 5.3.1.9) is encoded by three loci in all salmonids (Allendorf *et al.*, 1977; Donelly *et al.*, 1979; May *et al.*, 1979b; Grant *et al.*, 1980; May, 1980; Okazaki, 1982b, 1986b; Shaklee *et al.*, 1990b). The isozyme controlled by the locus *GPI-3** (synonym *GPI-A**; in our work on pink salmon designated *GPI-2**, see Kartavtsev *et al.*, 1990), which demonstrates the highest mobility in gel, is active in the liver, heart, muscle and eye tissues. Two other loci, encoding isozymes with a lower mobility (GPI-B, mainly in muscle), either are represented by the isolocus pair *GPI-B1.2**, as in pink salmon (Shaklee *et al.*, 1990b), or differ in their products, as in masu salmon (Okazaki, 1986b). Subunits controlled by all three loci interact and form heterodimers. Shaklee *et al.* (1990b) suggested that variation in each locus of the *GPI-B1,2** pair could be analysed by examining polymorphism in the heart and muscles simultaneously. The results showed that this variation in muscles is typical for isoloci, i.e. allozyme activities in heterozygous variants were in the ratios corresponding to one, two or three doses of the alternative alleles (approximately 9 : 6 : 1, 1 : 2 : 1, and 1 : 6 : 9). In the heart of the same individuals, the allozyme activities in some of these heterozygous variants were either at a ratio of 1 : 2 : 1 or other heterozygous variants were absent. Such a situation is only possible when one of the two isoloci is expressed in the heart; knowing its variation and the overall variation of both isoloci in muscle, it is possible to determine to which individual isolocus each variant belongs.

In chinook salmon, the locus *GPI-1** (or possibly *GPI-3**) proves to contain an allele that is not associated with changes in the mobility of its respective protein molecule but rather blocks the interaction between subunits of the loci *GPI-1** and *GP1-3** (Utter *et al.*, 1989).

Glutathione reductase

Glutathione reductase (GR, GSR; EC 1.6.4.2) is active in muscle and eye tissues and is probably encoded by one locus. GR polymorphism typical for dimeric molecules was described in chinook salmon (Shaklee *et al.*, 1990b, 1991).

α-Glycerophosphate dehydrogenase

α-Glycerophosphate dehydrogenase (G3PDH, EC 1.1.1.8) of different salmonid species is encoded by two or three loci (Engel *et al.*, 1971; Aspinwall, 1973; Allendorf *et al.*, 1977; Slynko, 1978; Donelly *et al.*, 1979; May *et al.*, 1979b; May, 1980;

Okazaki, 1982b, 1986b; Salmenkova & Omelchenko, 1983b; Semenova & Slynko, 1988). The most active locus in muscle is *G3PDH-1**, in the heart, loci *G3PDH-2** and *G3PDH-3** are sufficiently active to reveal polymorphism. Note that polymorphism at loci 1 and 2 is common for many species, and their pattern and segregation in the progeny from individual crosses demonstrate the absence of duplication at these loci.

Isocitrate dehydrogenase

Isocitrate dehydrogenase (IDHP, EC 1.1.142), as with MDH and AAT, has mitochondrial and cytoplasmic forms. The electrophoretically 'slower' *m*IDHP is mainly found in the heart and skeletal muscles and in the eye, where it is the least active. Relatively fast-moving *s*IDHP prevails in the liver, kidney, eye and stomach or pyloric caeca; it also occurs in other tissues but in low concentrations (Cross & Payne, 1971; Allendorf *et al.*, 1975, 1977; Kijima & Fujio, 1977; Reinitz, 1977; May *et al.*, 1979b; May, 1980; Salmenkova & Omelchenko, 1983a; Gagalchy, 1988; Shaklee & Phelps, 1992). The analysis of these and other studies shows that each form is encoded for by two loci. As a rule, the loci *mIDHP-1** and *mIDHP-2** encode subunits that differ in mobility and form three isozymes. In some cases, however, the activity of one subunit typically dominates, and only two zones of activity remain in zymograms (e.g. in muscles of chum salmon or rainbow trout).

In some species (e.g. chum salmon), the expression of *sIDHP-1** and *sIDHP-2** in the liver is also manifested in a 'three-band' phenotype resembling that of *m*IDHP (Altukhov *et al.*, 1980). In other species, such as chinook salmon, two loci apparently produce subunits of similar electric charge, as follows from the occurrence of heterozygous variants in which activities of liver *s*IDHP bands have a characteristic ratio of 9: 6: 1. In other words, these loci share the same common allele, i.e. they are isoloci *sIDHP-1,2** (formerly, they were designated *IDHP-3,4** or *IDHP-3**) (Shaklee & Phelps, 1992). It is interesting that some heterozygous variants observed in the chinook liver (or eye) were also revealed in the muscle tissue of the same fishes, but the ratio of activities in the latter case was 1 : 2: 1. This means that only one isolocus is active in muscle cells, and, hence, the simultaneous analysis of different tissues provides the possibility of estimating the variability of each isolocus separately.

Malate dehydrogenase

Malate dehydrogenase (MDH, EC 1.1.1.37) as an isozyme system has been studied in several salmonid species (Bailey *et al.*, 1969; Slynko, 1971, 1976; Altukhov *et al.*, 1972; Aspinwall, 1974b; Allendorf *et al.*, 1977; May *et al.*, 1979b). This enzyme has two forms: a slowly migrating mitochondrial *m*MDH and a faster migrating cytoplasmic *s*MDH, which are controlled by independent loci whose products do not interact. In turn, *s*MDH is usually represented by isozyme A, which is less mobile and prevails in the liver, brain and eye tissues; the more mobile isozyme B form which prevails in skeletal muscles; and the heterodimeric AB form, which is the product of their interaction. The genetic analysis of this polymorphism and the ratio of isozyme activities in heterozygotes showed that isozymes AA and BB are encoded by independent loci (Bailey *et al.*, 1970). It is known that in pink, sockeye and chinook salmon and in rainbow and cutthroat trout both these loci are duplicated; moreover, the duplicate loci are fixed for the same allele, i.e. they are isoloci: *sMDH-A1,2** and *sMDH-B1,2** (previous designations *MDH-1,2* and *MDH-3,4*, respectively) (Aspinwall, 1974b; Allendorf, 1975; Slynko, 1976; May, 1980; Salmenkova & Omelchenko, 1983b; Shaklee & Phelps, 1990). In brown trout and Atlantic salmon, the loci *sMDH-A1** and *sMDH-A2** are fixed at different alleles (Allendorf *et al.*, 1977; Cross & Ward, 1980). The analysis of crosses in pink salmon provided evidence for polymorphism at each of the *sMDH-B1,2** isoloci (our data). In zymograms it is sometimes difficult to distinguish between variation at loci controlling isozymes A and B. In such cases, it is recommended that electrophoresis be undertaken of the same samples in different buffer systems, choosing the optimal conditions for separating the corresponding protein variants (Allendorf, 1975).

Peptidases

Peptidases (PEP, EC 3.4.—. —.) used in the allozyme analysis of salmonids cleave dipeptides glycyl–leucine (peptidases A and C), phenylalanyl–proline (peptidase D), and leucyl–tyrosine (peptidase LT), and tripeptide leucyl–glycyl–glycine (peptidase B). Polymorphic variants provide evidence that peptidase molecules are dimeric. All these molecules are controlled by independent loci, and the number of loci for certain peptidases in different species ranges from one to three (Utter *et al.*, 1978; Okazaki, 1982b; Beacham *et al.*, 1985a,b; Varnavskaya, 1992; Bartley *et al.*, 1992b,c). The peptidase cleaving the Gly–Leu dipeptide is encoded by at least two loci: *PEPA**, which is expressed in all the tissues studied (skeletal and heart muscle, liver, eye), and *PEPC**, with expression confined to the eye. The peptidase controlled by the locus *PEP-LT**, is mainly active in skeletal and heart muscle. *PEPB* and *PEPD* loci are active in different tissues, including eye tissues. Virtually all salmonid species exhibit polymorphism of some genes encoding peptidases.

6-Phosphogluconate dehydrogenase

6-Phosphogluconate dehydrogenase (PGDH, EC 1.1.1.44) is encoded in salmonids by one non-duplicated locus. This enzyme is active in erythrocytes, muscles, heart and liver, and its polymorphism is observed in many salmonid species (Utter *et al.*, 1978; Donelly *et al.*, 1979; May, 1980; Salmenkova & Omelchenko, 1982, 1983b; Okazaki, 1982b, 1986b).

Superoxide dismutase

Superoxide dismutase (SOD; synonym tetrazolium oxidase; EC 1.15.1.1) is encoded by at least two loci, *mSOD** and *sSOD**, controlling the synthesis of mitochondrial and cytoplasmic forms of the enzyme, respectively, which have no distinct tissue specificity in most of the species studied (Utter *et al.*, 1976, 1984; Allendorf *et al.*, 1977; May, 1980; Okazaki, 1982b, 1986b; Shaklee *et al.*, 1990b). Some sympatric salmonid species, e.g. chum and pink salmon, differ markedly in the electrophoretic pattern of this enzyme. SOD is often revealed spontaneously in gels stained for many dehydrogenases, and this property can be used during electrophoretic analysis of fry or young fish for determining the species to which they belong, as visual determination is sometimes difficult.

Triose phosphate isomerase

To date, up to four genes have been found to encode the triose phosphate isomerase (TPI, EC 5.3.1.1) isozyme system in the muscle, heart and eye tissues of chum, chinook and pink salmon (Shaklee *et al.*, 1990b; Wilmot *et al.*, 1992) and in the liver of Atlantic salmon (Galvin *et al.*, 1994, Wilson *et al.*, 1995). One or two of the corresponding loci are usually polymorphic.

3.1.3 Enzymes and other proteins with monomeric structure

Adenosine deaminase

Adenosine deaminase (ADA, EC 3.5.4.4) is encoded by two loci, *ADA-1** and *ADA-2** (Cross & Ward, 1980; May *et al.*, 1980, 1982; Stoneking *et al.*, 1981; Kobayashi *et al.*, 1984; Utter *et al.*, 1984; Shaklee *et al.*, 1990b). In chinook salmon, they are expressed almost identically in skeletal muscle, liver, eye and heart tissues. Polymorphism was revealed at both loci (Kobayashi *et al.*, 1984).

Esterases

Esterases (EST, EC 3.1.1.–) are usually determined by incubating gels with artificial substrates containing ester bonds and manifesting their activity in a variety of tissues. These enzymes are controlled by several independent loci. In brown trout, for example, three loci are expressed in muscles (Allendorf *et al.*, 1977). One of them was described as polymorphic on the basis of its expression in muscles and blood serum (Nyman, 1970; Semenova & Slynko, 1988). In Arctic charr, one of several loci (designated *EST-2**) also proved to be polymorphic (Nyman, 1972; Anderson *et al.*, 1983; Dempson *et al.*, 1988; Osinov *et al.*, 1996). An original pattern of esterase polymorphism was observed in the liver of sockeye and pink salmon (Muske, 1988). Note, however, that even the authors of early studies (Utter *et al.*, 1974) emphasized that esterase molecules are ontogenetically labile and predisposed to modifications changing their electrophoretic properties, and, hence, the data on their polymorphism should be interpreted with caution.

Phosphoglucomutase

Phosphoglucomutase (PGM, EC 5.4.2.2), as with LDH, was one of the first enzymes for which genetic polymorphism was described in salmonids (Roberts *et al.*, 1969; Utter & Hodgins, 1970). To date, its variation has been analysed in many salmonid species (Utter *et al.*, 1972, 1984; Altukhov *et al.*, 1975a; May *et al.*, 1975; Allendorf *et al.*, 1977; Salmenkova & Omelchenko, 1978; Donelly *et al.*, 1979; Grant *et al.*, 1980; Kijima & Fujio, 1980; May, 1980; Okazaki, 1986b; Koljonen, 1989; Shaklee *et al.*, 1990b). The enzyme is active in the liver, muscles, eye, heart and apparently in some other tissues. Two bands of activity, corresponding to the loci *PGM-1** and *PGM-2**, are the most prominent in skeletal and heart muscles, and the third band (locus *PGM-3**) is mainly observed in the eye and liver tissues. In chinook salmon, the latter locus is designated as the isolocus pair *PGM-3,4**, according to the type of its variation (Shaklee *et al.*, 1990b).

At the locus *PGM-1**, represented in zymograms by the band with the highest cathodic mobility at approximately neutral pH, a null allele was discovered in sockeye (Utter *et al.*, 1984) and chum salmon (Winans *et al.*, 1994; Salmenkova & Omelchenko, unpublished). In masu (Okazaki, 1986b) and chinook salmon (Bartley *et al.*, 1992b), all the three loci proved to be polymorphic.

Allendorf *et al.* (1982) discovered the gene controlling expression of the *PGM* locus in rainbow trout liver (according to the last classification, *PGM-3**). The results of experimental crossings confirmed the existence of this regulatory gene, designated by the authors *PGM1-t**, with the additive type of inheritance, which ensures individual differences in the expression of the locus encoding liver PGM. An analogous regulatory locus for this enzyme was also revealed in Atlantic salmon (Verspoor, 1988). Subsequent studies showed that the allele *PGM1-t(b)* (authors' designation), which provides for a high PGM activity in the liver, has a pleiotropic effect: its carriers demonstrated more stable development and rapid growth than individuals characterized by zero or very low expression of the locus encoding liver PGM (Allendorf *et al.*, 1983; Ferguson *et al.*, 1987).

Transferrin

Transferrin (TF) as a polymorphic system has received more attention than other proteins of non-enzymatic nature. Transferrin is a blood serum protein that carries iron. Its polymorphism was described in rainbow trout (Utter *et al.*, 1972), coho salmon (Utter *et al.*, 1970), Atlantic salmon (Moller, 1970), and brook charr (Wright & Atherton, 1970).

The data on all the protein loci studied and the levels of their variation in different salmonid species are summarized in Tables 3.1 to 3.10 (see Appendix at end of chapter).

3.1.4 Inheritance and linkage

Numerous experiments on the inheritance of electrophoretically identifiable variants of biochemical loci (Table 3.11, see Appendix) demonstrated the following:

(1) These variants are determined genetically, and their occurrence is associated with allelic polymorphism of genes (loci) coding for the synthesis of corresponding enzymes.
(2) Mendelian codominant inheritance of alleles is typical for virtually all loci, except for the regulatory locus *PGM-r1** controlling the expression of *PGM** in the liver: relationships between (+) and (−) alleles of this locus are characterized by dominance–recessiveness.
(3) In all the salmonid species studied, the type of inheritance is disomic, and the published data on the tetrasomic inheritance of some loci have not been confirmed (Allendorf & Utter, 1973; Engel *et al.*, 1975; May, 1980). This is evidence for diploidization of the tetraploid salmonid genome owing to the divergence of homologous genes and chromosomes. However, some facts indicate that this process has not yet been completed, as revealed in deviations from random segregation, and the presence of unexpected genotypes in progenies from some crossings (May *et al.*, 1980). At the duplicated loci sharing the same alleles (i.e. at isoloci), so-called pseudolinkage is often observed, which is manifested in the excess of non-parental (recombinant) genotypes in the progeny from backcrosses with the involvement of diheterozygous males. This does not occur in the progeny of diheterozygous females (Morrison & Wright, 1966; Davisson *et al.*, 1973; Wright *et al.*, 1975, 1980; May *et al.*, 1979b, 1980; Taggart & Ferguson, 1984b). Deviations from disomic inheritance and pseudolinkage are regarded as manifestations of 'residual tetrasomy' resulting from the pairing of homologous multivalents in meiosis (Wright *et al.*, 1983).

The linkage of enzyme loci (also referred to as joint segregation, or association) has been studied in many salmonid species and their hybrids. Such investigations are difficult to perform because karyotypes of salmonid fishes consist of many chromosomes; the number of chromosome arms in different species varies from 50 to 100 (Vasiliev, 1985). To facilitate investigations, it is expedient to use fertile hybrids because their higher heterozygosity provides for a sharp increase in the number of loci that can be tested for joint segregation.

Studies on gene linkage in salmonids have been reviewed by several authors (May *et al.*, 1979a,b, 1982, 1989; May, 1980; Wright *et al.*, 1983; Johnson *et al.*, 1987; May & Johnson, 1993). Some works focus on the genera *Oncorhynchus* and *Salmo* (Donelly *et al.*, 1979; Salmenkova & Omelchenko, 1983b; Taggart & Ferguson, 1984b; Seeb &

Seeb, 1986; Kartavtsev *et al.*, 1990). In the most extensively studied species – *S. trutta* and *O. mykiss* – and interspecific hybrids from the related salmonid genus *Salvelinus fontinalis* × *S. namaycush* and *S. fontinalis* × *S. alpinus*, 18 linkage groups for enzyme loci were revealed (May & Johnson, 1993) (Fig. 3.4). It is noteworthy that most of these linkage groups proved to be conserved among all salmon species and hybrids under study. Moreover, different groups of classic or true linkage demonstrated 'pseudolinkage', i.e. non-random assortment with one another. This is probably explained by facultative formation of multivalent pairing and directional chromosome disjunction. It is also evident that some linkage groups proper are duplicated (e.g. groups 3 and 4, 7 and 8, and, partially, 13 and 14), and it is known that only pseudolinkage (rather than true linkage) can exist between duplicated loci.

These facts confirm once again that the genome of salmonids originated via tetraploidization, rather than from tandem duplications. They also show that Robertsonian fusions occurring after tetraploidization affected non-homologous chromosomes, as fusion of homologues would have resulted in true linkage of duplicated loci. In addition, the conservative pattern of linkage and pseudolinkage groups in different salmonid species provides evidence that the corresponding Robertsonian fusions occurred prior to species divergence from the common ancestral form. Note, however, that species-specific linkage groups have also been identified. Of many genes analysed in Arctic charr (*Salvelinus alpinus*), three –

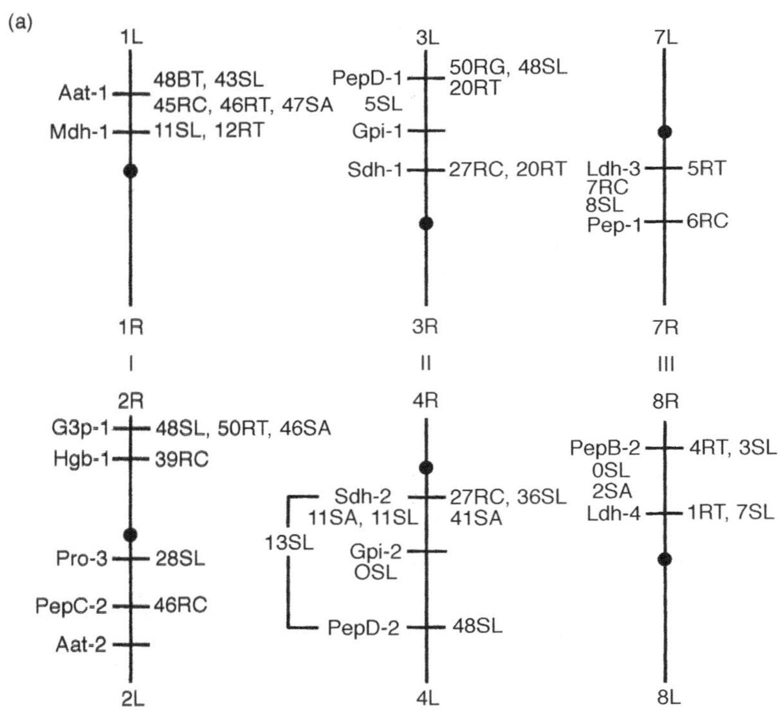

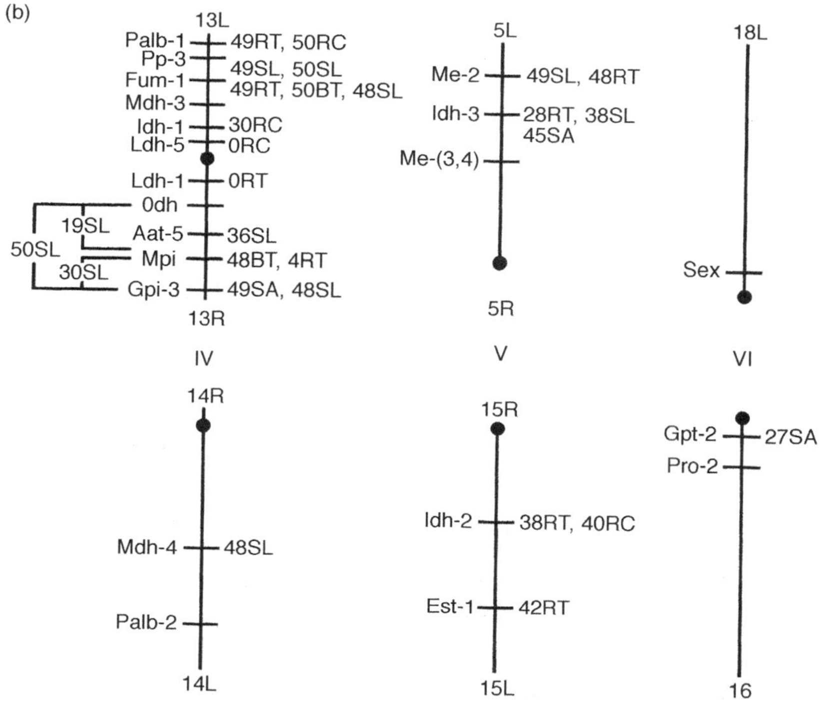

Fig. 3.4 Composite linkage map in salmonids. (a,b) Classical and pseudolinkage groups, (c) classical linkage groups (6, 9–12, 16–18) (from May & Johnson, 1993; reproduced with permission from the authors and Cold Spring Harbor Laboratory Press). Classical linkage groups (chromosomes) are given Arabic numerals and are represented by vertical lines with dots for centromeres. Pseudolinkage groups are given Roman numerals between certain classical linkage groups. Chromosome arms are marked as L and R. Estimated map distances in centimorgans are given for each species based on female intergenic recombination frequencies (values on the left side of the chromosomes) and based on female gene–centromere recombination frequencies (values on the right side of the chromosomes). Species abbreviations are as follows: RT – *Oncorhynchus mykiss*, CT – *O. clarki*, RC – *O. mykiss* × *O. clarki*, BT – *Salmo trutta*, ST – *Salvelinus fontinalis*, SL – *S. fontinalis* × *S. namaycush*, SA – *S. fontinalis* × *S. alpinus*. Assigning authors' locus designation to the recent nomenclature: AAT-1 and AAT-2 = *sAAT-1,2**, Mdh-1 = *sMDH-A1**, G3p-1 and G3p-3 = *G3PDH-1** and *G3PDH-3**, Pro-3 = Protein (?), PepC = *PEPC**, PepD = *PEPD**, Gpi-1 and Gpi-2 = *GPI-B1,2**, Gpi-3 = *GPI-A**, Sdh-1 and -2 = *SDH-1** and *-2**, Ldh-1 = *LDH-A1**, Ldh-3 = *LDH-B1**, Ldh-4 = *LDH-B2**, Ldh-5 = *LDHC**, Pep1 = *PEPA-1**, PepB = *PEPB**, Palb1 and -2 = *PALB1** and *-2**, Pp = *PP* (inorganic pyrophosphatase), Mdh-3 and -4 = *sMDH-B1,2**, Fum1 = *FH**, Odh = *ODH**, Me-(3,4) = *sMEP-1,2**, Idh-2 = *mIDHP-2**, Est1 = *EST1**, Gpt-2 = *ALAT2**, Ada-1 = *ADA-1**, Aat-4 = *AAT-4**, Gam-1 = *bGALA**, Ha = *bGLUA**, Ak-2 = *AK-2**, Sod = *SOD**; Sex is the sex-determining locus.

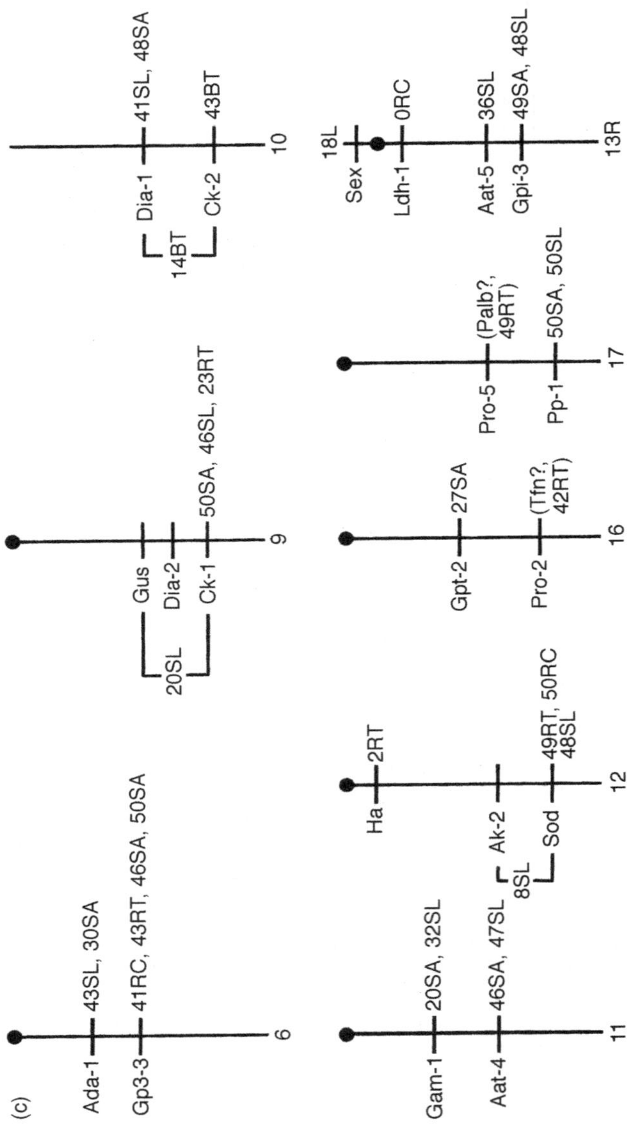
(c)
Ada-1
43SL, 30SA
Gp3-3
41RC, 43RT, 46SA, 50SA
6
20SL
Gus
Dia-2
Ck-1
50SA, 46SL, 23RT
9
14BT
Dia-1
41SL, 48SA
Ck-2
43BT
10
Gam-1
20SA, 32SL
Aat-4
46SA, 47SL
11
Ha
2RT
8SL
Ak-2
Sod
49RT, 50RC
48SL
12
Gpt-2
27SA
Pro-2
(Tfn?,
42RT)
16
Pro-5
(Palb?,
49RT)
Pp-1
50SA, 50SL
17
18L
Sex
Ldh-1
0RC
Aat-5
36SL
Gpi-3
49SA, 48SL
13R

Fig. 3.4 Cont.

*LDH-1**, *AAT-5** and *GPI-3** – proved to be strongly linked with the locus *SEX-1**, which is responsible for the phenotypic expression of sex. In rainbow trout (*Oncorhynchus mykiss*), *SEX-1** demonstrated linkage with the loci *HEX-2** and *SOD-1** (Allendorf *et al.*, 1994). Taking into account that parallel experiments on other salmonid species failed to reveal sex linkage of any enzyme locus, it appears that such linkage is only specific for rainbow trout and Arctic charr. In the latter, Robertsonian fusion of the chromosome arm responsible for sex determination with the corresponding autosome probably occurred in the course of species divergence.

As mentioned above (see Chapter 2), the question concerning an auto- or allotetraploid origin of salmonids still remains open. According to current views, the common ancestral form of salmonids could be a segmental tetraploid carrying in its genome both auto- and allotetrasomes. Indeed, segmental allotetraploidy can explain why some duplicate loci have diverged to a greater or lesser extent, whereas other duplicate loci are similarly expressed in all species. Initial differences of ancient diploid species upon their hybridization and subsequent evolutionary divergence could have contributed to the differentiation of duplicate loci (Johnson *et al.*, 1987). Note also that auto- and allotetraploidy have different consequences for the protein-encoding gene systems. In allotetraploids (and, to a lesser extent, segmental tetraploids), because of genetic differences between the initial species, polypeptides significantly diverging in their primary structure can associate into multimeric complexes, thus creating a specific type of gene interaction. In autotetraploids, the multiplicity of protein products is accounted for by genetic polymorphism, i.e. allelic gene interaction, whereas in allotetraploids, in addition to this system, a very important effect arises from interactions between loci, which lead to new products. The general level of gene multiplicity and constant heterozygosity in allotetraploids is especially high.

3.1.5 Species specificity of multiple proteins

Several multi-gene protein systems of salmonids proved to have two important properties: they are (1) invariant (monomorphic) throughout the vast ranges of these fish and (2) species specific within certain genera. An example of such monomorphism was provided by studies on chum and pink salmon haemoglobins, performed in 1968 by the method of agar gel electrophoresis (Fig. 3.5a,c) (Altukhov, 1969b). As follows from this figure, all conspecific individuals proved to have the same constantly heterozygous genotype. Such a situation was observed in all chum and pink salmon populations studied, suggesting a situation opposite to the phenomenon of genetic polymorphism (Altukhov, 1969b; Altukhov & Rychkov, 1972).

These observations and other data published in the late 1960s and early 1970s provided evidence that the genetic content of a species is not limited to variation alone, and that the biochemical approach to population analysis should always

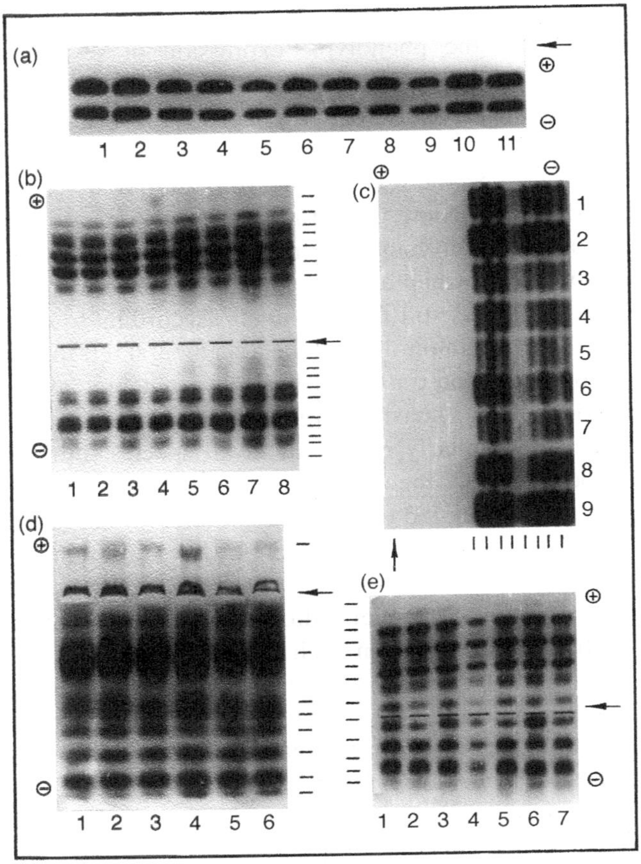

Fig. 3.5 Multiple molecular forms of proteins in Pacific salmon. (a) Haemoglobin of 11 specimens of chum salmon (*Oncorhynchus keta*). Electrophoresis in agar gel, medinal-veronal buffer, pH 8.6. (b) Haemoglobin of eight chum salmon specimens. Electrophoresis in a starch gel, borate buffer system, pH 8.5. (c, e) The same as (b) for several specimens of pink salmon (*O. gorbuscha*). The multiplicity of haemoglobins in salmon is determinated by the presence of seven or eight different subunits in them which form tetrameric molecules (Tsuyuki & Ronald, 1970; Wilkins, 1970). (d) Water-soluble proteins of the eye lens (cristallyne) in chum salmon. Electrophoresis in agar gel. All the photographs show the absence of individual variability in the number or position of separate protein components (monomorphism).

result in revealing not only polymorphic genetic markers, but also monomorphic, invariant proteins, marking at least $\frac{2}{3}$ structural genes.

As such traits are not expedient for studying the genetic divergence of populations, they are largely ignored by researchers. However, each protein has a certain functional role in the organism, and the fact that the greater part of the genome is invariant deserves special emphasis. We performed such analysis (Altukhov, 1969b, 1974; Altukhov & Rychkov, 1972; Altukhov *et al.*, 1972; Omelchenko, 1974) and, on this basis, formulated a hypothesis that appears to explain adequately the genetic diversity of natural populations with respect to different proteins (either with or without enzymatic activity). We postulate dualism in the structural–functional organization of the eukaryotic genome. According to this concept, protein polymorphism is a relatively neutral variation associated with 'minor' adaptive properties of the species, whereas genetically monomorphic proteins are regarded as markers of the key functions whose variation in the norm is biologically inadmissible; mutations in the corresponding part of the genome lead to pathologies and, hence, should be immediately eliminated by selection, especially at the early stages of ontogeny.

We defined genetic monomorphism as *the absence of variation in an a priori hereditary trait throughout the species range or the occurrence within a species of rare discrete variants with such a frequency that their maintenance by repeated mutations is not excluded* (Altukhov & Rychkov, 1972; Altukhov, 1989a). This definition is conditional, essentially being more limited in application than Ford's definition of polymorphism. However, we consider it relevant because it emphasizes that genetic monomorphism is an actual natural phenomenon characterizing the species as a whole and suggests the detection of such invariability at each structural level of life organization.

There is no doubt about the phenomenology of biochemical polymorphism and its reality. But as far as genetic monomorphism is concerned, recognition of its reality requires special reasoned argument. One cannot exclude the possibility of invariance simply being a reflection of inadequate sampling, a charactristic of an individual population, or a technical lack of resolving capacity. Finally, the thesis of the absolute connection between monomorphic proteins and viability needs to be substantiated.

Owing to the work carried out by our own group and by other researchers in recent years, these difficulties have now been overcome.

(1) It may be regarded as an established fact that the genetic invariance recorded during protein electrophoresis cannot be ascribed to an inadequate technique in all cases. Indeed, employing more improved methods of analysis, for example electrophoresis in starch gel, does not change the conclusion on monomorphism of one or other protein. There are the zymograms of the same haemoglobins of chum and pink salmon relatively, analysed by the starch electrophoresis method on Fig. 3.5 (b,e). Certainly the number of haemoglobin bands here is significantly more than in an agar gel, but there is no individual variation that could be interpreted as genetic polymorphism. Electrophoresis in a polyacrylamide gel gives a similar picture of monomorphism, but it reveals fewer components (Omelchenko, 1974). It has been known for a long time that during the complete analysis of the primary structure of enzymes, their catalytic centre is often invariable, whereas a molecule's functionally less significant areas may vary. For instance, lactate dehydrogenase has a different

amino acid composition in different vertebrate species, but the enzyme's catalytic centre, composed of 12 amino acids, is identical in all the cases researched (Kaplan, 1965). Histidine residues, located respectively in the functionally important 58th and 87th, and the 63rd and 92nd positions, respectively, of the α and β chains of the human haemoglobin molecule, are completely invariant; in many groups of vertebrates this monomorphism has been conserved for the 500 million years of their natural history (Jukes, 1971). The substitution of tyrosine for any of these histidines causes haemoglobinopathy in humans (see Kimura & Ohta, 1973; for other proteins see also Berger & Weber, 1974).

In addition the reality of the monomorphic state of several protein loci has been reaffirmed recently, despite the discovery of heat-sensitive alleles. Lewontin, for example, showed at the 14th International Genetics Congress that further refinements of methods for detecting hereditary variations at a molecular level do not alter the conclusion about the existence of two groups of gene loci that are responsible for protein synthesis. The rule that emerges is that loci which are highly variable under one method of analysis also show increased allelic diversity when more improved methods are employed, whereas loci which appear to be monomorphic in routine electrophoresis remain monomorphic even when finer analysis is carried out (Lewontin, 1978b, p. 465).

Singh (1979, p. 1014), when analysing *Drosophila pseudoobscura* proteins electrophoretically, concluded that monomorphism was a reality. The author indicated that the so-called monomorphic loci 'form a separate group and are not just the tail of the same distribution covering the polymorphic loci'.

Studies of tissue proteins of insects and mammals by means of two-dimensional electrophoresis revealed such low levels of genetic variability that a revision of previous estimates of polymorphism and heterozygosity in natural populations was required (Brown & Langley, 1979; Lee *et al.*, 1979; McConkey *et al.*, 1979; Walton *et al.*, 1979 Racine & Langley, 1980). According to these new data, the proportion of polymorphic genes does not exceed 11%, while average heterozygosity per locus is only 1–4%.

However, it should be recalled that a very similar result was obtained in 1971 (Altukhov *et al.*, 1972) when we deliberately included several groups of proteins not previously investigated by electrophoretic analysis, not confining ourselves to enzymes alone. Lewontin, unfortunately, considered this to be a defect of our work, whereas in fact this method enabled us to reveal for the first time protein systems characterized by extremely low levels of genetic variability.

(2) It may be regarded as proven that genetic monomorphism, as had been postulated, is not only a population phenomenon but also a phenomenon that characterizes a species as a whole. This has been established with particular

reliability for *Homo sapiens*, representatives of all the main races of whom have been studied up to the present from an adequately large number of gene protein markers and at least some of which are monomorphic throughout the species' range (Nei & Roychoudhury, 1974).

Taking into account the findings of American, Canadian and Japanese colleagues, our laboratory has obtained evidence of monomorphism of certain protein systems in Pacific salmon. This conclusion is based on electrophoretic analysis of several thousand specimens gathered from large areas of the Asian as well as American coasts of the North Pacific (Koch *et al.*, 1964; Tsuyuki *et al.*, 1965; Wilkins, 1968; Yamanaka *et al.*, 1967; Altukhov, 1969b, 1974, 1977; Altukhov & Rychkov, 1972; Altukhov *et al.*, 1972; Omelchenko, 1975; Salmenkova *et al.*, 1986). As research into different species using a variety of gene protein markers is extended, the reality of the existence of genetically monomorphic systems becomes increasingly obvious (see also Coulthart *et al.*, 1984).

(3) Direct evidence has been obtained of the link between genetically monomorphic systems and the viability of organisms. It has been found that only a small proportion of the total number of gene mutations arising in a given generation appear within the reproductive section of a population and are transmitted in generations as relatively neutral. The direct evidence that has been obtained of the negative selective value of rare electrophoretic protein variants makes it possible to explain in a new way the excess within populations of different species of rare alleles when approximating the empirical distributions of their frequencies, by means of a mathematical function which hypothesizes the neutrality of protein polymorphism. Dubrova (1980) performed the appropriate analysis, and the approximation proved to be more satisfactory when our model was adopted, according to which the existence of rare alleles in a population reflects to a considerable degree the balance between selection and spontaneous mutation.

Data obtained recently not only corroborate the conclusion of a species' genetic monomorphism as a natural phenomenon, but also shed light on the mechanisms which maintain phenotypic stability at the protein level. The following are among the most important mechanisms.

The effects of purifying selection

A protein with a quaternary structure that is responsible for essential biochemical functions will mean the vast majority of mutations should be categorized as 'forbidden', which means that they are eliminated by natural selection at the earliest stages of ontogenesis. These are mutations impinging on the active centre of an enzyme, or else sections of a molecule that are important to the process of association of subunits.

The organization of genetic material

If there is a mechanism for correcting and removing mutations in protein synthesis, then it is evident that its particular function lies in maintaining the stability of primary protein structure. The analysis of the connection between the fine structure of eukaryotic genes and their levels of variability may lead to an important area of research. At this point, however, attention should be paid to yet another factor which, from our viewpoint, is of major importance for elucidating the biological significance of the genetic monomorphism of a species: considerable invariance often reveals protein systems that are characterized by exceptional multiplicity of structural components. By way of example, we can indicate several zymograms of monomorphic protein systems described by us for Pacific salmon, which are allotetraploid (Fig. 3.5). It is still not clear what mechanism is involved in such tolerance to spontaneous mutations of duplicated genes that encode the synthesis of isofunctional protein families.

This phenomenon has been explained on a purely functional plane: the maximum duplication of genetic information in eukaryotes is in proportion to its biological significance. In other words, the most important functions may be encoded by multiple cistrons, promoting broad adaptation in a normally fluctuating environment (Altukhov & Rychkov, 1972; Altukhov, 1974, 1977, 1982).

If Pacific salmon are in fact amphidiploid (Altukhov, 1974; Altukhov *et al.*, 1972; Bushuyev, 1973; Bushuyev *et al.*, 1975; Omelchenko & Gerasimenko, 1981), then with regard to homologous genes which differ in primary structure and are obtained from different ancestors, they, like other groups of the same origins, are effectively fixed heterozygotes. Correspondingly, in the case of multimer proteins, their formation is effected by the interaction of noticeably diverged subunits, encoded by different gene loci and not by alleles of the same gene.

It is possible that selection places particularly harsh demands on precisely these complexes, so constraining the levels of polymorphism in allopolyploids (Altukhov *et al.*, 1972; Salmenkova & Volokhonskaya, 1973; Salmenkova & Omelchenko, 1978).

Harris *et al.* (1977) have noted reduced variability of multimer proteins in humans, being concerned not so much with multimer proteins themselves, as much as with which of them are encoded by different loci and form hybrid molecules.

It is undeniable that the multiplicity of isofunctional proteins, characterizing the monomorphism of a species as a whole, results from the duplication of genetic material caused by unequal crossing-over and polyploidy.

Monomorphic multi-gene protein systems may therefore be expected to behave like true species-specific traits (Fig. 3.6). In all the studied groups of salmonids, except for several problematic species of charr, species within genera are easy to

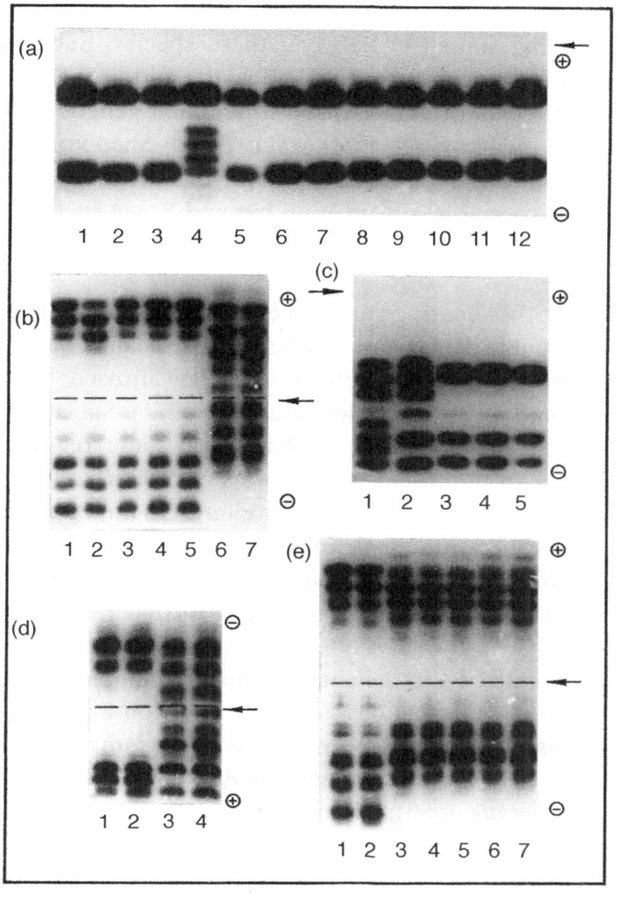

Fig. 3.6 Species specificity discovered in electrophoretic research of multiple haemoglobins in fish belonging to the family Salmonidae. (a) 1–3 and 5–12 – *Oncorhynchus keta*, 4 – *O. kisutch* (electrophoresis in agar gel); (b) 1–5 – *Salvelinus leucomaenis*, 6, 7 – *O. gorbuscha* (electrophoresis in starch gel); (c) 1 – *O. gorbuscha*, 2 – hybrid (F_1) *O. gorbuscha* × *O. masu*, 3–5 – *O. masu* (electrophoresis in agar gel); (d) 1, 2 – *O. keta*, 3, 4 – *O. gorbuscha*; (e) 1, 2 – *Salvelinus leucomaenis*, 3–7 – *O. keta* (electrophoresis in starch gel).

identify by multiple monomorphic protein systems, which confirms the typological concepts of species. It is interesting that intraspecies conservatism along with interspecies differences were evident not only for multiple proteins. Such a phenomenon was described for electrophoretic patterns of repeated nucleotide sequences of nuclear DNA, which were obtained through restriction analysis (so named 'taxonoprints') (Mednikov *et al.*, 1995).

Applying monomorphic proteins allowed researchers to clarify several controversial issues of species identification (taxonomy) of salmonids. For example, analysing monomorphic water-soluble proteins of muscle tissue and several enzymes of *Salmo gairdneri* and *S. mykiss*, Ryabova & Goncharova (1981) demonstrated a striking similarity between these species. Subsequent research, including studies on biological, ecological, allozyme and karyological parameters and the analysis of mitochondrial DNA, confirmed their genetic similarity (Frolov, 1981; Gorshkova & Gorshkov, 1985; Okazaki, 1986a; Shed'ko, 1991a; Osinov & Pavlov, 1993), indicating the possibility of unifying them into one species. As mentioned above, the species involving *S. mykiss* and *S. gairdneri* was included into the genus of Pacific salmon and received the name *Oncorhynchus mykiss* (Smith & Stearley, 1989).

The taxonomically problematic group of the genus *Salvelinus* inhabiting the lower reaches of the Kamchatka River and southern Kuril Islands had been attributed either to Arctic charr *S. alpinus* (Savvaitova, 1973; Savvaitova & Movchan, 1973) or to Dolly Varden *S. malma* (Berg, 1948b; Behnke, 1972). Specific electrophoretic types of monomorphic haemoglobins in charr from these regions proved to be similar to those in anadromous and brook forms of *S. malma* from Sakhalin, differing from the haemoglobin patterns of Arctic charr. On this basis, the group under consideration was classified as *S. malma* (Omelchenko, 1975). Karyological differences between *S. alpinus* from Sweden and different forms of Far Eastern *S. malma* confirm this conclusion (Nygren *et al.*, 1971; Viktorovsky, 1975; Viktorovsky & Glubokovsky, 1977).

Later genetic differentiation in the charr group was demonstrated by comparing different protein systems. On this basis, species isolation has been supported for sympatric populations of lake charr *Salvethymus svetovidovi* (Chereshnev & Skopetz), *Salvelinus elgyticus* (Viktorovsky & Glubokovsky), and *Salvelinus boganidae* (Berg) (Glubokovsky *et al.*, 1993); *Salvelinus levanidovi* (Chereshnev, Skopetz et Gudkov) and *Salvelinus malma* (Walbaum) (Omelchenko *et al.*, 1996); and *Salvelinus taranetzi* (Kaganovski) and *Salvelinus malma* (Omelchenko *et al.*, 1998). All species except *S. boganidae* and *S. malma* were studied from the localities of their first description. Species independence of allopatric populations of *Salvelinus alpinus* L. from the Arctic Ocean coast and *Salvelinus malma* from the Okhotsk sea basin (Omelchenko & Salmenkova, 1998; Politov *et al.*, 1998) was also supported through analysis of these sympatric forms in Alaska (Gharrett *et al.*, 1991, cited by Crane *et al.*, 1994).

3.2 DNA

3.2.1 DNA polymorphism and methods of its analysis

Although allozymes and other proteins are very convenient for analysing genetic variation, they have certain drawbacks when used as markers. These drawbacks, determined by both natural properties of protein molecules and technical constraints for detecting polymorphism, are as follows:

(1) The variability of these markers in some species and populations is very low.
(2) Protein mark variation is only a small, coding part of the genome.
(3) Synonymous and many non-synonymous substitutions that do not lead to changes in the electric charge, size or shape of protein molecules are impossible to reveal by electrophoresis.
(4) It is usually impossible to take samples of different tissues from living organisms in an amount sufficient for comprehensive analysis of genetic variation.
(5) In samples stored for a long time, protein molecules degrade and lose enzymatic activity.

As will be shown in this chapter, genetic markers directly associated with nucleotide sequence polymorphism in different components of genomic DNA are virtually free from such drawbacks and, in this sense, have an advantage over proteins (for a review, see Carvalho & Pitcher, 1995).

DNA polymorphism arises because of point mutations (single nucleotide substitutions) and certain errors in DNA replication (insertions or deletions ranging in size from one to hundreds and thousands of nucleotides). Investigations into this variation became possible owing to achievements in the technique of DNA analysis, primarily the discovery of restriction enzymes, which cleave specifically the DNA molecule into fragments, and the development of the polymerase chain reaction (PCR) technique, which allows amplification of fragments of DNA from minimal quantities of tissue, including those from partially degraded DNA molecules.

In the cell, DNA is located in the nucleus (nuclear DNA) and in mitochondria (mitochondrial DNA). In bony fishes, the size of the nuclear genome ranges from 0.3 to 4 billion base pairs (Ohno, 1974), and that of the mitochondrial genome, from 16 000 to 18 000 base pairs (Park & Moran, 1994). All DNA molecules consist of coding and non-coding regions. In the mammalian genome, only 1% of total DNA codes for protein synthesis (Park & Moran, 1994) and, hence, is generally highly conserved. The non-coding component is much more variable because largely it has no selective constraints, rendering it more convenient for analysing hereditary variation. DNA contains both unique ('single copy') sequences and those repeated several-fold (from a few to thousands of copies); in the mammalian genome, the proportion of the latter is about 30% (Alberts *et al.*, 1983).

Molecules of *mitochondrial DNA* (mtDNA) in vertebrates are circular (closed). The complete sequence analysis of human and other vertebrate mtDNA has revealed a conservative arrangement of mitochondrial genes (Anderson *et al.*, 1981; Beckenbach, 1991). The mtDNA molecule contains more than 30 different genes, including 2 ribosomal RNA (rRNA) genes, 22 transfer RNA (tRNA) genes, and 13 protein-coding sequences, and has a non-coding regulatory region, called the D-loop, which is involved in replication (Ferris & Berg, 1987; Beckenbach, 1991). The same organization of mitochondrial genes proved to be characteristic of fishes, e.g. several species of *Salmo* and *Oncorhynchus* genera (Davidson *et al.*, 1989a; Thomas & Beckenbach, 1989; Zardoya *et al.*, 1995; Carney *et al.*, 1997).

As mitochondria are transmitted to the progeny with the cytoplasm, mtDNA is inherited maternally (Gyllensten *et al.*, 1985; Avise & Vrijenhoek, 1987) and is haploid, i.e. all the mitochondria usually contain only one type of DNA, with the exception of cases referred to as heteroplasmy (Arnason & Rand, 1992). Indeed, recombination is not observed in mitochondrial genomes of most organisms, and mtDNA is clonally inherited (Avise, 1994). It was shown that the greatest number of mutations occur in the D-loop, and the fewest in tRNA genes (Cann *et al.*, 1984; Ferris & Berg, 1987); however, this probably is not the case with some fish, particularly salmonid species (Bernatchez *et al.*, 1992; Park *et al.*, 1993; Apostolidis *et al.*, 1996b). In studies on different organisms, the rate of nucleotide substitution in mtDNA was estimated at 1–2% per million years (Brown *et al.*, 1979; Ferris *et al.*, 1983), whereas in *S. salar* it may be higher, reaching 2–3% (Shed'ko, 1991b).

A considerable fraction of *repetitive nuclear DNA*, referred to as satellite DNA, consists of tandem repeats of certain 'core' sequences with lengths ranging from two to several thousand base pairs. Mutations such as insertions and deletions resulting from DNA slippage and unequal crossing-over change the number of repeats, i.e. the total length of this highly repetitive sequence (Park & Moran, 1994). This variation, observed in different parts (chromosomes) of the genome and in different individuals, is known as *variable number of tandem repeats* (VNTR). In the families of tandem repeats, those exhibiting valuable characteristics for use as genetic markers attracted special attention. These are *minisatellites* consisting of repeated copies with a size of 9–10 to 100 nucleotides long each (Jeffreys *et al.*, 1985a,b; Taggart & Ferguson, 1990), and *microsatellites* in which repeat length varies from one to four or, rarely, six nucleotides (Tautz, 1989; Wright & Bentzen, 1994). A minisatellite locus may comprise two to several hundred repeats, and a microsatellite locus from ten to a hundred repeats. Individual alleles of these loci differ from one another in the number of tandemly repeated copies. Numerous loci of both types are distributed throughout the genome, and there is a hypothesis concerning the evolutionary origin of minisatellites from microsatellites (Wright, 1994).

Other promising genetic markers are also repeated but non-tandem sequences spread over the genome and designated *short* and *long interspersed repetitive elements* (SINE and LINE) (Singer, 1982).

In addition, high variability is characteristic of *intergenic spacers* that separate from one another repeated gene pairs in multi-gene families coding for ribosomal 28S and 18S RNAs. Non-transcribed and transcribed spacers, unlike conservative ribosomal genes, accumulate mutations and can be used successfully for analysing DNA variation (Arnheim, 1983).

For a detailed review of corresponding methods, we refer the reader to the paper by Park & Moran (1994) or special manuals (e.g. Taylor, 1997) and the lists of references they provide. Here, we will discuss briefly only the basic approaches to such an analysis.

According to Park & Moran (1994), most of the methods are based on two technologies: (1) DNA treatment with different enzymes and (2) separation of a mixture of DNA fragments in order to isolate and visualize them. The most important enzymes are:

(1) Bacterial restriction enzymes (endonucleases), which cleave DNA molecules into fragments of different length at sites where specific recognition sequences are located (Fig. 3.7).
(2) Polymerases (especially thermostable), which synthesize new DNA strands using a DNA template.
(3) Ligases, which are used for joining the ends of two DNA strands (for example, in the cloning process).

DNA fragments are usually separated by electrophoresis in agarose or polyacrylamide gels and unspecifically visualized by staining with ethidium bromide or silver oxide. Thus, individual variation of DNA, resulting from nucleotide substitutions or insertions and deletions, is manifested as differences between individuals in the size and pattern of restriction fragments. This phenomenon is referred to as *restriction fragment length polymorphism* (RFLP) (for details, see Dowling *et*

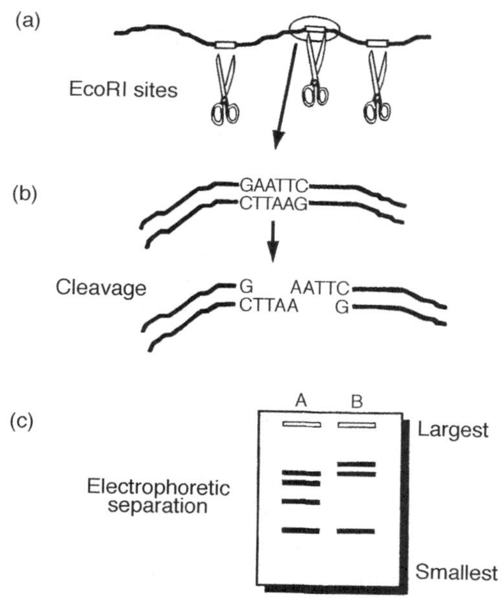

Fig. 3.7 Restriction enzyme digestion and electrophoretic separation of resulting DNA fragments. (a) This DNA molecule has three *Eco*RI sites. At every position in the DNA molecule where the sequence GAATTC occurs (open rectangles), the restriction enzyme will cleave (b), producing, in this case, four fragments separated on an agarose gel (lane A in (c)). If an alternative allele exists in the population that lacks the circled restriction site in (a), the two middle fragments in the A profile will not be cleaved and will be seen as the single larger fragment in the B profile (from Park & Moran, 1995. Reproduced with permission from the authors and Kluwer Academic Publishers).

al., 1990). The size of fragments is determined by comparing them to a standard reference kit, which contains DNA fragments of known length and is separated in the same gel. It should be noted, however, that unspecific visualization is inconvenient for analysing complex mixtures of restriction fragments obtained from total nuclear DNA. This method is generally used in experiments with individual fragments of nuclear DNA or with mtDNA.

Selective vizualization of DNA fragments in gels is performed by means of hybridization with a specific *probe*, usually by the method proposed by Southern (1975) and known as *Southern blotting*. The probe is a cloned DNA sequence, or PCR product, or synthetic oligonucleotide complementary to the specific sequence of the restriction fragment(s) and carrying a radioactive or fluorescent label, allowing subsequent visualization of the target fragment(s).

VNTRs are analysed by hybridization with probes of this type, i.e. complementary to the sequence of the repeated motif. Initially, universal *multilocus probes* were used, which recognized sequences widely represented in the genome as families of corresponding repeats (Jeffreys *et al.*, 1988). Such probes reveal in the same gels numerous similar loci, which essentially have multiple alleles and, hence, are in the heterozygous state. These complex DNA profiles, usually unique for each individual, were named *DNA fingerprints*. The minisatellite loci of different species proved to be extremely variable (Jeffreys *et al.*, 1988; Wright, 1993). DNA fingerprinting, being an efficient method for revealing relatedness or parentage, is less applicable to the analysis of interpopulation differences because it is virtually impossible to attribute multiple allele variants to a certain locus and to determine allele frequencies. Nevertheless, methods for estimating heterozygosity, genetic distances and gene diversity were developed (Lynch, 1990; Jin & Chakraborty, 1994).

To date, numerous *single-locus probes* (SLP) have already been constructed, including those for salmonids (Taggart & Ferguson, 1990; Bentzen *et al.*, 1991; Heath *et al.*, 1994; Prodohl *et al.*, 1994; Taggart *et al.*, 1995a). They simplified the final analysis significantly since variation (DNA profile) revealed by hybridization with such a probe reveals variation at a single locus. Such patterns can be interpreted on the basis of Mendelian segregation (Fig. 3.8a,b), allowing direct estimation of allele frequencies, levels of heterozygosity and other population parameters (Goldstein *et al.*, 1995; Slatkin, 1995; Zhivotovsky & Feldman, 1995).

One of the most important modern methods for analysing polymorphism at the DNA level is PCR, the most rapid and simple technique for amplifying target DNA sequences. The principle of this technique is that the DNA region to be amplified (more precisely, each of its two strands) serves as a template for synthesizing the complementary sequence *in vitro*. The reaction is catalysed by a thermostable DNA polymerase in the presence of two primers, i.e. synthetic oligonucleotides complementary to sequences surrounding the amplified fragment (one for each strand of the DNA molecule), and involves several steps proceeding at different temperatures (for details, see Mullis *et al.*, 1986; Saiki *et al.*, 1988; Innis *et al.*, 1990).

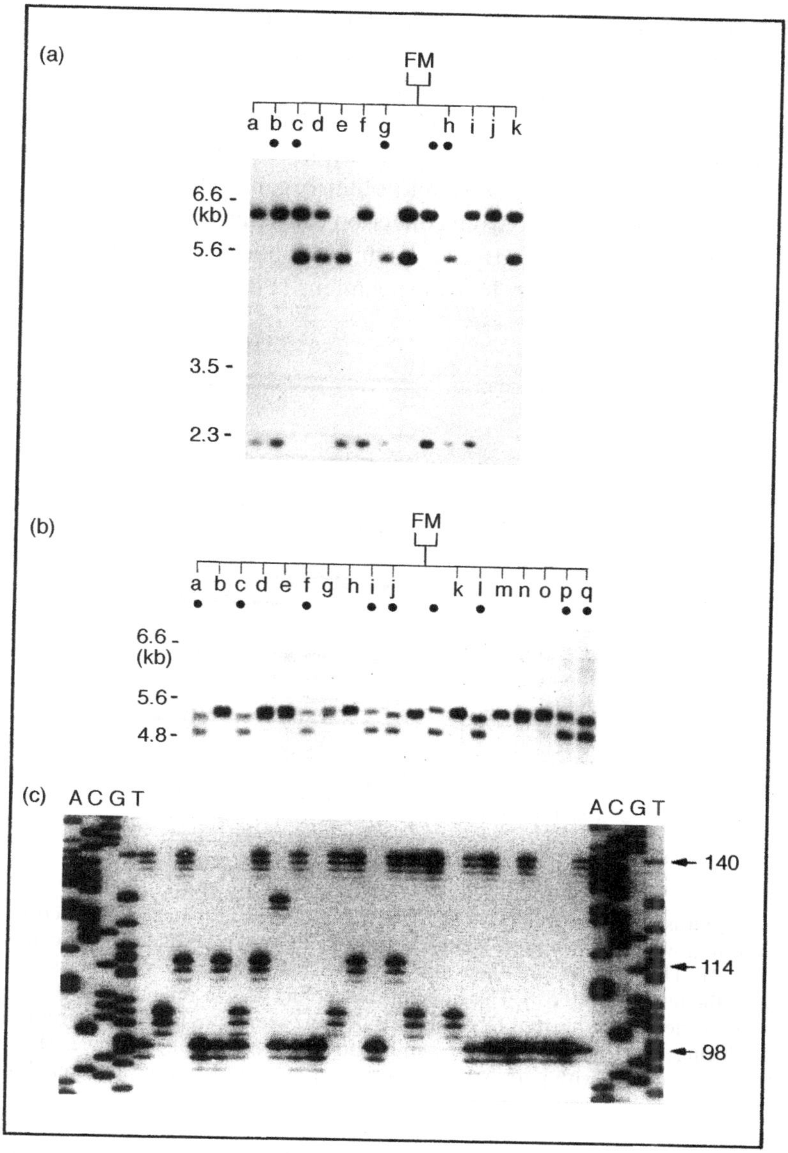

Fig. 3.8 Genetic variability for microsatellite DNA loci and inheritance of minisatellite DNA loci in salmonid fishes. (a, b) Examples of DNA profiles detected with brown trout minisatellite DNA single-locus probes (a) – p*Str*-A3, (b) – p*Str*-A9). In both cases a brown trout family comprising male parent (M), female parent (F) and 11 to 17 progeny are shown. Male progeny are denoted by black points. DNA fragment size scale is given in kilobase pairs (kb). (From Prodohl *et al.*, 1994. Reproduced with permission from Blackwell Science Ltd, Oxford). (c) PCR-amplified microsatellite alleles from several individuals of *Oncorhynchus mykiss* at microsatellite locus *Omy77*. The 'size' of each allele was estimated by reference to comigrating M13 sequence fragments (A, C, G and T) run on either side of the trout samples. (From Morris *et al.*, 1996. Reproduced with permission from the National Research Council of Canada).

The procedure is automated, and reaction cycles are repeated many times. As each cycle doubles the amount of DNA synthesized in the previous cycle, after 30 cycles the initial amount is amplified by more than a million-fold (Fig. 3.9). It is especially important that the PCR technique allows amplification of fragments from minimal amounts of the initial DNA (even partially degraded) present in drops of blood, scales, fragments of skin, bones, fins, and other organs. The method proved to be applicable also to DNA isolated from collection (museum) samples stored for a long time in a dry state (Paabo, 1989; Beckenbach, 1991; Nielsen *et al.*, 1997; Hutchinson *et al.*, 1999), including palaeontological specimens (Poinar, 1993).

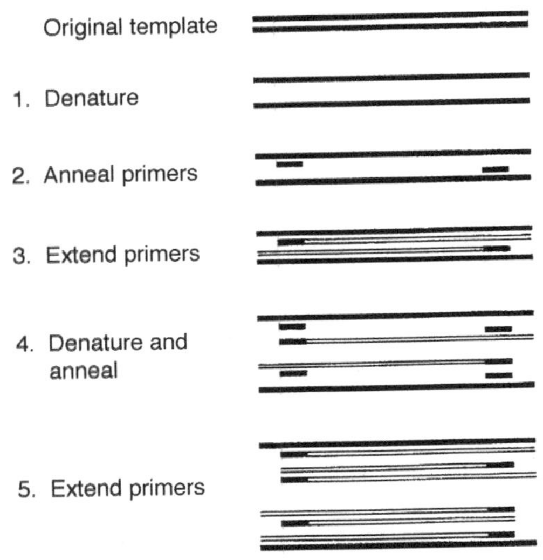

Fig. 3.9 Polymerase chain reaction (PCR). The DNA template is heat-denatured (step 1) followed by cooling, which allows short synthetic primers to anneal to each of the complementary strands (step 2). The primers are then extended by a thermostable polymerase, resulting in duplication of the template (step 3). This heating and cooling cycle is repeated 20–40 times (step 4). In each cycle, the newly synthesized DNA strands become templates for the next replication (step 5), thus producing over one million-fold amplification of the target sequence. (From Park & Moran, 1995. Reproduced with permission from the authors and Kluwer Academic Publishers).

DNA amplification is also performed by the method of cloning, which, in contrast to the PCR technique, allows operations with large DNA fragments (thousands of base pairs) (for details, see Datta *et al.*, 1988). The fragment to be cloned is inserted into the self-reproducing 'construction', termed a vector, by means of cutting with restriction nucleases and subsequent 'sealing' of resulting molecules with the DNA ligase enzyme. Vectors are DNA molecules that penetrate bacterial or yeast cells (or are introduced into them) and replicate as these cells grow and divide, producing a great number of copies containing the inserted DNA fragment. Bacterial plasmids and phages are used as such molecules (Fig. 3.10).

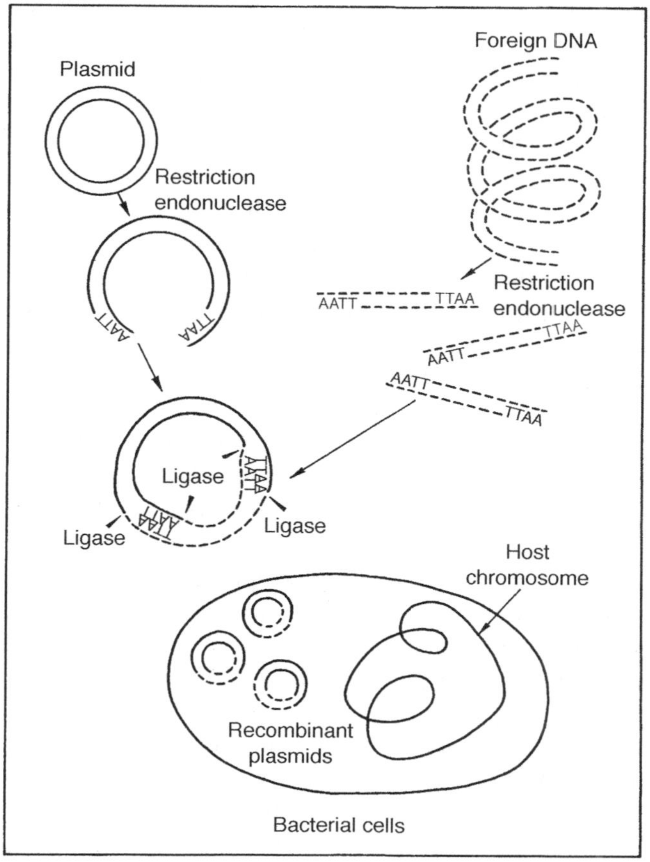

Fig. 3.10 Scheme of DNA cloning. A restriction fragment of a DNA molecule is joined by a ligase with a preliminarily cleaved plasmid. The cleavage of the plasmid and foreign DNAs is carried out by the same restriction enzyme, that provides complementarity of end sequences of the plasmid and the fragment DNAs and joining their 'sticky ends'. The recombinant construction enters the bacterial host and multiples along with the host. (This figure is adapted from Vosberg (1977) by Vogel & Motulsky, 1986. Reproduced with permission from the author and Springer-Verlag).

Amplification of fragments (mainly by means of PCR) is used for analysing both nuclear and mitochondrial DNAs (Beckenbach, 1991). In the latter case, primers necessary for the procedure are easier to prepare: mtDNA contains evolutionarily conservative genes, and its nucleotide sequences in many organisms are already known. In addition, specific amplification of individual mtDNA fragments makes unnecessary the tedious procedures of separating mtDNA from nuclear DNA.

Polymorphism of amplified DNA fragments is analysed either by means of sequencing (Sanger *et al.*, 1977) and direct comparison of individual nucleotide sequences, or by indirect methods. The latter include electrophoretic fractionation of DNA fragments based on differences in their length (RFLP) or conformation (single strand conformation polymorphism, SSCP). The method of SSCP analysis is based on the fact that even single nucleotide substitutions resulting from point mutations change the conformation (i.e. shape and spatial structure) of relatively short single-strand PCR fragments, producing associated changes in electrophoretic mobility (Hayashi, 1985; Sheffield *et al.*, 1993).

Each microsatellite locus, i.e. a tandem repeat of two to six nucleotides, is flanked by a unique sequence. If the latter is known, it is possible to synthesize a complementary primer and to amplify this repeat by means of PCR. This provides an opportunity to analyse this locus alone further by determining the 'length' of its alleles by means of electrophoresis in polyacrylamide gel with a set of standard DNA fragments of known size used as markers (Fig. 3.8c). The size of microsatellite loci in some salmonids and other fish species is almost twice as great as in mammals (Slettan *et al.*, 1993; Brooker *et al.*, 1994).

An efficient approach to the assessment of DNA polymorphism consists in the analysis of so-called random amplified polymorphic DNA (RAPD). The principle of this method is that short (10–20 base pairs) primers with random nucleotide sequences are used in the PCR for amplifying anonymous DNA regions, and amplification products are analysed by electrophoresis. The number and size of amplified fragments depend on the size and sequence of these arbitrarily chosen primers. Priming sites are randomly distributed over the genome, and polymorphism at such sites is displayed in the presence (or absence) of corresponding fragments on gels.

To date, data on nucleotide sequences of many genes in numerous different taxa have accumulated, allowing the possibility of synthesizing gene-specific primers and analysing polymorphism in different fragments of certain genes with known functions. For example, studies are being performed on polymorphism in different segments of the growth hormone gene (Gross & Nilsson, 1995) and in mtDNA segments including portions of individual genes and even several neighbouring genes (Hansen & Loeschcke, 1996; Burger *et al.*, 1997; Hansen *et al.*, 1997; Radchenko *et al.*, 1997).

3.2.2 *Inheritance and linkage of DNA markers*

As noted in the previous section, mtDNA is inherited maternally. Maternal inheritance of mtDNA in salmonid fishes was demonstrated for different interspecific hybrids, such as masu × pink salmon obtained artificially (Ginatulina & Maksimovich, 1994), brown trout × Atlantic salmon (McGowan & Davidson, 1992; Hartley, 1996) and others. In the majority of organisms, the mitochondrial genome does not recombine. Hence, although mtDNA contains up to 30 different genes, from the standpoint of formal genetics it is regarded as a single locus, and mtDNA haplotypes as alleles of this locus and as individual clones.

Codominant Mendelian inheritance in families was demonstrated for minisatellites (revealed with the aid of single-locus probes) in Atlantic salmon, brown trout and chinook salmon (Prodohl *et al.*, 1994; Withler *et al.*, 1994; Taggart *et al.*, 1995a) and for microsatellites in Atlantic salmon, brown trout, rainbow trout and pink salmon (Estoup *et al.*, 1993; Sakamoto *et al.*, 1994; McConnell *et al.*, 1995; Slettan *et al.*, 1997; Seeb *et al.*, 1998). SINE markers are inherited similarly, as observed in families of pink salmon (Greene & Seeb, 1997) and other salmonids (Spruell & Thorgaard, 1996). The Mendelian inheritance with dominance is typical for RAPD markers; in particular, this was experimentally demonstrated in *Salmo salar* and *S. trutta* (Elo *et al.*, 1997).

The tetraploid origin of salmonids also shows up at the level of DNA markers. Thus, the analysis of segregation in Atlantic salmon and brown trout revealed duplication of several mini- and microsatellite loci and the disomic type of their inheritance (Prodohl *et al.*, 1994; Taggart *et al.*, 1995a; Slettan *et al.*, 1997). Young *et al.* (1998) found that in rainbow trout many microsatellite loci exhibited disomic inheritance, though at one locus tetrasomic inheritance was evident, showing support for residual chromosome pairing. As noted above (section 3.1.4), cases of such 'residual tetrasomy' have been observed previously in studies on the inheritance of allozyme loci (Allendorf & Thorgaard, 1984).

As mini- and microsatellites and some other DNA loci are highly polymorphic and widely distributed over the genome, they are very convenient markers for gene linkage research and gene mapping.

Linkage between molecular markers in salmonids and other fish species is usually analysed in haploids (Lie *et al.*, 1994) or double haploids (Young *et al.*, 1998) produced by methods based on the use of gynogenesis and androgenesis. In the haploid progeny, these markers can be studied at embryonic stages of development, and this facilitates and accelerates the process of obtaining experimental families necessary for the analysis of linkage (for details, see Slettan *et al.*, 1997; Young *et al.*, 1998).

Recently Young *et al.* (1998) used double haploids obtained in *Oncorhynchus mykiss* lines for the detailed analysis of linkage between 475 molecular markers such as multi-locus minisatellites, microsatellites, SINE, RAPD and amplified fragment length polymorphism (AFLP). The resulting genetic map included 42 linkage groups (31 large and 11 small). Markers of different types differed in their preferential location on chromosomes: AFLP markers clustered mainly in the regions corresponding to centromeres, multi-locus minisatellites localized closer to telomeres, and SINE occupied an intermediate position. In general, the minimal length of the rainbow trout genome was estimated at 2627.5 centimorgans (cM), which is close to the value calculated from the data on crossing-over frequency in females of this species (2600 cM). It is noteworthy that the number of large linkage groups (31) proved to be similar to the haploid chromosome number (30) in the rainbow trout lines used in these experiments. Thus, this genetic map represented the *Oncorhynchus mykiss* genome fairly adequately, although many gaps remained, suggesting the presence of small linkage groups and unlinked markers.

Linkage maps of DNA markers are an important tool for analysing linkage or association of these markers with different quantitative phenotypic traits. By revealing joint segregation of certain quantitative traits and alleles of molecular markers, it is possible to locate chromosome regions that contain the loci controlling these traits (quantitative trait loci, QTL). Identification of molecular markers strongly linked with QTL offers the possibility of directly observing the effect of QTL on a certain adaptive trait, estimating the magnitude of such effects, and determining the number of chromosome regions involved (Paterson *et al.*, 1988). Such studies in *Oncorhynchus mykiss* found linkage between two microsatellite loci and QTL determining resistance to high temperature (Jackson *et al.*, 1998). Previously, associations of minisatellite markers with visible mutations were observed in the same species (Sakamoto *et al.*, 1994). Some DNA markers of salmonid fishes proved to be sex-linked (Devlin *et al.*, 1991; Du *et al.*, 1993; Prodohl *et al.*, 1994; Clifton & Rodriguez, 1997; Young *et al.*, 1998).

Thus, DNA technologies are progressing rapidly. Although the analysis of DNA markers can be relatively tedious, time-consuming and expensive, the amount of genetic variation revealed in this way is far greater than in allozyme analysis. Currently, the main methodological trend in this field is the development of more rapid and less labour intensive experimental procedures and, especially, their automation. For example, oligonucleotide synthesis (including that of labelled products used as primers), amplification (PCR) and sequencing have already been automated. Computer databases on nucleotide sequences of many genes, DNA markers, primers and other related information have been created, and are readily accessible to researchers. Special methods of statistical analysis and corresponding computer software have been developed for processing the data obtained with the aid of DNA markers (for a review, see Ferguson *et al.*, 1995). This creates conditions for effectively applying the analysis of DNA polymorphism to a broad range of population and evolutionary studies, including those related to fisheries.

In the next section, we shall compare the attributes of DNA and allozyme analyses.

3.2.3 Application of DNA loci

Mitochondrial DNA

The rate of nucleotide substitution in mtDNA is five to ten times as high as in nuclear DNA (Moritz *et al.*, 1987), and, as the mitochondrial genome is haploid, the effective population size for mtDNA is only one-fourth of that for nuclear genes (Nei & Tajima, 1981). Hence, the degree of mtDNA divergence should be higher than that of nuclear allozyme genes because of the effects of higher mutation rates and genetic drift. Moreover, if males of some species migrate during reproduction more often than females, the divergence of populations may be not manifested in allozymes but will occur with respect to mtDNA inherited maternally. If the intensity of migration is equal in both sexes, the exchange of migrants necessary to prevent divergence in mtDNA should occur at a rate four times as great as in the case of nuclear genes (Billington & Hebert, 1991).

Another important feature of mtDNA study is that mtDNA haplotypes, in contrast to allozyme and VNTR alleles, can be interconnected with one another by tracing the network of consecutive evolutionary events. Relationships between haplotypes are established with reference to the minimal number of stepwise mutational changes necessary for transformations of haplotypes into one another, which are determined from the appearance or loss of restriction sites. Taking into account the results obtained for all the restriction enzymes used, it is possible to construct the phylogeny of complex haplotypes using special computer programs developed for this purpose (McElroy *et al.*, 1991; Felsenstein, 1992). Molecular distances between mtDNA haplotypes may be included in estimations of genetic differentiation (Excoffier *et al.*, 1992). The analysis of haplotype phylogeny can also provide additional data on genetic relatedness of populations. Such an analysis proved to be especially informative in studies on revealing large phylogenetic groups in brown trout (Osinov & Bernatchez, 1996; Apostolidis *et al.*, 1996b).

Billington & Hebert (1991) noted that the use of mtDNA for analysing spatial genetic divergence provides the possibility for discriminating stocks that had been isolated in the past but subsequently came in contact and became indistinguishable by allozyme analysis. The point is that mtDNA can 'fix' the past history of isolation in a way that is impossible for the nuclear genome. If different alleles and clones of mtDNA have been fixed in populations during the period of isolation, their secondary contact results in the formation of a cline, which gradually levels off until the eventual disappearance of differences in allozyme alleles. The erosion of divergence in mtDNA proceeds much slower, and clones fixed during the period of isolation are retained in populations. As such processes accompany the evolution of populations, the analysis of the spatial pattern of mtDNA differentiation can also provide evidence of the allopatric origin of forms that are currently sympatric (Bernatchez & Dodson, 1990).

Thus, mtDNA is an efficient genetic marker of population divergence, and it has been used actively in the population analysis of salmonid fishes in recent years (see Chapter 6).

Spatial or temporal variation in populations for mtDNA often proves to be greater than that estimated by allozyme analysis. In one of the lakes located in British Columbia, for example, the ecotypes of kokanee (*Oncorhynchus nerka*) spawning in the tributaries and the shore waters of the lake differed in mtDNA, whereas their allozyme and minisatellite allele frequencies were similar (Taylor *et al.*, 1997). Hansen & Loeschcke (1996), analysing variation between generations in a small brown trout population, revealed it only with respect to mtDNA and observed no variation in allozyme allele frequencies. Adams *et al.* (1994) estimated the values of genetic identity (according to Nei, 1978) for different chinook salmon populations and stocks entering the rivers at different times: these values for mtDNA ranged from 0.454 to 0.998, and those for allozyme loci from 0.994 to 0.999. Some researchers, however, observed an inverse situation: thus, North American chum salmon populations differing in allozyme loci (Winans *et al.*, 1994) proved to be similar in mtDNA (Park *et al.*, 1993). This may be associated with the fact that the set of allozymes is encoded by a group of loci, whereas mtDNA represents only one locus which, in addition, was investigated only in part. In some brown trout and Atlantic salmon populations, the analysis of mtDNA and minisatellite DNA revealed no additional divergence compared to that in allozyme loci (Ferguson *et al.*, 1995; O'Connell *et al.*, 1995).

As mtDNA is more sensitive to the effect of evolutionary factors leading to the decrease of genetic variation in a population (e.g. to genetic drift, since the effective population size for mtDNA is smaller), it is especially valuable as a marker for determining genetic changes in the cultural brood stocks (Ferguson, M.M., 1995).

Significant geographic variation and maternal (clonal) inheritance make mtDNA an efficient genetic marker of different hatchery stocks and strains. Thus, several mtDNA haplotypes proved to be specific for several hatchery strains of lake and brook trout, which were used for restoring populations of these species in Lake Ontario (Danzmann *et al.*, 1991; Grewe *et al.*, 1993). Unique patterns of restriction fragments or haplotype frequencies sharply differing from those in natural populations were found in hatchery (brood) stocks and strains of rainbow trout (Palva & Palva, 1987), Atlantic salmon (Palva *et al.*, 1989) and brown trout (Gyllensten & Wilson, 1987; Hynes *et al.*, 1989; Hansen *et al.*, 1997). Such markers allow specialists to distinguish these stocks and strains reliably, to identify their origin, to estimate the efficiency of their use for introduction and in aquaculture, and to monitor the effects of their interaction with natural stocks (Billington & Hebert, 1991; Knox & Verspoor, 1991; Ferguson, 1995; Fontaine *et al.*, 1997). Thus, mtDNA analysis performed to evaluate the result of long-term introductions of brown trout fry from hatcheries into natural populations showed that the genetic contribution of introduced fishes was only 2%, compared to an expected 52–70% (Hansen *et al.*, 1995).

An interesting result was obtained in studies on mtDNA from nine hatchery populations of several salmon species and 22 neighbouring populations of the same species in California (Nielsen *et al.*, 1994). The diversity of hatchery populations proved to be significantly greater than that of natural populations, thus confirming the data on their origin from numerous and geographically remote sources.

The combined analysis of mtDNA and allozyme markers can be used successfully for revealing interspecific hybridization or introgression of genes from introduced to local species (Gyllensten *et al.*, 1985; Gyllensten & Wilson, 1987; Allendorf & Leary, 1988). In its course, the analysis of mtDNA provides information about directionality of hybridization (with respect to sex) or allows researchers to reveal certain asymmetry in gene exchange between the parental species (e.g. between introduced and local species; see Forbes & Allendorf, 1991). In studies on hybrids of Atlantic salmon and brown trout from Canadian rivers, which were identified by diagnostic allozyme markers, the analysis of mtDNA showed that hybridization was always unidirectional, i.e. female brown trout × male Atlantic salmon (McGowan &

Davidson, 1992); in other regions, however, reverse combinations were also found (Hartley, 1996). Unidirectional hybridization was observed in some cases of interaction between cultivated strains and natural populations (Ferguson, 1994).

Owing to a high rate of mtDNA evolution, its analysis is an efficient way to investigate phylogenetic relationships between close species. Thus, in a series of studies (Berg & Ferris, 1984; Thomas *et al.*, 1986; Gyllensten & Wilson, 1987), the analysis of mtDNA has demonstrated that Pacific *Salmo gairdneri* and *S. clarki* are closer to species of the genus *Oncorhynchus* than to Atlantic *Salmo* species. These results served as one of arguments for including Pacific *Salmo* species to the genus *Oncorhynchus* (Smith & Stearley, 1989). Subsequently, these relationships between the above taxa were confirmed in studies on individual fractions of nuclear DNA, including ribosomal (Phillips *et al.*, 1992), on retroposons, or SINEs (Murata *et al.*, 1993, 1996), and on the growth hormone gene (Phillips & Oakley, 1997). It should be noted, however, that other researchers who performed cladistic analysis with the use of different markers, such as mtDNA, allozymes and various morphological and biological traits, disagree with the aforementioned interpretation of taxonomic relationships between these groups (Oleinik, 1994; Glubokovsky, 1995) (see Chapter 2).

Nuclear DNA

Here, we shall concentrate on the use of mini- and microsatellite loci, the properties of which make them effective genetic markers: (Wright, 1993; O'Reilly & Wright, 1995, O'Connell & Wright 1997).

(1) The genome is very rich in such loci. Thus, according to a rough estimation, the numbers of $(GT)_n$ and $(CT)_n$ microsatellites in the *Salmo trutta* genome are 109 000 and 33 000, respectively (Estoup *et al.*, 1993).
(2) These loci are non-coding and, hence, selectively neutral (except for the cases of strong linkage with adaptively significant genes).
(3) They can be analysed using very small quantities of blood or tissues, and samples for this purpose can be taken from living animals (in fishes, dry scales or otoliths can be used as well).
(4) These loci evolve rapidly. The spontaneous mutation rate for mini- and microsatellite loci is about 10^{-2} to 10^{-4} per locus per generation, compared to 10^{-6} for allozyme genes (Nei, 1987). Hence, the divergence at mini- and microsatellite loci is influenced by both genetic drift and mutations, whereas that at allozyme genes (selectively neutral) is influenced by drift only (Ward & Grewe, 1994). Heterozygosity for minisatellites may exceed that for allozyme genes by more than an order of magnitude, reaching almost 100% (Jeffreys *et al.*, 1988; Huang *et al.*, 1992); the degree of polymorphism in microsatellites is variable (O'Connell & Wright, 1997).

(5) Micro- and minisatellites (single locus) usually demonstrate Mendelian codominant inheritance.
(6) Microsatellite loci of closely related species are similar, enabling use of the same primers and analytical procedures (e.g. see Scribner *et al.*, 1996).
(7) The analysis of microsatellites can be automated (Olsen *et al.*, 1996).

Both mini- and microsatellites have already been used in the population analysis of salmonid fishes (see Chapter 6). Owing to a high level of polymorphism, these markers have proven useful for studying species with low allozyme variation such as coho salmon (Small *et al.*, 1998), Atlantic salmon (McConnell *et al.*, 1995; Tessier *et al.*, 1995; Galvin *et al.*, 1996; Nilsson, 1997) and Arctic charr (Brunner *et al.*, 1998). In Atlantic salmon, for example, several microsatellite loci with very high levels of polymorphism (up to 38 alleles per locus) and heterozygosity (up to 0.9) were identified (McConnell *et al.*, 1995).

Moreover, as the PCR technique allows the analysis of microsatellite DNA extracted from scales (Nielsen *et al.*, 1997) or otoliths (Hutchinson *et al.*, 1999), it is possible to trace variation in certain populations retrospectively on the basis of archived collections.

It is highly probable that these markers will prove their efficiency even in studies on inbred stocks (e.g. in conditions of aquaculture) and sharply declining, geographically close, or recently separated populations (Wright & Bentzen, 1994).

Indeed, a level of genetic differentiation similar to that typical for populations of river basins was demonstrated with the aid of minisatellite loci for *Salmo salar* from the system of tributaries of the Shannon River in Ireland (Galvin *et al.*, 1996). This population suffered a drastic decline after the Shannon hydroelectric scheme was constructed in the 1920s, but the subpopulation inhabiting one of the tributaries remained largely undamaged. Its intensive breeding over the past 30 years provided for the recovery of population size, and salmon populated other tributaries. This was accompanied by genetic differentiation, which was traced using minisatellite variation.

Owing to very high allelic diversity, many micro- and minisatellites allow successful identification of individuals and their family relations. This concerns both multi-locus minisatellites (DNA fingerprints) and combinations of several multi-allelic microsatellites or single-locus minisatellites, which can produce a great number of different genotypes. Ferguson *et al.* (1995) noted that five loci with ten alleles per locus can form more than 500 million combinations of genotypes and, hence, are sufficient for identifying individual families in most of the corresponding studies.

Thus, these markers allow researchers to identify progenies of certain parents not only in the first, but also in subsequent generations, providing an opportunity for analysis of reproductive success and fitness of individuals in the wild differing in their biological and ecological parameters. It is important that extensive variation ensures feasibility of these studies in both experimental and natural populations

(Ferguson *et al.*, 1995). Thus, in studies on reproductive success of Atlantic salmon mature parr spawning with adult females, six single-locus minisatellites were used in the paternity assessment of the progeny. The experiments showed that such males are able to fertilize 25–89% of the total eggs (Moran *et al.*, 1996).

Based on the possibility of determining the degree of kinship among experimental progeny with the aid of minisatellite markers, Mousseau *et al.* (1998) proposed a new method for assessing heritability of quantitative traits. This method makes unnecessary the long-term controlled breeding in long-lived species, such as salmonids, because studies are performed on the progeny of spawners taken from natural populations. Estimates of heritability indices for chinook salmon from such populations – 0.20 for body weight, 0.38 for body length, 0.67 for early sexual maturation (jacking), and 0.76 for muscle colour – corresponded to those obtained by the classic methods of quantitative genetics.

It is apparent that mini- and microsatellite loci will be used increasingly in aquaculture as genetic markers associated with other genomic loci controlling commercially important traits (see Jackson *et al.*, 1998).

Takezaki & Nei (1996) demonstrated that microsatellite loci are very useful for analysing evolutionary relationships of closely related populations. However, both mini- and microsatellite loci are less effective in phylogenetic studies because of broad intrapopulation variation (Morris *et al.*, 1996), the mechanisms of which still remain unclear (Wright & Bentzen, 1994).

Conclusion

The data discussed demonstrate the significant progress made in studying biochemical and molecular genetic variation in salmonid fishes. We may assert without exaggeration that today this information reveals unique, previously inaccessible opportunities for a comprehensive analysis of genetic structure and dynamics of salmon populations as units of evolution and commercial activity.

In this book, we shall not be concerned with the interesting, but in many respects contradictory, aspects of speciation and evolution. In accordance with the title of the book, our principal objective is to demonstrate how a genetic approach has contributed to the biology and conservation of salmon populations.

In the following chapters, in addition to the data presented in Chapter 2, we will review evidence that at least one of the significant features of native populations of Pacific and Atlantic salmon is their temporal genetic stability, inferred from their patterns of spatial and temporal organization.

Appendix

Table 3.1 Enzymes and coding genetic loci studied in pink salmon (*Oncorhynchus gorbuscha*) populations.

Enzyme, EC number	Locus	Number of alleles	Tissue	Level of variation
Aspartate aminotransferase, 2.6.1.1	*mAAT-1**	4	*mh*	R
	*mAAT-2**	2	*ml*	R
	*sAAT-1,2**	4	*mh*	R
	*sAAT-3**	4	*e*	P
	*sAAT-4**	4	*l*	P
Acid phosphatase, 3.1.3.2	*ACP-1**	1	*l*	M
	*ACP-2**	1	*l*	M
Adenosine deaminase, 3.5.4.4	*ADA-1**	3	*m*	R
	*ADA-2**	5	*mhe*	P
Alcohol dehydrogenase, 1.1.1.1	*ADH**	1	*l*	M
Aconitate hydratase, 4.2.1.3	*mAH-1**	3	*mh*	R
	*mAH-2**	2	*mh*	R
	*mAH-3**	3	*mh*	R
	*mAH-4**	5	*mh*	P
	*sAH-1**	4	*lmh*	P
	*sAH-2**	1	*mhl*	M
	*sAH-3**	1	*mhl*	M
Adenylate kinase, 2.7.4.3	*AK**	2	*me*	R
Alanine aminotransferase, 2.6.1.2	*ALAT**	6	*mh*	P
Creatine kinase, 2.7.3.2	*CK-A1**	4	*m*	R
	*CK-A2**	5	*m*	R
	*CK-B**	2	*e*	R
	*CK-C1**	2	*e*	P
	*CK-C2**	3	*e*	R
Cytochrome-b reductase, 1.6.2.2	*CYBR**	4	*e*	P
Enolase, 4.2.1.11	*ENO**	1	*m*	M
Esterase D, 3.1.1.–	*ESTD**	1	*m*	M
Fructose-1, 6-bisphosphate aldolase, 4.1.2.13	*FBALD-3**	1	*e*	M
	*FBALD-4**	1	*e*	M
Formaldehyde dehydrogenase, 1.2.1.1	*FDHG**	3	*lm*	P
Fumarate hydratase, 4.2.1.2	*FH**	2	*m*	R

Cont.

Table 3.1 Cont.

Enzyme, EC number	Locus	Number of alleles	Tissue	Level of variation
β-Acetylgalactose aminidase, 3.2.1.53	*bGALA**	1	*l*	M
Guanine deaminase, 3.5.4.3	*GDA**	8	*l*	P
Glyceraldehyde-3-phosphate dehydrogenase, 1.2.1.12	*GAPDH-1**	1	*m*	M
	*GAPDH-2**	1	*h*	M
	*GAPDH-3**	1	*h*	M
	*GAPDH-4**	1	*e*	M
	*GAPDH-5**	1	*e*	M
Glucose-6-phosphate isomerase, 5.3.1.9	*GPI-A**	5	*m*	R
	*GPI-B1**	5	*m*	R
	*GPI-B2**	3	*hm*	P
Glycerol-3-phosphate dehydrogenase, 1.1.1.8	*G3PDH-1**	6	*mh*	P
	*G3PDH-2**	1	*h*	M
	*G3PDH-3**	1	*h*	M
Glutathione reductase, 1.6.4.2	*GR**	3	*el*	R
β-Glucuronidase, 3.2.1.31	*bGUS**	1	*m*	M
Hydroxyacylglutathione hydrolase, 3.1.2.6	*HAGH**	4	*mh*	P
L-Iditol dehydrogenase, 1.1.1.14	*sIDDH**	1	*l*	M
Isocitrate dehydrogenase, 1.1.1.42	*mIDHP-1**	4	*mh*	R
	*mIDHP-2**	2	*mh*	R
	*sIDHP-1**	6	*le*	P
	*sIDHP-2**	1	*l*	M
Lactate dehydrogenase, 1.1.1.27	*LDH-A1**	2	*ml*	P
	*LDH-A2**	1	*m*	M
	*LDH-B1**	4	*mhe*	R
	*LDH-B2**	4	*mle*	R
	*LDH-C**	3	*e*	R
Malate dehydrogenase, 1.1.1.37	*mMDH-1**	2	*mh*	R
	*mMDH-2**	1	*h*	M
	*sMDH-A1,2**	7	*lhem*	R
	*sMDH-B1,2**	7	*mhe*	R
Malic enzyme ($NADP^+$), 1.1.1.40	*mMEP-1**	3	*mh*	R
	*mMEP-2**	3	*mhle*	R
	*sMEP**	1	*mhle*	M
Mannose-6-phosphate isomerase, 5.3.1.8	*MPI**	3	*mhe*	P

Cont.

Table 3.1 Cont.

Enzyme, EC number	Locus	Number of alleles	Tissue	Level of variation
Nucleoside triphosphate pyrophosphatase, 3.6.1.19	*NTP**	2	*ml*	R
Peptidases, 3.4.–.–				
Gly–Leu	*PEPA**	3	*mhle*	R
	*PEPC**	1	*e*	M
Leu–Gly–Gly	*PEPB-1**	4	*mhel*	R
	*PEPB-2**	1	*mhe*	M
Phe–Pro	*PEPD-1**	2	*he*	R
	*PEPD-2**	5	*mhle*	P
Leu–Tyr	*PEP-LT1**	1	*mh*	M
	*PEP-LT2**	4	*mhl*	P
Phosphogluconate dehydrogenase, 1.1.1.44	*PGDH**	6	*mhe*	P
Phosphoglycerate kinase, 2.7.2.3	*PGK-1**	2	*lmhe*	R
	*PGK-2**	3	*lmhe*	R
Phosphoglucomutase, 5.4.2.2	*PGM-1**	1	*mhe*	M
	*PGM-2**	5	*hme*	P
Pyruvate kinase, 2.7.1.40	*PK-1**	1	*h*	M
	*PK-2**	1	*mhe*	M
Purine nucleoside phosphorylase, 2.4.2.1	*PNP-1**	1	*e*	M
	*PNP-2**	1	*e*	M
Superoxide dismutase, 1.15.1.1	*mSOD**	3	*h*	R
	*sSOD-1**	2	*lm*	R
	*sSOD-2**	2	*hl*	R
Triose phosphate isomerase, 5.3.1.1	*TPI-1**	3	*leh*	R
	*TPI-2**	4	*he*	R
	*TPI-3**	2	*mhe*	R
	*TPI-4**	2	*he*	R
Xanthine oxidase, 1.1.3.22	*XO**	1	*l*	M

M – monomorphic, R – polymorphic locus with frequency of common allele > 0.95, P – polymorphic locus with frequency of common allele < 0.95. Tissue: *m* – muscle, *h* – heart muscle, *e* – eye retina, *l* – liver.

Data from Salmenkova & Omelchenko, 1983a,b; Gharrett *et al.*, 1988; Shaklee *et al.*, 1990b, 1991; Oleinik, 1994; Shaklee & Varnavskaya, 1994.

Table 3.2 Enzymes and coding genetic loci studied in chum salmon (*Oncorhynchus keta*) populations.

Enzyme, EC number	Locus	Number of alleles	Tissue	Level of variation
Aspartate aminotransferase, 2.6.1.1	*mAAT-1**	2	*mhl*	P
	*mAAT-2**	2	*lh*	P
	*sAAT-1,2**	4	*mh*	P
	*sAAT-3**	4	*e*	P
	*sAAT-4**	?	*l*	R
Acid phosphatase, 3.1.3.2	*ACP-1**	2	*l*	R
Adenosine deaminase, 3.5.4.4	*ADA-1**	?	*ehl*	R
	*ADA-2**	1	*meh*	M
Alcohol dehydrogenase, 1.1.1.1	*ADH**	1	*l*	M
Aconitate hydratase, 4.2.1.3	*mAH-1,2**	3	*hl*	R
	*mAH-3**	2	*hl*	P
	*mAH-4**	3	*h*	R
	*sAH**	3	*l*	R
Adenylate kinase, 2.7.4.3	*AK**	1	*m*	M
Alanine aminotransferase, 2.6.1.2	*ALAT**	4	*m*	P
Creatine kinase, 2.7.3.2	*CK-A1*	3	*m*	R
	*CK-A2**	3	*m*	R
	*CK-B**	2	*e*	R
	*CK-C1**	3	*e*	R
	*CK-C2**	2	*e*	R
Diaphorase, 1.6.4.3	*DIA**	1	*h*	M
Enolase, 4.2.1.11	*ENO-1**	1	?	M
	*ENO-2**	1	?	M
Esterase, 3.1.1.–	*EST**	1	*l*	M
Esterase D, 3.1.1.–	*ESTD**	3	*mh*	P
Fructose-1, 6-bisphosphate aldolase, 4.1.4.13	*FBALD-3**	1	?	M
	*FBALD-4**	1	?	M
Formaldehyde dehydrogenase, 1.2.1.1	*FDHG**	2	*l*	R
Fumarate hydratase, 4.2.1.2	*FH**	3	*mel*	P
β-Galactosidase, 3.2.1.23	*bGAL**	?	*l*	P
β-*N*-Acetyl galactosaminidase, 3.2.1.53	*bGALA**	4	*l*	P

Cont.

Table 3.2 Cont.

Enzyme, EC number	Locus	Number of alleles	Tissue	Level of variation
Glyceraldehyde-3-phosphate dehydrogenase, 1.2.1.12	*GAPDH-1**	1	*m*	M
	*GAPDH-2**	2	*h*	P
	*GAPDH-3**	2	*hm*	P
	*GAPDH-4**	1	*e*	M
	*GAPDH-5**	3	*e*	R
Guanine deaminase, 3.5.4.3	*GDA**	?	*l*	P
N-Acetyl-β-glucosaminidase, 3.2.1.30	*bGLUA**	2	*l*	P
Glucose-6-phosphate isomerase, 5.3.1.9	*GPI-A**	3	*mhe*	R
	*GPI-B1,2**	3	*mh*	R
Glycerol-3-phosphate dehydrogenase, 1.1.1.8	*G3PDH-1**	3	*mh*	R
	*G3PDH-2**	3	*hm*	P
	*G3PDH-3**	2	*hm*	R
	*G3PDH-4**	1	*hm*	M
Glutathione reductase, 1.6.4.2	*GR**	?	*eh*	M
β-Glucuronidase, 3.2.1.31	*GUS**	1	*l*	M
Hydroxyacylglutathione hydrolase, 3.1.2.6	*HAGH**	1	*m*	M
L-Iditol dehydrogenase, 1.1.1.14	*IDDH-1**	2	*l*	R
	*IDDH-2**	3	*l*	R
Isocitrate dehydrogenase, 1.1.1.42	*mIDHP-1**	4	*mh*	P
	*mIDHP-2**	1	*mh*	M
	*sIDHP-1**	4	*l*	R
	*sIDHP-2**	7	*l*	P
Lactate dehydrogenase, 1.1.1.27	*LDH-A1**	4	*m*	P
	*LDH-A2**	1	*m*	M
	*LDH-B1**	3	*meh*	R
	*LDH-B2**	4	*lme*	P
	*LDH-C**	2	*e*	R
Lactoylglutathione lyase, 4.4.1.5	*LGL**	2	*m*	R
α-Mannosidase, 3.2.1.24	*aMAN**	1	*leh*	M
Malate dehydrogenase, 1.1.1.37	*mMDH-1**	1	*mhel*	R
	*mMDH-2**	2	*mh*	R
	*mMDH-3**	3	?	R
	*sMDH-A1**	4	*leh*	P
	*sMDH-A2**	3	*leh*	R
	*sMDH-B1,2**	6	*hml*	P
Malic enzyme ($NADP^+$), 1.1.1.40	*mMEP-1**	?	*m*	R
	*mMEP-2**	3	*mh*	P
	*sMEP-1**	3	*mlh*	P
	*sMEP-2**	1	*m*	M

Cont.

Table 3.2 Cont.

Enzyme, EC number	Locus	Number of alleles	Tissue	Level of variation
Mannose-6-phosphate isomerase, 5.3.1.8	*MPI**	4	*leh*	P
Nucleoside triphosphate pyrophosphatase, 3.6.1.19	*NTP**	2	*m*	P
Peptidases, 3.4.–.–				
Gly–Leu	*PEPA**	3	*me*	R
Leu-Gly–Gly	*PEPB-1**	6	*lme*	P
	*PEPB-2**	2	*hm*	P
Phe–Pro	*PEPD-1**	2	*mh*	R
Leu-Tyr	*PEP-LT**	4	*ml*	P
Phosphogluconate dehydrogenase, 1.1.1.44	*PGDH**	5	*lme*	P
Phosphoglycerate kinase, 2.7.2.3	*PGK-1**	2	*meh*	R
	*PGK-2**	1	*meh*	M
Phosphoglucomutase, 5.4.2.2	*PGM-1**	2	*meh*	P
	*PGM-1r**	2	*meh*	P
	*PGM-2**	4	*meh*	R
	*PGM-3,4**	?	*l*	P
Pyruvate kinase, 2.7.1.40	*PK-1**	1	*h*	M
	*PK-2**	2	*h*	R
Purine nucleoside phosphorylase, 2.4.2.1	*PNP**	1	*e*	M
Superoxide dismutase, 1.15.1.1	*mSOD**	1	?	M
	*sSOD-1**	2	*lm*	R
	*sSOD-2**	1	?	M
Triose phosphate isomerase, 5.3.1.1	*TPI-1**	3	*ehm*	R
	*TPI-2**	2	*ehm*	R
	*TPI-3**	3	*ehm*	R
	*TPI-4**	3	*ehm*	R
Xanthine dehydrogenase, 1.2.3.2	*XDH**	1	*e*	M

M – monomorphic, R – polymorphic locus with frequency of common allele > 0.95, P – polymorphic locus with frequency of common allele < 0.95. Tissue: *m* – muscle, *h* – heart muscle, *e* – eye retina, *l* – liver.

Data from Okazaki, 1982b; Salmenkova *et al.*, 1992; Kondzela *et al.*, 1994; Phelps *et al.*, 1994b; Wilmot *et al.*, 1994; Winans *et al.*, 1994.

Table 3.3 Enzymes and coding genetic loci studied in sockeye salmon (*Oncorhynchus nerka*) populations.

Enzyme, EC number	Locus	Number of alleles	Tissue	Level of variation
Aspartate aminotransferase, 2.6.1.1	*mAAT-1**	2	*mh*	P
	*mAAT-2**	2	*l*	R
	*sAAT-1,2**	3	*mh*	R
	*sAAT-3**	2	*e*	R
	*sAAT-4**	1	*l*	M
Acid phosphatase, 3.1.3.2	*ACP-1**	1	*l*	M
	*ACP-2**	1	*l*	M
Adenosine deaminase, 3.5.4.4	*ADA-1**	2	*m*	R
	*ADA-2**	3	*m*	R
Aconitate hydratase, 4.2.1.3	*mAH-1,2**	2	*mhl*	R
	*mAH-3**	2	*mh*	P
	*mAH-4**	1	*mh*	M
	*sAH-1**	1	*mhl*	M
	*sAH-2**	3	*mhl*	R
	*sAH-3**	1	*em*	R
Adenylate kinase, 2.7.4.3	*AK**	1	*em*	R
Alanine aminotransferase, 2.6.1.2	*ALAT-1**	?	*h*	R
	*ALAT-2**	5	*mh*	P
Aldolase, 4.1.2.7	*ALD-4**	1	*e*	M
Creatine kinase, 2.7.3.2	*CK-A1*	2	*m*	R
	*CK-A2**	1	*m*	M
	*CK-B**	2	*e*	R
	*CK-C1**	1	*e*	M
	*CK-C2**	1	*e*	M
Enolase, 4.2.1.11	*ENO**	1	*l*	M
Esterase, 3.1.1.–	*EST**	1	*m*	M
Esterase D, 3.1.1.–	*ESTD**	1	*m*	M
Fructose-1,6-bisphosphate aldolase, 4.1.2.13	*FBALD-3**	1	*e*	M
	*FBALD-4**	2	*e*	P
Formaldehyde dehydrogenase, 1.2.1.1	*FDHG**	1	*m*	M
Fumarate hydratase, 4.2.1.2	*FH**	2	*m*	R
β-*N*-Acetylgalactosaminidase, 3.2.1.53	*bGALA**	1	*l*	M
Glyceraldehyde-3-phosphate dehydrogenase, 1.2.1.12	*GAPDH-1**	1	*m*	M
	*GAPDH-2**	1	*h*	M
	*GAPDH-3**	1	*h*	M
	*GAPDH-4**	2	*e*	R
	*GAPDH-5**	1	*e*	M

Cont.

Table 3.3 Cont.

Enzyme, EC number	Locus	Number of alleles	Tissue	Level of variation
Guanine deaminase, 3.5.4.3	*GDA-1**	1	*l*	M
	*GDA-2**	1	*l*	M
N-Acetyl-β-glucosaminidase, 3.2.1.30	*bGLUA**	1	*me*	M
Glyoxalase I, 4.4.1.5	*GLO I**	1	*l*	M
Glyoxalase II, 3.1.2.6	*GLO II**	1	*l*	M
Glucose-6-phosphate isomerase, 5.3.1.9	*GPI-A**	4	*m*	P
	*GPI-B1,2**	3	*m*	P
Glycerol-3-phosphate dehydrogenase, 1.1.1.8	*G3PDH-1**	2	*m*	R
	*G3PDH-2**	1	*h*	M
	*G3PDH-3**	1	*h*	M
Glutathione reductase, 1.6.4.2	*GR**	1	*e*	M
Hydroxyacylglutathione hydrolase, 3.1.2.6	*HAGH**	1	*mh*	M
Isocitrate dehydrogenase, 1.1.1.42	*mIDHP-1**	2	*mh*	R
	*mIDHP-2**	2	*mh*	R
	*sIDHP-1**	3	*el*	R
	*sIDHP-2**	3	*el*	P
Inosine triphosphatase, 3.6.1.19	*ITP-1,2**	1	*m*	M
Lactate dehydrogenase, 1.1.1.27	*LDH-A1**	1	*m*	M
	*LDH-A2**	2	*m*	R
	*LDH-B1**	3	*hme*	R
	*LDH-B2**	3	*mle*	P
	*LDH-C**	3	*e*	R
Malate dehydrogenase, 1.1.1.37	*mMDH-1**	2	*h*	R
	*mMDH-2**	1	*h*	M
	*sMDH-A1**	3	*leh*	R
	*sMDH-A2**	1	*lhe*	M
	*sMDH-B1,2**	4	*mhe*	P
Malic enzyme ($NADP^+$), 1.1.1.40	*mMEP-1**	2	*m*	P
	*mMEP-2**	1	*mhle*	M
	*sMEP-1**	2	*m*	P
	*MEP-3**	1	*m*	M
Mannose-6-phosphate isomerase, 5.3.1.8	*MPI**	2	*hm*	P
Nucleoside triphosphate pyrophosphatase, 3.6.1.19	*NTP**	1	*m*	M

Cont.

Table 3.2 Cont.

Enzyme, EC number	Locus	Number of alleles	Tissue	Level of variation
Peptidases, 3.4.–.–				
Gly–Leu	*PEPA**	2	*mhel*	R
	*PEPC**	2	*e*	P
Leu–Gly–Gly	*PEPB-1**	1	*mhe*	M
	*PEPB-2**	1	*mhe*	M
Phe–Pro	*PEPD-1**	1	*he*	M
	*PEPD-2**	2	*mhel*	R
Leu–Gly–Leu	*PEP-LGL**	1	*m*	M
Leu–Tyr	*PEP-LT1**	1	*mh*	M
	*PEP-LT2**	3	*mh*	P
Phosphogluconate dehydrogenase, 1.1.1.44	*PGDH**	3	*mh*	P
Phosphoglycerate kinase, 2.7.2.3	*PGK-1**	2	*mhe*	R
	*PGK-2**	1	*mhe*	M
Phosphoglucomutase, 5.4.2.2	*PGM-1**	4	*mhe*	P
	*PGM-2**	4	*mhe*	P
	*PGM-3,4**	4	*mh*	P
Pyruvate kinase, 2.7.1.40	*PK-1**	1	*h*	M
	*PK-2**	2	*mhe*	R
Purine nucleoside phosphorylase, 2.4.2.1	*PNP-1**	1	*e*	M
	*PNP-2**	1	*e*	M
Superoxide dismutase, 1.15.1.1	*mSOD**	1	*h*	R
	*sSOD-1**	1	*l*	M
	*sSOD-2**	1	*l*	M
Triose phosphate isomerase, 5.3.1.1	*TPI-1**	1	*he*	M
	*TPI-2**	1	*he*	M
	*TPI-3**	2	*he*	R
	*TPI-4**	1	*he*	M
Xanthine oxidase, 1.1.3.22	*XO**	1	*l*	M

M – monomorphic, R – polymorphic locus with frequency of common allele > 0.95, P – polymorphic locus with frequency of common allele < 0.95. Tissue: *m* – muscle, *h* – heart muscle, *e* – eye retina, *l* – liver.
Data from Utter *et al.*, 1984; Wood *et al.*, 1994; Varnavskaya *et al.*, 1994a; Oleinik, 1994.

Table 3.4 Enzymes and coding genetic loci studied in chinook salmon (*Oncorhynchus tshawytscha*) populations.

Enzyme, EC number	Locus	Number of alleles	Tissue	Level of variation
Aspartate aminotransferase, 2.6.1.1	*mAAT-1**	3	*mh*	R
	*mAAT-2**	3	*mhl*	P
	*mAAT-3**	2	*mh*	R
	*sAAT-1,2**	3	*mh*	P
	*sAAT-3**	2	*e*	R
	*sAAT-4**	3	*l*	P
Acid phosphatase, 3.1.3.2	*ACP-1**	1	*ml*	M
	*ACP-2**	1	*m*	M
Adenosine deaminase, 3.5.4.4	*ADA-1**	3	*m*	P
	*ADA-2**	1	*m*	M
Alcohol dehydrogenase, 1.1.1.1	*ADH**	2	*l*	R
Aconitate hydratase, 4.2.1.3	*mAH-1**	2	*eh*	P
	mAH-2	2	*el*	R
	*mAH-3**	2	*mh*	R
	*mAH-4**	4	*mh*	P
	*sAH-**	3	*lme*	P
Alanine aminotransferase, 2.6.1.2	*ALAT**	2	*mh*	P
Creatine kinase, 2.7.3.2	*CK-A1*	1	*m*	M
	*CK-A2**	?	*m*	P
	*CK-B**	1	*e*	M
	*CK-C1**	1	*e*	M
	*CK-C2**	4	*e*	R
Diaphorase, 1.6.4.3	*DIA**	1	*m*	M
Esterase, 3.1.1.–	*EST-1**	1	*m*	M
	*EST-2**	1	*m*	M
Esterase D, 3.1.1.–	*ESTD**	3	*me*	R
Fructose-1,6-bisphosphate aldolase, 4.1.2.13	*FBALD-4**	1	*e*	M
Formaldehyde dehydrogenase, 1.2.1.1	*FDHG**	1	*m*	M
Fumarate hydratase, 4.2.1.2	*FH**	1	*mh*	M
β-Galactosidase, 3.2.1.23	*bGAL**	1	*l*	M
β-*N*-Acetylgalactosaminidase, 3.2.1.53	*bGALA**	4	*l*	P
Glyceraldehyde-3-phosphate dehydrogenase, 1.2.1.12	*GAPDH-5**	1	*e*	M
	*GAPDH-6**	1	*e*	M
N-Acetyl-β-glucosaminidase, 3.2.1.30	*bGLUA**	1	*l*	M

Cont.

Table 3.4 Cont.

Enzyme, EC number	Locus	Number of alleles	Tissue	Level of variation
Glucose-6-phosphate isomerase, 5.3.1.9	*GPI-A**	1	*meh*	M
	*GPI-B1**	2	*m*	R
	*GPI-B2**	3	*m*	P
	*GPI-B1r**	2	*m*	P
Glycerol-3-phosphate dehydrogenase, 1.1.1.8	*G3PDH-1**	1	*m*	M
	*G3PDH-2**	1	*m*	M
	*G3PDH-3**	1	*m*	M
	*G3PDH-4**	1	*m*	M
Glutathione reductase, 1.6.4.2	*GR**	2	*mel*	P
β-Glucuronidase, 3.2.1.31	*bGUS**	1	*m*	M
Hydroxyacylglutathione hydrolase, 3.1.2.6	*HAGH**	3	*lme*	R
L-Iditol dehydrogenase, 1.1.1.14	*IDDH-1**	2	*l*	P
	*IDDH-2**	3	*l*	P
Isocitrate dehydrogenase, 1.1.1.42	*mIDHP-1**	1	*meh*	M
	*mIDHP-2**	2	*m*	P
	*sIDHP-1**	7	*mel*	R
	*sIDHP-2**	4	*el*	P
Lactate dehydrogenase, 1.1.1.27	*LDH-A1**	2	*m*	P
	*LDH-A2**	1	*m*	M
	*LDH-B1**	?	*he*	P
	*LDH-B2**	4	*le*	R
	*LDH-C**	3	*e*	P
α-Mannosidase, 3.2.1.24	*MAN**	1	*l*	M
Malate dehydrogenase, 1.1.1.37	*mMDH-1**	2	*me*	P
	*mMDH-2**	2	*mh*	P
	*sMDH-A1,2**	4	*em*	P
	*sMDH-B1,2**	4	*me*	P
Malic enzyme ($NADP^+$), 1.1.1.40	*mMEP*	1	*m*	M
	*sMEP-1**	3	*hlm*	P
	*sMEP-2**	2	*melh*	P
Mannose-6-phosphate isomerase, 5.3.1.8	*MPI**	2	*eml*	P
Peptidases, 3.4.–.–				
Gly–Leu	*PEPA**	2	*meh*	P
	*PEPC**	1	*e*	M
Leu–Gly–Gly	*PEPB-1**	4	*mehl*	P
	*PEPB-2**	1	*e*	M
Phe–Pro	*PEPD-2**	3	*me*	R
Leu–Tyr	*PEP-LT**	2	*mh*	P
Phosphogluconate dehydrogenase, 1.1.1.44	*PGDH**	3	*mel*	P

Cont.

Table 3.4 Cont.

Enzyme, EC number	Locus	Number of alleles	Tissue	Level of variation
Phosphoglycerate kinase, 2.7.2.3	*PGK-1**	1	*l*	M
	*PGK-2**	2	*mel*	P
Phosphoglucomutase, 5.4.2.2	*PGM-1**	3	*meh*	P
	*PGM-2**	4	*mel*	P
	*PGM-3**	2	*elm*	P
	*PGM-4**	6	*elm*	P
Pyruvate kinase, 2.7.1.40	*PK-1**	1	*m*	M
	*PK-2**	1	*m*	M
Superoxide dismutase, 1.15.1.1	*mSOD-1**	2	*hml*	P
	*sSOD-1**	4	*lm*	P
Triose phosphate isomerase, 5.3.1.1	*TPI-1**	1	*hm*	M
	*TPI-2**	3	*hm*	P
	*TPI-3**	3	*e*	P
	*TPI-4**	4	*melh*	P

M – monomorphic, R – polymorphic locus with frequency of common allele > 0.95, P – polymorphic locus with frequency of common allele < 0.95. Tissue: *m* – muscle, *h* – heart muscle, *e* – eye retina, *l* – liver.
Data from Bartley *et al.*, 1992b; Wilmot *et al.*, 1992.

Table 3.5 Enzymes and coding genetic loci studied in coho salmon (*Oncorhynchus kisutch*) populations.

Enzyme, EC number	Locus	Number of alleles	Tissue	Level of variation
Aspartate aminotransferase, 2.6.1.1	*AAT-1**	1	*ml*	M
	*AAT-2**	2	*ml*	P
	*AAT-3**	2	*ml*	P
Alcohol dehydrogenase, 1.1.1.1	*ADH**	2	*l*	P
Aconitate hydratase, 4.2.1.3	*AH-1**	3	*l*	P
	*AH-2**	1	*m*	M
	*AH-3**	1	*m*	M
Adenylate kinase, 2.7.4.3	*AK-1**	1	*me*	M
	*AK-2**	1	*me*	M
Creatine kinase, 2.7.3.2	*CK-A1*	1	*me*	M
	*CK-A2**	2	*me*	P
	*CK-C2**	2	*me*	P
	*CK-4**	1	*me*	M
	*CK-5**	1	*me*	M
Diaphorase, 1.6.4.3	*DIA**	1	*m*	M

Cont.

Table 3.5 Cont.

Enzyme, EC number	Locus	Number of alleles	Tissue	Level of variation
Esterase, 3.1.1.–	*EST-1**	1	*m*	M
	*EST-2**	1	*m*	M
Esterase D, 3.1.1.–	*ESTD**	1	*m*	M
Fructose-1,6-bisphosphate aldolase, 4.1.2.13	*FBALD**	1	*e*	M
Formaldehyde dehydrogenase, 1.2.1.1	*FDHG**	1	*m*	M
β-*N*-Acetylgalactosaminidase, 3.2.1.53	*bGALA**	2	*l*	P
Glucose-6-phosphate isomerase, 5.3.1.9	*GPI-A**	4	*m*	P
	*GPI-B1**	1	*m*	M
	*GPI-B2**	3	*m*	P
Glycerol-3-phosphate dehydrogenase, 1.1.1.8	*G3PDH-1**	1	*m*	M
	*G3PDH-2**	1	*m*	M
	*G3PDH-3**	1	*m*	M
L-Iditol dehydrogenase, 1.1.1.14	*IDDH-1**	2	*l*	P
	*IDDH-2**	1	*l*	M
Isocitrate dehydrogenase, 1.1.1.42	*mIDHP-1**	2	*mel*	P
	*mIDHP-2**	2	*mel*	P
	*sIDHP-1**	4	*mel*	P
	*sIDHP-2**	4	*mel*	P
Lactate dehydrogenase, 1.1.1.27	*LDH-A1**	1	*mel*	M
	*LDH-A2**	1	*mel*	M
	*LDH-B1**	2	*mel*	R
	*LDH-B2**	2	*mel*	P
	*LDH-C**	1	*e*	M
Malate dehydrogenase, 1.1.1.37	*sMDH-A1**	1	*ml*	M
	*sMDH-A2**	1	*ml*	M
	*sMDH-B1,2**	3	*ml*	P
Malic enzyme (NADP$^+$), 1.1.1.40	*MEP-1**	1	*m*	M
	*MEP-2**	1	*m*	M
	*MEP-3**	1	*m*	M
Mannose-6-phosphate isomerase, 5.3.1.8	*MPI**	3	*mel*	P
Peptidases, 3.4.–.–				
Gly–Leu	*PEPA**	4	*m*	P
	*PEPC**	3	*e*	P
Leu–Gly–Gly	*PEPB-1**	1	*m*	M
Phe–Pro	*PEPD**	3	*m*	P
Phosphogluconate dehydrogenase, 1.1.1.44	*PGDH**	2	*ml*	P

Cont.

Table 3.5 Cont.

Enzyme, EC number	Locus	Number of alleles	Tissue	Level of variation
Phosphoglycerate kinase, 2.7.2.3	*PGK-1**	1	*ml*	M
Phosphoglucomutase, 5.4.2.2	*PGM-1**	2	*ml*	P
	*PGM-2**	2	*ml*	P
Superoxide dismutase, 1.15.1.1	*SOD**	1	*l*	M
Transferrin	*TF**	4	*s*	P

M – monomorphic, R – polymorphic locus with frequency of common allele > 0.95, P – polymorphic locus with frequency of common allele < 0.95. Tissue: *m* – muscle, *h* – heart muscle, *e* – eye retina, *l* – liver, *s* – blood serum.
Data from Bartley *et al.*, 1992c; Oleinik, 1994.

Table 3.6 Enzymes and coding genetic loci studied in masu salmon (*Oncorhynchus masou*) populations.

Enzyme, EC number	Locus	Number of alleles	Tissue	Level of variation
Aspartate aminotransferase, 2.6.1.1	*mAAT-1**	2	*m*	?
	*mAAT-2**	1	*m*	M
	*sAAT-1**	3	*l*	?
	*sAAT-2**	1	*l*	M
	*sAAT-3**	1	*e*	M
Alcohol dehydrogenase, 1.1.1.1	*ADH**	2	*l*	P
Aconitate hydratase, 4.2.1.3	*AH-1**	1	*m*	M
	*AH-2**	2	*m*	P
	*AH-3**	2	*m*	P
Adenylate kinase, 2.7.4.3	*AK**	1	*m*	M
Creatine kinase, 2.7.3.2	*CK-A1*	1	*m*	M
	*CK-A2**	1	*m*	M
Diaphorase, 1.6.4.3	*DIA**	1	*m*	M
Esterase, 3.1.1.–	*EST-1**	1	*m*	M
	*EST-2**	1	*m*	M
Esterase D, 3.1.1.–	*ESTD**	6	*mh*	P
Formaldehyde dehydrogenase, 1.2.1.1	*FDHG**	1	*m*	M

Cont.

Table 3.6 Cont.

Enzyme, EC number	Locus	Number of alleles	Tissue	Level of variation
Glyceraldehyde-3-phosphate dehydrogenase, 1.2.1.12	*GAPDH-1**	1	*m*	M
	*GAPDH-2**	1	*m*	M
Glucose-6-phosphate isomerase, 5.3.1.9	*GPI-B1**	3	*m*	R
	*GPI-B2**	2	*m*	R
	*GPI-A**	3	*me*	R
	*GPI-4**	2	*m*	?
Glycerol-3-phosphate dehydrogenase, 1.1.1.8	*G3PDH-1**	3	*m*	R
	*G3PDH-2**	1	*m*	M
	*G3PDH-3**	2	*m*	?
L-Iditol dehydrogenase, 1.1.1.14	*IDDH-1**	3	*l*	P
	*IDDH-2**	1	*l*	M
Isocitrate dehydrogenase, 1.1.1.42	*mIDHP-1**	2	*m*	?
	*mIDHP-2**	1	*m*	M
	*sIDHP-1**	2	*l*	P
	*sIDHP-2**	6	*m*	P
Lactate dehydrogenase, 1.1.1.27	*LDH-A1**	1	*m*	M
	*LDH-A2**	1	*m*	M
	*LDH-B1**	2	*m*	R
	*LDH-B2**	2	*ml*	?
	*LDH-C**	2	*e*	?
Malate dehydrogenase, 1.1.1.37	*mMDH-1**	1	*m*	M
	*sMDH-A1,2**	5	*l*	P
	*sMDH-B1,2**	4	*m*	P
Malic enzyme (NADP$^+$), 1.1.1.40	*MEP-1,2**	4	*m*	P
	*MEP-1**	3	*m*	P
	*MEP-2**	1	*lm*	M
	*MEP-3**	2	*m*	?
Mannose-6-phosphate isomerase, 5.3.1.8	*MPI**	3	*em*	P
Octanol dehydrogenase, 1.1.1.73	*ODH**	2	*l*	?
Peptidases, 3.4.–.–				
Gly–Leu	*PEPA**	1	*em*	M
	*PEPC**	1	*em*	M
Leu–Gly–Gly	*PEPB-1**	3	*e*	P
Phosphogluconate dehydrogenase, 1.1.1.44	*PGDH**	2	*lm*	R

Cont.

Table 3.6 Cont.

Enzyme, EC number	Locus	Number of alleles	Tissue	Level of variation
Phosphoglucomutase, 5.4.2.2	*PGM-1**	4	*m*	O
	*PGM-2**	2	*m*	R
	*PGM-3**	3	*l*	?
	*PGM-4**	1	*m*	M
Superoxide dismutase, 1.15.1.1	*SOD**	2	*l*	P

M – monomorphic, R – polymorphic locus with frequency of common allele > 0.95, P – polymorphic locus with frequency of common allele < 0.95. Tissue: *m* – muscle, *h* – heart muscle, *e* – eye retina, *l* – liver.
Data from Nakajima *et al.*, 1986; Okazaki, 1986b; Severtsov (unpublished data), Omelchenko *et al.*, 1991; Oleinik, 1994.

Table 3.7 Enzymes and coding genetic loci studied in Atlantic salmon (*Salmo salar*) populations.

Enzyme, EC number	Locus	Number of alleles	Tissue	Level of variation
Aspartate aminotransferase, 2.6.1.1	*mAAT-1**	1	*m*	M
	*mAAT-2**	1	*l*	M
	*sAAT-1**	1	*m*	M
	*sAAT-2**	1	*m*	P
	*sAAT-3 (4)**	4	*l*	P
Acid phosphatase, 3.1.3.2	*ACP**	1	*l*	M
Adenosine deaminase, 3.5.4.4	*ADA-1**	1	*m*	M
	*ADA-2**	2	*l*	R
	*ADA-3**	1	*ml*	M
Alcohol dehydrogenase, 1.1.1.1	*ADH**	2	*l*	R
Adenylate kinase, 2.7.4.3	*AK-1**	1	*m*	M
	*AK-2**	1	*m*	M
Aminopeptidase, 2.6.1.2	*AP-1**	1	*ml*	M
	*AP-2**	1	*ml*	M
Creatine kinase, 2.7.3.2	*CK-AL**	3	*m*	P
	*CK-A2**	1	*m*	M
	*CK-A3**	1	*m*	M
Diaphorase, 1.6.4.3	*DIA**	2	*l*	?
Esterase, 3.1.1.–	*EST-1**	1	*m*	M
	*EST-2**	1	*l*	M
	*EST-3**	1	*lm*	M
	*EST-4**	1	*l*	M
Esterase D, 3.1.1.–	*ESTD**	2	*m*	P

Cont.

Table 3.7 Cont.

Enzyme, EC number	Locus	Number of alleles	Tissue	Level of variation
Glyceraldehyde-3-phosphate dehydrogenase, 1.2.1.12	*GAPDH-1**	1	*m*	M
	*GAPDH-2**	1	*e*	M
Glyoxalase, 4.4.1.5	*GLO**	2	*m*	R
Glycerol-3-phosphate dehydrogenase, 1.1.1.8	*G3PDH-1**	1	*m*	M
	*G3PDH-2**	2	*mcf*	?
	*G3PDH-3**	1	*l*	M
	*G3PDH-4**	1	*l*	M
Glucose-6-phosphate isomerase, 5.3.1.9	*GPI-A**	6	*mcf*	P
	*GPI-B1,2**	5	*mf*	P
L-Iditol dehydrogenase, 1.1.1.14	*IDDH-1,2**	4	*l*	P
	*IDDH-1**	3	*l*	P
	*IDDH-2**	2	*l*	P
Isocitrate dehydrogenase, 1.1.1.42	*IDHP-1**	1	*m*	M
	*IDHP-2**	1	*l*	M
	*IDHP-3**	3	*lcf*	P
	*IDHP-4**	2	*l*	M
Lactate dehydrogenase, 1.1.1.27	*LDH-A1**	1	*m*	M
	*LDH-A2**	1	*me*	M
	*LDH-B1**	1	*me*	M
	*LDH-B2**	1	*lme*	M
	*LDH-C**	1	*e*	M
Malate dehydrogenase, 1.1.1.37	*mMDH-1**	1	*ml*	M
	*mMDH-2**	1	*ml*	M
	*sMDH-A1**	2	*l*	P
	*sMDH-A2**	1	*l*	M
	*sMDH-B1,2**	6	*m*	P
	*sMDH-B1**	3	*m*	P
	*sMDH-B2**	1	*ml*	M
Malic enzyme ($NADP^+$), 1.1.1.40	*mMEP-1**	1	*mh*	M
	*mMEP-2**	3	*mlh*	P
	*sMEP**	1	*ml*	M
Muscle myogens	*MP-1**	1	*m*	M
	*MP-2**	1	*m*	M
	*MP-3**	1	*m*	M
Mannose-6-phosphate isomerase, 5.3.1.8	*MPI**	1	*m*	M
Octanol dehydrogenase, 1.1.1.73	*ODH**	1	*l*	M
Phosphogluconate dehydrogenase, 1.1.1.44	*PGDH**	2	*ml*	R

Cont.

Table 3.7 Cont.

Enzyme, EC number	Locus	Number of alleles	Tissue	Level of variation
Phosphoglucomutase, 5.4.2.2	*PGM-1**	2	*m*	P
	*PGM-2**	2	*lm*	?
	*PGM-1r**	2	*l*	P
Peroxidase, 1.11.1.7	*PX-2**	3	*h*	P
Superoxide dismutase, 1.15.1.1	*mSOD**	1	*l*	M
	*sSOD**	1	*lmcf*	M
Transferrin	*TF**	2	*s*	P
Triose phosphate isomerase, 5.3.1.1	*TPI-3**	2	*l*	P
Trypsin, 3.4.4.4	*TRP-1**	3	*p*	P
	*TRP-2**	2	*p*	P
	*TRP-3**	1	*p*	M
Xanthine dehydrogenase, 1.2.3.2	*XDH**	1	?	M

M – monomorphic, R – polymorphic locus with frequency of common allele >0.95, P – polymorphic locus with frequency of common allele <0.95. Tissue: *m* – muscle, *h* – heart muscle, *e* – eye retina, *l* – liver, *s* – blood serum, *c* – scale, *f* – fin, *p* – pyloric caeca.

Data from Cross & Ward, 1980; Verspoor, 1988; Koljonen, 1989; Vuorinen & Berg, 1989; Jordan & Youngson, 1991; McElligot & Cross, 1991; Kazakov & Titov, 1991, 1993; 1995; Torrisen *et al.*, 1993; Elo *et al.*, 1994; Galvin *et al.*, 1994; Skaala & Jorstad, 1994; Wilson *et al.*, 1995.

Table 3.8 Enzymes and coding genetic loci studied in rainbow trout (*Oncorhynchus mykiss*) populations.

Enzyme, EC number	Locus	Number of alleles	Tissue	Level of variation
Aspartate aminotransferase, 2.6.1.1	*mAAT-1**	2	*mh*	R
	*sAAT-1,2**	5	*mh*	R
	*sAAT-3**	3	*e*	R
Adenosine deaminase, 3.5.4.4	*ADA-1**	6	*me*	R
	*ADA-2**	5	*m*	R
Alcohol dehydrogenase, 1.1.1.1	*ADH**	4	*l*	P
Aconitate hydratase, 4.2.1.3	*mAH-1**	2	*hm*	R
	*mAH-2**	2	*hme*	R
	*mAH-3**	3	*hm*	R
	*mAH-4**	2	*h*	R
	*sAH**	4	*l*	P
Alanine aminotransferase, 2.6.1.2	*ALAT**	3	*mh*	P

Cont.

Table 3.8 Cont.

Enzyme, EC number	Locus	Number of alleles	Tissue	Level of variation
Creatine kinase, 2.7.3.2	*CK-A1**	3	*m*	R
	*CK-A2**	2	*m*	R
	*CK-C1**	3	*e*	R
	*CK-C2**	3	*e*	R
Fructose bisphosphatase, 3.1.3.11	*FBP**	2	?	P
Fumarate hydratase, 4.2.1.2	*FH**	3	*m*	R
Glyceraldehyde-3-phosphate dehydrogenase, 1.2.1.12	*GAPDH-3**	3	*hm*	R
N-Acetyl-β-glucosaminidase, 3.2.1.30	*bGLUA**	4	*l*	R
Glycerate dehydrogenase, 1.1.1.29	*GLYDH-1,2**	3	*m*	P
Glycerol-3-phosphate dehydrogenase, 1.1.1.8	*G3PDH-1**	4	*m*	P
	*G3PDH-2**	2	*m*	R
	*G3PDH-3**	2	*mh*	P
Glucose-6-phosphate isomerase, 5.3.1.9	*GPI-B1**	3	*m*	R
	*GPI-B2**	3	*m*	R
	*GPI-A**	4	*m*	P
Hydroxyacylglutathione hydrolase, 3.1.2.6	*HAGH**	4	*mh*	P
L-Iditol dehydrogenase, 1.1.1.14	*IDDH-1**	4	*l*	R
	*IDDH-2**	3	*l*	P
Isocitrate dehydrogenase, 1.1.1.42	*mIDHP-1**	2	*mh*	R
	*mIDHP-2**	4	*mh*	P
	*sIDHP-1,2**	12	*le*	P
Lactate dehydrogenase, 1.1.1.27	*LDH-A1**	2	*m*	R
	*LDH-A2**	1	*m*	M
	*LDH-B1**	3	*em*	R
	*LDH-B2**	3	*em*	P
	*LDH-C**	2	*e*	P
Malate dehydrogenase, 1.1.1.37	*mMDH-1,2**	3	*hm*	R
	*mMDH-3**	3	*hml*	R
	*sMDH-A1,2**	6	*lhm*	R
	*sMDH-B1,2**	8	*hm*	P
Malic enzyme (NAD^+), 1.1.1.38	*ME**	2	*hm*	R
Malic enzyme ($NADP^+$), 1.1.1.40	*mMEP-1**	4	*m*	R
	*sMEP-1**	7	*mh*	P
	*sMEP-2**	4	*l*	R

Cont.

Table 3.8 Cont.

Enzyme, EC number	Locus	Number of alleles	Tissue	Level of variation
Mannose-6-phosphate isomerase, 5.3.1.8	*MPI**	4	*helm*	P
Nucleoside triphosphate pyrophosphatase, 3.6.1.19	*NTP**	4	*m*	P
Para-albumin	*PALB-1,2**	2	*s*	P
Peptidases, 3.4.–.–				
Gly–Leu	*PEPA**	5	*mhle*	P
	*PEPC**	3	*e*	P
Leu–Gly–Gly	*PEPB-1**	4	*mhle*	P
	*PEPB-2**	3	*m*	P
Phe–Pro	*PEPD*	3	*mhe*	P
Leu–Tyr	*PEP-LT**	3	*m*	P
Leu–Ala	*PEP-LA3**	2	*mh*	R
Phosphogluconate dehydrogenase, 1.1.1.44	*PGDH**	1	*lme*	M
Phosphoglycerate kinase, 2.7.2.3	*PGK-2**	6	*mhe*	P
Phosphoglucomutase, 5.4.2.2	*PGM-1**	4	*me*	R
	*PGM-1r**	2	*e*	R
	*PGM-2**	4	*me*	P
Purine nucleoside phosphorylase, 2.4.2.1	*PNP**	2	*e*	P
Superoxide dismutase, 1.15.1.1	*mSOD**	2	*hm*	P
	*sSOD-1**	3	*lmh*	P
Triose phosphate isomerase, 5.3.1.1	*TPI-1**	2	*mhe*	R
	*TPI-2**	2	*mhe*	R
	*TPI-3**	4	*mhe*	P
	*TPI-4**	3	*mhe*	R

** M – monomorphic, R – polymorphic locus with frequency of common allele >0.95, P – polymorphic locus with frequency of common allele <0.95. Tissue: *m* – muscle, *h* – heart muscle, *e* – eye retina, *l* – liver, *s* – blood serum.

*** Data from Campton & Johnston, 1985; Okazaki, 1986a; Krueger & May, 1987b; Berg & Gall, 1988; Reisenbichler & Phelps, 1989; Phelps *et al.*, 1994a.

Table 3.9 Enzymes and coding genetic loci studied in brown trout (*Salmo trutta*) populations.

Enzyme, EC number	Locus	Number of alleles	Tissue	Level of variation
Aspartate aminotransferase, 2.6.1.1	*sAAT-1,2**	3	*mh*	P
	*sAAT-4(3)**	3	*l*	P
Acid phosphatase, 3.1.3.2	*ACP-1**	1	?	M
	*ACP-2**	1	?	M
Adenosine deaminase, 3.5.4.4	*ADA-1**	1	*mhe*	M
	*ADA-2**	1	*mhe*	M
Alcohol dehydrogenase, 1.1.1.1	*ADH**	1	*l*	M
Aconitate hydratase, 4.2.1.3	*AH-1**	1	*h*	M
	*AH-2**	1	*h*	M
Adenylate kinase, 2.7.4.3	*AK**	1	*m*	M
Alanine aminotransferase, 2.6.1.2	*ALAT-1**	1	*m*	M
Diaphorase, 1.6.4.3	*DIA-1**	3	*be*	P
Creatine kinase, 2.7.3.2	*CK-A1(2)**	3	*m*	P
	*CK-C1**	2	*e*	P
Esterase, 3.1.1.–	*EST-1**	1	*l*	M
	*EST-2**	3	?	P
	*EST-3**	2	?	?
	*EST-5**	2	*m*	P
Fructose bisphosphatase, 3.1.3.11	*FBP-1**	3	*l*	P
Fumarate hydratase, 4.2.1.2	*FH-1,2**	5	*m*	P
β-Galactosidase, 3.2.1.23	*bGAL-2**	3	*l*	P
N-Acetyl-β-glucosaminidase, 3.2.1.30	*bGLUA*	2	*l*	P
Guanine deaminase, 3.5.4.3	*GDA**	1	*l*	M
Glycerate dehydrogenase, 1.1.1.29	*GLYDH-1,2**	3	*l*	P
Glucose-6-phosphate dehydrogenase, 1.1.1.49	*G6PDH-2**	2	?	?
Glucose-6-phosphate isomerase, 5.3.1.9	*GPI-B1**	2	*m*	P
	*GPI-B2**	5	*m*	P
	*GPI-A**	4	*me*	P
Glycerol-3-phosphate dehydrogenase, 1.1.1.8	*G3PDH-1(2)**	4	*m*	P
	*G3PDH-3**	1	?	M
L-Iditol dehydrogenase, 1.1.1.14	*IDDH-1**	3	*l*	P
	*IDDH-2**	2	*l*	P

Cont.

Table 3.9 Cont.

Enzyme, EC number	Locus	Number of alleles	Tissue	Level of variation
Isocitrate dehydrogenase, 1.1.1.42	*sIDHP-1**	5	*ehm*	P
	*sIDHP-2**	2	*le*	P
	*IDHP-3,4**	1	*l*	M
Lactate dehydrogenase, 1.1.1.27	*LDH-A1**	3	*m*	P
	*LDH-A2**	1	*m*	M
	*LDH-B1**	2	*l*	P
	*LDH-B2**	1	*ml*	M
	*LDH-C**	4	*e*	P
Glyoxalase 1, 4.4.1.5	*GLO1**	1	?	M
α-Mannosidase, 3.2.1.24	*aMAN**	3	*l*	P
Malate dehydrogenase, 1.1.1.37	*sMDH-A1**	2	*hel*	R
	*sMDH-A2**	3	*hle*	P
	*sMDH-B1**	6	*meh*	P
	*sMDH-B2**	2	*me*	P
Malic enzyme ($NADP^+$), 1.1.1.40	*MEP-1**	3	*m*	P
	*MEP-2**	2	*m*	P
	*MEP-3**	2	*mh*	?
	*MEP-4**	2	*mh*	?
Mannose-6-phosphate isomerase, 5.3.1.8	*MPI-2(1)**	3	*mlh*	P
Para-albumin	*PALB-1,2**	1	*s*	M
Peptidases, 3.4.–.–				
Leu–Gly–Gly	*PEPB**	2	*m*	R
Leu–Tyr	*PEP-LT**	2	*mle*	R
Phosphogluconate dehydrogenase, 1.1.1.44	*PGDH-1**	1	*h*	M
	*PGDH-2**	2	*m*	P
Phosphoglucomutase, 5.4.2.2	*PGM-1**	2	*m*	P
	*PGM-2**	2	*m*	P
Purine nucleoside phosphorylase, 2.4.2.1	*PNP**	1	?	M
Superoxide dismutase, 1.15.1.1	*SOD-1**	3	*ml*	P
	*SOD-2**	3	?	P
Transferrin	*TF**	3	*s*	P
Xanthine dehydrogenase, 1.2.3.2	*XDH**	1	*l*	M

M – monomorphic, R – polymorphic locus with frequency of common allele > 0.95, P – polymorphic locus with frequency of common allele < 0.95. Tissue: *m* – muscle, *h* – heart muscle, *e* – eye retina, *l* – liver, *s* – blood serum, *b* – brain.

Data from Taggart & Ferguson, 1984b; Vuorinen & Piironen, 1984; Krueger & May, 1987a; Ferguson, 1989; Jorde *et al.*, 1991; Skaala, 1992; Hansen *et al.*, 1993; Jorde, 1994; Presa *et al.*, 1994; Riffel *et al.*, 1995.

Table 3.10 Enzymes and coding genetic loci studied in cutthroat trout (*Oncorhynchus clarki*) populations.

Enzyme, EC number	Locus	Number of alleles	Tissue	Level of variation
Aspartate aminotransferase, 2.6.1.1	*sAAT-1,2**	3	*m*	P
	*sAAT-3**	1	*e*	M
Alcohol dehydrogenase, 1.1.1.1	*ADH**	2	*l*	P
Creatine kinase, 2.7.3.2	*CK-1,2**	1	*m*	M
Glucose-6-phosphate isomerase, 5.3.1.9	*GPI-A**	1	*ml*	M
	*GPI-B1**	3	*m*	P
	*GPI-B2**	4	*m*	R
Glycerol-3-phosphate dehydrogenase, 1.1.1.8	*G3PDH-1**	2	*m*	R
	*G3PDH-2**	1	*m*	M
L-Iditol dehydrogenase, 1.1.1.14	*IDDH-1,2**	3	*l*	P
Isocitrate dehydrogenase, 1.1.1.42	*mIDHP-1,2**	1	*m*	M
	*sIDHP-1,2**	4	*l*	P
Lactate dehydrogenase, 1.1.1.27	*LDH-A1,2**	1	*m*	M
	*LDH-B1**	3	*me*	R
	*LDH-B2**	2	*mle*	R
	*LDH-C**	1	*e*	M
Malate dehydrogenase, 1.1.1.37	*MDH-A1,2**	3	*l*	R
	*MDH-B1,2**	4	*m*	P
Malic enzyme ($NADP^+$), 1.1.1.40	*MEP-1,2**	1	*m*	M
	*MEP-3**	1	*ml*	M
	*MEP-4**	2	*l*	R
Mannose-6-phosphate isomerase, 5.3.1.8	*MPI**	2	*e*	P
Peptidases, 3.4.–.–				
Gly–Leu	*PEPA**	3	*me*	P
Leu–Gly–Gly	*PEPB**	1	*e*	M
Phosphogluconate dehydrogenase, 1.1.1.44	*PGDH**	3	*ml*	P
Phosphoglucomutase, 5.4.2.2	*PGM**	2	*m*	P
Superoxide dismutase, 1.15.1.1	*SOD**	2	*l*	P

M – monomorphic, R – polymorphic locus with frequency of common allele > 0.95, P – polymorphic locus with frequency of common allele < 0.95. Tissue: *m* – muscle, *e* – eye retina, *l* – liver.

Data from Campton & Utter, 1987.

Table 3.11 Polymorphic proteins for which Mendelian inheritance of electrophoretic variants with codominant expression of corresponding genes was demonstrated.

Species	Protein	Coding loci	References
Oncorhynchus mykiss	ADA, CK, FBP, G3PDH, IDDH, IDHP, LDH, MDH, MEP, PGDH, PGI, PGK, PGM, SOD, TF	ADA-1*, CK-1*, FBP*, G3PDH-1*, -2(3)*, IDDH-1*, IDHP-1,2*, IDHP-3,4*, LDH-B1*, -B2*, LDH-C*, LDH-A1*, MDH-B1,2*, MEP-2*, PGDH*, PGI-2*, PGK*, PGM-2*, PGMr-1*, SOD-2*, TF*	Utter *et al.*, 1972, 1979; Ropers *et al.*, 1973; Allendorf, 1975; Wright *et al.*, 1975; Reinitz, 1977; May, 1980; Allendorf & Thorgaard, 1984; Danzmann *et al.*, 1985
O. gorbuscha	G3PDH, IDHP, MDH, MEP, PGDH, PGI, PGM	MDH-A1,2*, MDH-B1,2*, IDHP-3,4*, G3PDH-2(1)*, MEP-2*, PGDH*, PGI-3*, PGM-2*	Aspinwall, 1973, 1974a; Donelly *et al.*, 1979; Johnson, 1979; Salmenkova & Omelchenko, 1983a; Gagalchy, 1988; Kartavtsev *et al.*, 1990
O. keta	AAT, EST-D, LDH, MDH, MEP, PGDH	AAT-1,2*, ESTD*, LDH-A1*, MDH-B1,2*, MEP-2*, PGDH*	Slynko, 1971; Sachko, 1973; Altukhov, 1974; Allendorf *et al.*, 1975; May *et al.*, 1975; Kartavtsev *et al.*, 1985; authors' own unpublished data
O. nerka	LDH, PGM	LDH-B2*, PGM-2*	Utter *et al.*, 1973; Kirpichnikov, 1987
O. tshawytscha	AAT, ADA, IDHP, IDDH, MDH, MPI, SOD	AAT-1,2*, AAT-3*, ADA*, IDHP-1*, -2*, IDDH-2*, MDH-B1,2*, MPI*, SOD-1*	May, 1980; Kobayashi *et al.*, 1984; Shaklee & Phelps, 1992
O. kisutch	CK, IDDH, LDH, MDH, PGI	CK-1*, IDDH-1*, LDH-B2*, MDH-A1(2)*, MDH-B1,2*, PGI-2*	Bailey *et al.*, 1970; May, 1980
Salmo salar	AAT, IDDH, IDHP, MDH, MEP, PGI, PGM, TF	AAT-3*, IDDH-1,2*, IDHP-1*, MDH-B1,2*, MEP-2*, PGI-1,2*, PGI-3*, PGM-2*, TF*	Moller, 1970; Crozier & Moffett, 1990; Skaala & Jorstad, 1994
S. trutta	AAT, CK, DIA, G3PDH, IDDH, IDHP, LDH, MDH, MPI, PGI, SOD	AAT-1,2*, AAT-4*, CK-2*, DIA*, G3PDH-2(1)*, IDDH-1*, IDHP-1*, -2*, LDH-A1*, LDH-C*, MDH-A1*, -A2*, MDH-B1,2*, MPI*, PGI-2*, -3*, SOD-1*	Bailey *et al.*, 1970; May, 1980; Ståhl & Ryman, 1982; Guyomard & Krieg, 1983; Taggart & Ferguson, 1984a; Semenova & Slynko, 1988
Salvelinus fontinalis	ADA, AAT, ADH, CK, G3PDH, GUS, IDDH, IDHP, DIA, GPT, LDH, MEP, MDH, MPI, ODH, PGI, TF	AAT-1,2*, AAT-4*, ADA-1*, ADH*, CK-1*, DIA*, G3PDH-1*, -2*, GPT*, GUS*, IDDH*, IDHP-3*, LDH-B1*, LDH-B2*, MDH-1*, MDH-B1,2*, MEP-1*, -2*, MPI*, PGI-2*, SOD*, TF*	Morrison & Wright, 1966; Wright & Atherton, 1970; May *et al.*, 1979b, 1980; Stoneking *et al.*, 1979; Wright *et al.*, 1980

Chapter 4
Genetic Dynamics and Stability of a Population System

The natural subdivision of salmon populations into different hierarchical levels (see Chapter 2) requires a special approach to studying their subpopulation structures. It is impossible to understand genetic processes in a species without clear knowledge of what is the hierarchical level in the data, whether dealing with the system as a whole or a subpopulation as its structural component. For the information to be appreciated, it is necessary to fulfil at least three conditions.

(1) A population should be well defined in a historical–geographical sense as a natural system that constantly perpetuates itself through several generations.
(2) It is essential to have a reasonably complete record of the characteristics of its subpopulation structure.
(3) The collection of data should be so organized as to provide the maximum possible complete coverage of a system's subpopulation structure in space and time.

At present, such information is available only for an isolated system of sockeye salmon subpopulations, spawning in Lake Azabachye (Kamchatka). Let us turn to the analysis of these data.

4.1 Temporal stability of a native population system

4.1.1 Subpopulation structure and mean effective size of local subpopulations

Isolated populations of sockeye salmon, spawning in a small area of $9 \times 14\,\text{km}^2$ in Lake Azabachye situated in the River Kamchatka basin (Fig. 4.1), have been investigated by the authors and their colleagues since 1971, using allozyme polymorphisms in the autosomal loci of lactate dehydrogenase and phosphoglucomutase (Figs 3.1 and 3.2). Simultaneously, within the overall programme, a group of scientific workers at the Institute of Marine Biology, USSR Academy of Sciences'

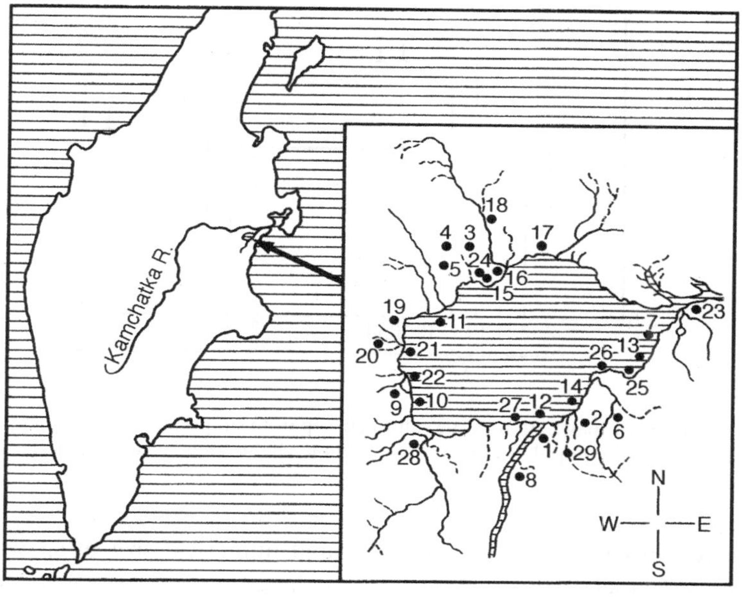

Fig. 4.1 Location of Lake Azabachye in the Kamchatka River system (arrow) and the spatial location of spawning subpopulations (1–29) of sockeye salmon.

Far Eastern Scientific Centre, led by S.M. Konovalov, assessed the numbers and sex ratio on the spawning grounds, thus enabling us to approach the matter of determining the local effective population size (Altukhov, 1974; Altukhov et al., 1975 a, b; Konovalov, 1980).

In addition, on the basis of the data of Hartman & Raleigh (1964) and Il'in *et al.* (1983), we obtained possibly the highest estimate of gene migration rate in the subdivided population of sockeye salmon ($\simeq 2\%$; Chapter 2).

The structure of the sockeye assemblage in Lake Azabachye is a community of subpopulations which reproduce both in streams and 'ponds' (the 'spring race') connected to the lake and in its coastal, littoral zone (the 'summer race' in Fig. 4.1; for further details, see Konovalov, 1980).

During the spawning season there may be 30–40 elementary populations in the lake. In 11 years of research, 29 different spawning sites were investigated, but because many of them have been studied repeatedly we shall present data about the biological and genetic structure of 183 such communities, isolated by time or space.

Despite subdivision, spawning sockeye are characterized by considerable unity and stability of biological parameters such as the sex ratio and mean age. A sex ratio close to one is characteristic of the assemblage as a whole, though there are frequently observed excesses of males or females in different subpopulations. This is clearly seen in the example of the most detailed analyses of generations that arrive at a spawning ground without noticeable traces of the effects of excessive fishing (Altukhov, 1974; see also Fig. 4.2).

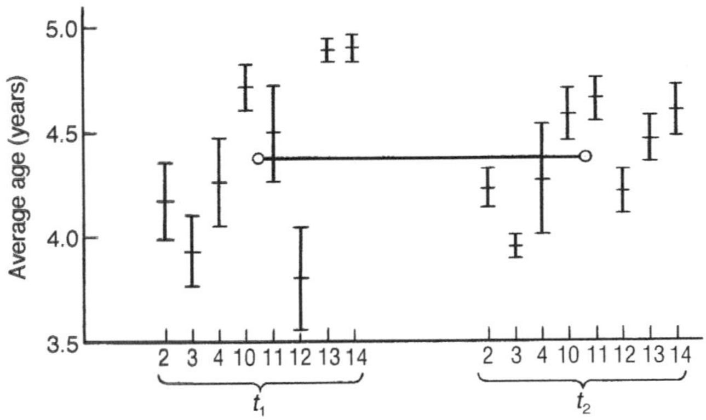

Fig. 4.2 Average age of spawning sockeye salmon in separate subpopulations (dashes with double standard error) and in the population system as a whole (light circles) for two successive generations, t_1 and t_2.

As indicated in Chapter 2, a relatively small numerical size of spawning subpopulations is characteristic of all species of the genus *Oncorhynchus*, a feature which is quite strongly marked in sockeye salmon. These breeding units, which reproduce especially in shallow water streams and ponds or lake littorals, are accessible to direct observation; it is possible to calculate the actual number of spawning fish and to determine the sex ratio. Control mechanisms dependent on density may also exercise a certain influence on the N_e value of sockeye subpopulations in addition to the effect of the sex ratio. Thus, for example, in years of large numbers, the fish that arrive later from the sea spawn by excavating the gravel that has been used already for the spawning nests of fishes which had reached the spawning grounds earlier (Krogius *et al.*, 1969). Although the role of this phenomenon has so far not been investigated in detail, we believe that its influence upon the reproductive structure of the Azabachye sockeye salmon is likely to be significant. During the years of our most detailed research, the numbers of the spawning run were small and overcrowding of spawning grounds with mature fish was not observed because of intensive sea fishing. Large returns were noticed only after the international agreement establishing the 200 nautical mile exclusive economic zone in 1977. In addition it was observed that the effects of fishing had impaired the sexual structure of several generations that returned to the spawning grounds.

Concerning the equilibrium between numbers of the sexes – an important indicator of the 'nativeness' of the population system (Altukhov, 1974) – an evaluation of the reproductive part of the sockeye populations in Lake Azabachye should be made on the basis of an appropriate analysis on only those generations that as a whole are characterized by an equilibrium sex ratio.

If one takes this factor into consideration in constructing the distribution of numbers observed throughout the research period, then a series is formed with a mode at an interval of 0–500 individuals and with a clearly expressed left-sided asymmetry (Fig. 4.3a). The geometric mean of such a series is 266.5 for spring subpopulations, 350.2 for summer ones, and 332.4 for the entire population. The population's effective size, found as a harmonic mean with correction for the sex ratio, is around 200 individuals. In view of the properties of N_e as harmonic mean value and the large number of subpopulations studied ($n = 183$), the high reliability of the estimate is evident.

A similar processing of data on the number of fish in the stream spawning grounds of the Wood River basin (Alaska) (Marriott, 1964) leads to a similar distribution (Fig. 4.3b) and an estimated value of $N_e = 174$ individuals; the geometric mean convention equals 1504. Throughout their exploitation period, the Wood River's sockeye populations suffered less from fishing than did the populations of Lake Azabachye and other water bodies in the Kamchatka River basin. Nevertheless, estimates of the effective size in both cases are extremely close; the differences refer only to the variance of reproductive size, which, for Alaskan populations, is roughly two orders of magnitude higher (2.193 against 0.118 for logarithmic values). The difference in these two distributions is of specific interest. First, one can see that the N_e value, estimated as the harmonic mean, depends mostly on small subpopulations; this results from the properties of the harmonic mean. Second, the fact that the N_e value for sockeye subpopulations of Lake Azabachye is similar to the corresponding value for the River Wood's subpopulations indicates that the population structure of sockeye salmon of Lake Azabachye has not yet been disturbed irreversibly by fishing. It suggests that if we were to stop or attenuate significantly such a fishing pressure, the total and reproductive size of the sockeye population in Lake Azabachye could increase greatly. This assumption, made in 1973 (Altukhov, unpublished), was confirmed after 1977 upon the introduction of the international

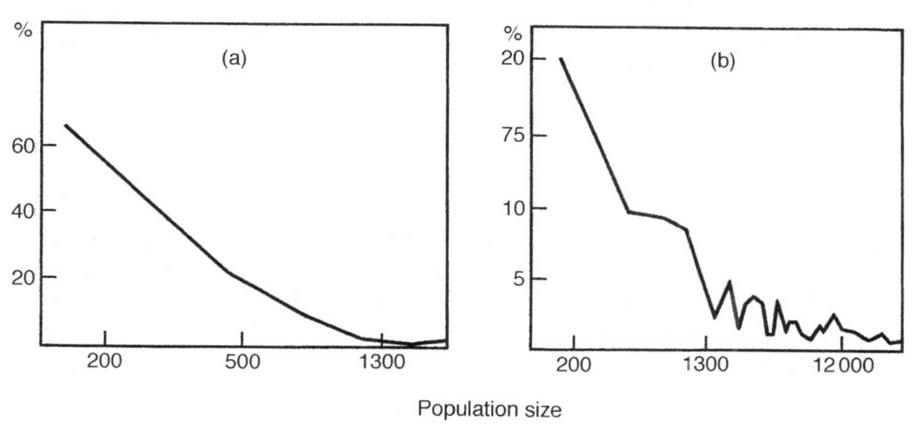

Fig. 4.3 The distribution of size of 121 spawning subpopulations of Lake Azabachye in Kamchatka (a) and 179 subpopulations in the Wood River basin in Alaska (b) (extracted from data by Marriott, 1964). The abscissa scale is logarithmic.

agreement on the 200 nautical mile exclusive economic zone, when sockeye populations of Lake Azabachye turned out to be prohibitive for Japanese sea fishing. If before 1977 the annual numbers of Azabachye sockeye were not more than 30 000–40 000 spawners, after 1977 the numbers increased by approximately one order of magnitude. This fact is of basic importance for management, because it implies that even removing 90% of the sockeye population by fishing may not threaten its biological well-being, provided the population structure was not destroyed (see also section 2.5).

Additionally, this important fact merely signifies that *fishing*, as one might expect, *has the greatest effect on large populations*, whereas the N_e value is more dependent on groups that contain few individuals with variability of this parameter in time or space (see section 1.3.1, Equation 1.2).

Our estimate of the average effective size of a local subpopulation of sockeye salmon from Lake Azabachye is in good agreement with the estimate obtained from Aro & Shepard's (1967) data for British Columbia (Table 2.4). Overall, there is strong evidence for the relative stability of mean effective size of local subpopulations in sockeye salmon in space and time.

4.1.2 Genetic processes in a native population system

Let us consider the data on genetic processes in separate subpopulations and in the subdivided population as a whole. Since dealing with an assemblage of spawning units, the collection of field material was organized so as to provide maximum coverage of the biological and ecological heterogeneity of the Azabachye sockeye salmon not only in space but also in time. Standard biological analyses were repeated annually on the same spawning grounds. Simultaneously in each sample, individual fishes were typed for the *LDH-B2** and *PGM-2** loci. As a result, unique information was obtained which makes the approximation of the empirical distribution of subpopulations in the gene frequency ranges of both loci, using Wright's mathematical functions for the island model of a population structure (see Chapter 1, and section 5.1).

It follows from Chapter 1, however, that Wright's F_{ST} statistics can be applied only if it is known for certain that the test situation is indeed genetically stable. Indirect arguments favouring the structural stability of the sockeye population in Lake Azabachye have been examined above. The stratigraphy of Kamchatkan quaternary deposits attests that Lake Azabachye was formed as a lake, differing little from that of today, at least 7000 years ago (Braitseva, 1968; Kuprina, 1970). If one estimates the average reproductive age of the Azabachye sockeye salmon as 5 years, it is clear that over 1000 generations have passed during the population system's existence in conditions of isolation, a period sufficient for the development of the stationary genetic process, taking into account the effective population size.

It is possible to obtain direct evidence of a population's genetic stability by comparing the distributions of gene frequencies in appropriate time 'intervals' that is, cross-sections through the subpopulation structure of a subdivided population (Altukhov & Rychkov, 1970). As already indicated, we have carried out an analysis of this sort, enabling an assessment of the distributions of genotypic and allelic frequencies in successive generations (Table 4.1).

Even with a cursory analysis of the tabular material one can see a similarity across generations of different years with respect to average values, as well as to the intergroup variance of the allelic frequencies of both loci, despite the overall genetic heterogeneity.

Heterogeneity is particularly well expressed for the *LDH-B2** locus. Statistically significant heterogeneity was not observed at the *PGM-2** locus in 1972, 1979, 1980 and 1982, but this finding does not influence the conclusion that heterogeneity need only be discovered at any polymorphic locus for it to be regarded as proven. However, where other genes are concerned, especially if polymorphism is not

Table 4.1 Stability of the main genetic parameters of a population system of sockeye salmon in Lake Azabachye.

Locus	Research year	K	n	$\bar{q}$	V_q	χ^2 test for variance homogeneity (Bartlett's criterion (d.f.))	Test for the population homogeneity
*LDH-B2**	1971	14	737	0.65	0.0074		50.18***
	1972	20	1022	0.61	0.0048		37.53***
	1973	21	1007	0.65	0.0078		65.11***
	1974	14	676	0.66	0.0076		45.59***
	1977	19	934	0.65	0.0085		69.34***
	1978	23	1244	0.65	0.0071	$2.69 < \chi^2_{0.05}$ (10) = 18.30	76.62***
	1979	14	846	0.65	0.0091		67.30***
	1980	17	843	0.67	0.0078		63.36***
	1981	11	2239	0.68	0.0073		137.22***
	1982	11	1380	0.67	0.0053		66.74***
	1984	19	1408	0.64	0.0061		79.22***
Totals		183	12336	0.66	0.0077		804.49***
*PGM-2**	1971	14	748	0.79	0.0028		22.57**
	1972	20	1004	0.77	0.0019		21.76
	1973	21	981	0.79	0.0041		41.36**
	1974	13	627	0.79	0.0039		45.01**
	1977	19	936	0.81	0.0032	$6.47 < \chi^2_{0.05}$ (9) = 16.90	39.12**
	1978	23	1241	0.77	0.0025		35.78*
	1979	14	846	0.78	0.0018		17.05
	1980	17	831	0.77	0.0015		24.90
	1982	11	1363	0.79	0.0035		14.53
	1984	19	1353	0.76	0.0019		45.94***
Totals		171	9950	0.78	0.0028		317.31***

* $P < 0.05$.
** $P < 0.01$.
*** $P < 0.001$.

K – number of subpopulations; n – number of fish studied; $\bar{q}$ – average gene frequency; V_q intergroup variance of the gene frequency.

expressed externally and there are no grounds for allowing the differential migration of genotypes, interpopulation variability may be stabilized by the effects of balancing selection.

Turning to Table 4.1, we see clear interlocus differences in variances: the allelic frequencies at the *LDH-B2** locus are more variable than those at the *PGM-2** locus, and this pattern persists throughout the period of study. Moreover, in all analyses of *PGM-2** genotype distributions one observes either agreement with the Hardy–Weinberg expectations or an excess of heterozygotes, as seen in 1973 and 1974 and in the total distribution for the entire period (Table 4.2). These observations are reminiscent of a typical picture of balanced polymorphism maintained through an adaptive advantage of heterozygotes. Selection of this type may influence the distribution of *LDH-B2** genotypes, but its pressure would be less, since a systematic heterozygote deficit for this locus was observed indicating a Wahlund effect.

Thus, the work here shows that despite clearly expressed intersubpopulation differences in gene frequencies, the system of subpopulations as a whole remains genetically stable in successive generations. Furthermore, the substantial interlocus differences in the variance of gene frequencies, repeated from year to year, indicate dissimilarity in the contributions made by random (drift) and systematic (selection) factors to maintaining the polymorphisms under study.

The role of natural selection in the maintenance of allozyme polymorphism will be considered separately (see Chapter 5). Here, attention will be focused on the genetic stability of a population system in time. The Lake Azabachye example is one of a few illustrations of this stability, which has been demonstrated for a variety of biological species, including humans (Altukhov & Rychkov, 1970; Rychkov, 1973; Altukhov, 1974, 1983a,b, 1990). We did not find any other studies in which the stability of genetic parameters of population systems is traced in a long time-series of generations. None the less, the feasibility of such an approach has been acknowledged by our colleagues (Campton & Utter, 1987) who demonstrated that local genetic differentiation in *Oncorhynchus* (*Salmo*) *clarki clarki* is as strong as that described above for sockeye salmon. Considerable subpopulation differentiation on a microgeographical scale has also been confirmed by means of DNA markers in *Salmo salar* (Tessier *et al.*, 1997), *O. clarki clarki* (Wenburg *et al.*, 1998) and *O. kisutch* (Small *et al.*, 1998). It is evident that when the subdivided population systems of these species are studied using the approach adopted here, their temporal genetic stability is revealed, in spite of variability in the separate subpopulations.

The phenomenon of *stability of average allele frequencies estimated in consecutive generations* of the population system should be distinguished from the spatial and temporal *uniformity* of gene frequencies resulting from balancing (stabilizing) selection pressure (see sections 5.2, 5.5). The main factor of genetic stability in the system of subpopulations is the balance between the processes of differentiation and integration of their gene pools (e.g. a balance of drift and migration or drift and selection, etc.).

Table 4.2 Observed and expected distribution of genotypes and gene frequencies (q) of *LDH-B2** and *PGM-2** loci in the population system of sockeye salmon in Lake Azabachye.

Year	*LDH-B2**						*PGM-2**					
	BB	*BB′*	*B′B′*	*n*	*qB*	χ^2	AA	AB	BB	*n*	*qA*	χ^2
1971	327	319	91	737	0.6601	0.92	465	258	25	749	0.7931	2.27
	321.1	330.7	85.2				471.1	245.8	32.1			
1972	386	467	169	1022	0.6061	1.88	589	366	49	1004	0.7689	0.67
	375.4	488.0	158.6				593.6	356.8	53.6			
1973	427	461	118	1007	0.6529	0.10	596	351	34	981	0.7864	4.14*
	429.3	456.4	121.3				606.7	329.6	44.7			
1974	296	302	78	676	0.6612	0.01	378	253	16	647	0.7797	12.38***
	295.5	302.9	77.6				393.3	222.3	31.4			
1977	409	410	115	934	0.6574	0.61	611	298	27	936	0.8119	1.70
	403.6	420.8	109.6				617.0	285.9	33.1			
1978	534	538	172	1244	0.6455	3.77	738	443	60	1241	0.7732	0.39
	518.3	569.4	156.3				741.9	435.3	63.8			
1979	373	349	124	846	0.6472	7.84**	517	284	45	846	0.7790	0.54
	354.3	386.4	105.3				513.3	291.4	41.3			
Totals for 7 years	2752	2846	868	6466	0.6457	9.35**	3894	2253	256	6403	0.7841	9.85**
	2695.8	2958.5	811.7				3936.6	2167.9	298.5			

* $P<0.05$.
** $P<0.01$.
*** $P<0.001$.
n – number of fish analysed. Here and further genotypes of *LDH-B2**: *BB* = 100/100, *BB′* = 100/120, *B′B′* = 120/120; genotypes of *PGM-2**: AA = 100/100, AB = 100/135, BB = 135/135.

Unfortunately, the effects of such interactions can rarely be revealed and studied in detail in natural populations of salmonids and other fishes. However, these genetic processes are possible to simulate in experiments with subdivided laboratory populations of *Drosophila melanogaster*, thus demonstrating the long-term genetic stability of population systems with the island or stepping-stone population model. As these data were discussed in detail previously (Altukhov, 1990), here we need only note an important fact, that the same result can now be obtained by applying mathematical methods to the analysis of interactions between different micro-evolutionary factors in the systems of semi-isolated populations.

4.2 Numerical modelling of the genetic dynamics of a population system

To date, many researchers have proven the validity of mathematical methods for analysing the genetic dynamics of subdivided populations represented by sets of migrationally interconnected subpopulations (Malecot, 1948; Kimura & Weiss, 1964; Kimura, 1968; Maruyama, 1971; Altukhov *et al.*, 1984; Nagylaki, 1989; Altukhov & Blank, 1991, 1992; Blank & Altukhov, 1992, 1995). Based on the corresponding theoretical data, we will present here the numerical results of the analytical modelling of genetic dynamics in the simplest one-locus diallelic population system with a circular stepping-stone structure, and in a panmictic population having the same size and gene pool.

As an illustration, let us consider the examples of genetic dynamics in ten panmictic (Fig. 4.4a) and ten subdivided (Fig. 4.4b) model populations of similar size (500 individuals each). The subdivided populations conform to the circular stepping-stone model and have the following structure: effective size $N_e = 20$, number of subpopulations $k = 25$, coefficient of short-range migration $m = 0.03$.

Comparing Figs 4.4a and 4.4b, it is seen that by the thousandth generation, eight out of ten model panmictic populations of 500 individuals each 'degraded' (i.e. became completely homozygous), whereas none of the subdivided populations of the same size ($N_e = 20$, $k = 25$, $m = 0.03$) lost genetic diversity. If we calculate the 'lifetime' of panmictic and subdivided populations of equal size (i.e. the number of generations necessary for a 99% decrease in heterozygosity from the initial level), this parameter will be 2301 generations in the first case and 5341 generations in the second ($N_e = 20$, $k = 25$, $m = 0.005$). In other words, the existence of the simplest subpopulation structure of the circular type with a limited intensity of gene migrations (about 0.5–1%) is all that is needed to slow down the loss of genetic diversity as if the effective size of the panmictic population is doubled. Table 4.3 shows examples of such comparisons.

Figure 4.5 represents this trend graphically, thus making the advantages of the population system in preserving genetic polymorphism over long time intervals more apparent. This fact is important for conservation biology, because the

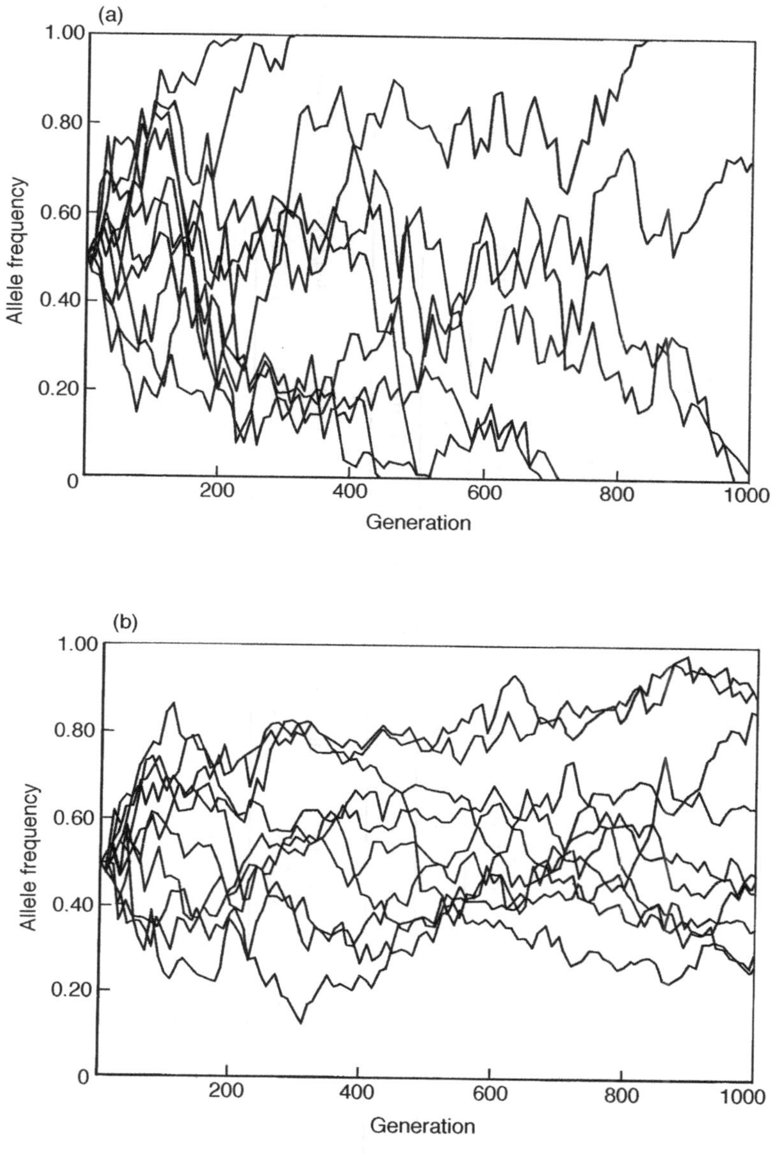

Fig. 4.4 Gene frequency dynamics in generations of ten panmictic populations consisting of 500 haploid individuals each (a) and ten subdivided populations (b). Subdivided population corresponds to a circular model with the stepping-stone structure of gene migrations. The structure of the circular model: number of subpopulations $k = 25$, effective size of a subpopulation $N_e = 20$, gene migration coefficient $m = 0.03$.

Table 4.3 'Lifetime' (in generations) of panmictic and subdivided populations of equal size until losing 95/99% of initial heterozygosity.

Subpopulation size (N)	Number of subpopulations (k)	Subdivision with a gene migration coefficient (m)						Panmixia
		0.000	0.0005	0.001	0.005	0.010	0.030	
10	10	29/44	33/1119	40/1997	420/1336	440/957	364/628	299/459
20	25	59/90	81/4977	314/8251	1731/5341	1772/3801	1437/2464	1497/2301

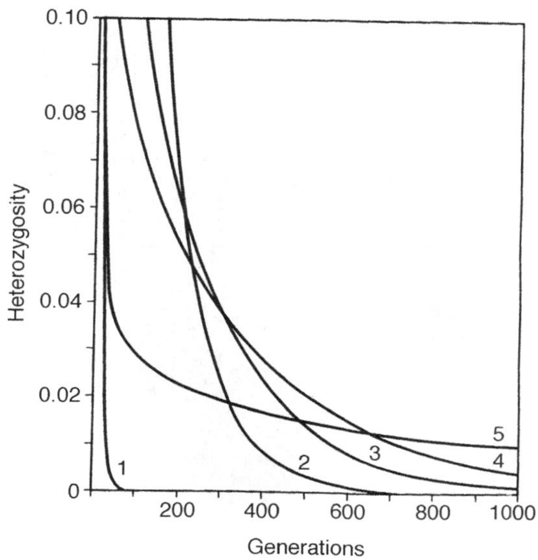

Fig. 4.5 The rate of loss of heterozygosity in successive generations of (1) ten isolates consisting of ten individuals each, (2) panmictic population of 100 individuals, and three subdivided populations of the same size ($N = 10$, $k = 10$) but differing in intensity of gene migrations: (3) $m = 0.01$, (4) $m = 0.005$, (5) $m = 0.001$.

resources (size of range, amount of food, etc.) required to achieve the same practical results are obviously prohibitive; in one case, the size of the *minimal viable population* (Soulé, 1987), due to inner fragmentation, can amount to no more than hundreds of organisms, whereas in another case it should be no less than several thousand.

Theoretically, it is interesting that the loss of genetic variation due to random genetic drift in the structured population is, as expected, inversely proportional to the intensity of gene migration and, in the first few dozen generations, higher than in the panmictic population (Fig. 4.5). Subsequently, the situation becomes fundamentally different, and the subdivided population of only 100 individuals ($N_e = 10$, $k = 10$, $m = 0.001$) retains a certain level of heterozygosity over the time interval corresponding to hundreds of generations after 'degradation' of the structureless panmictic community (Table 4.3, Fig. 4.5; see also Kimura & Ohta, 1971).

Stability of the subdivided population results from the formation of a non-random subpopulation structure whose rearrangements in 'biological' time resembles, at least superficially, a stationary (more precisely, quasi-stationary) process: the linearly ordered structures demonstrate the appearance of waves (Fig. 4.6). Introducing the long-term migration factor into the model (this can be gene migration from the 'continent' to the peripheral subpopulations; see Chapter 1) can result in the rapid stabilization of the process (Blank & Altukhov, 1992). The same effect, in accordance with the theory of Kimura & Weiss (1964; Chapter 1),

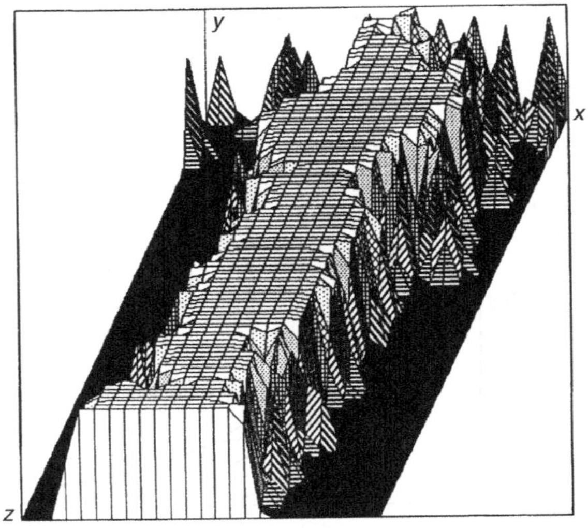

Fig. 4.6 Spatial–temporal dynamics of a circular population system with the stepping-stone structure of gene migration. Number of subpopulations $k = 25$, effective size of a subpopulation $N_e = 20$, and coefficient of gene migration $m = 0.03$. Axes: z – time in generations (1–1000); x – subpopulation number (1–25); y – gene frequency.

is observed even under slight pressure of balancing selection (Altukhov & Blank, 1992).

The facts discussed above confirm our previous conclusion that the population system is more stable than the structureless panmictic population (Altukhov, 1983a). It can be considered as proven that, both in nature and experimentally, the population system retains genetic characteristics of the ancestral population in a series of generations owing to the reciprocal balance of evolutionary factors.

Conclusion

The temporal stability of mean gene frequencies and their variances are an intrinsic emergent property of the population system, not following from the properties of subpopulations that are unstable structural components. Such a conclusion has been already drawn, when the concept of the population system was introduced (Altukhov & Rychkov, 1970), and since then has been demonstrated repeatedly for different biological species. The genetic stability of population systems was observed in time intervals commensurate with the time of their historical existence, measured in dozens, hundreds and thousands of generations (for details, see Altukhov, 1983a).

In this connection we should emphasize that the concept of the population system differs principally from theoretical views on a *subdivided population*. Although any population system is a subdivided population, its distinctive essential feature is the naturally formed hierarchy of structural levels, which reflects the history of its development (see Chapters 2 and 6). Conversely, a subdivided population as a conceived structure is not necessarily a system. This is why mathematical techniques for analysing subdivided populations, which are also applicable to the population system, reveal a previously unknown phenomenon: genetic stability is greater, the higher the level of the historical–geographical hierarchy analysed. Here is the principal difference of our conclusion about the genetic stability of population systems from the concept of subdivision as a factor accelerating the evolution of populations, developed by Sewall Wright (1951).

The concept of 'metapopulations', at least in the first stage of its formulation, also differs principally from our views. According to its authors (Levins, 1970; Gilpin, 1991; Hanski, 1991), metapopulations are plainly remnants of large panmictic populations that had high levels of genetic diversity. It is quite natural that the anthropogenic fragmentation of a habitat and the appearance of 'patchiness' lead to a decrease in effective size (N_e^{meta}, Gilpin, 1991) and, as a consequence, to the loss of heterozygosity. In other words:

> 'a metapopulation is an extreme form of spatial structure in which loosely coupled local populations "turn over", that is, suffer extinction followed by recolonization from elsewhere within the metapopulation' (Gilpin, 1991, p. 165).

The ideology of metapopulations admits that 'the classic conservation dilemma' is still unresolved, i.e. it remains unclear what is better for species preservation, 'a single large or several small patches' (the SLOSS controversy; Gliddon & Goudet, 1994).

Some scientists have already arrived at the extreme conclusion that it is inexpedient to maintain the subpopulation structure of the species while solving the problems of conservation biology. In their recent paper, Whitlock & Barton (1997, p. 440) claimed that

> '... a decrease in N_e with realistic population structures is a common phenomenon, and the increase in N_e with the island model is an artefact of the unnatural assumptions of that model'.

This reasoning is obviously based on the metapopulation viewpoint centred around the model in which extinction is caused by demographic stochasticity and chaos. The corresponding structure is by no means *realistic* but reflects only the result of excessive anthropogenic influences on the formerly stable population systems in which the processes of extinction and recolonization were in equilibrium.

Nowadays, such population systems are becoming increasingly difficult to find in nature. However, when this task is accomplished, we obtain results that are fundamental for an understanding of the natural mechanisms responsible for the maintenance of species' population structure and of the intra- and interpopulation components of gene diversity (Altukhov & Rychkov, 1970; Rychkov, 1973; Altukhov, 1974, 1983a, 1990). As demonstrated in this chapter, there are population systems of some salmonid fishes that provide this unique opportunity (see also Chapters 5 and 6).

It is very important to stress that in recent versions of the metapopulation concept we may find a systemic approach to the study of such communities, and many new conclusions of the authors (Hanski & Simberloff, 1997; Stacey *et al.*, 1997) are similar to those originally formulated for the concept of population systems (Altukhov & Rychkov, 1970).

Nevertheless, it is apparent that the population system, in contrast to the metapopulation, is not an artificial construction but a real natural entity characterized by distinct intra- and intersystemic relationships, specific territorial distribution (range), and a certain level of genetic diversity, which is maintained over long periods of time, and was initially characteristic of the ancestral population that progressively differentiated in a series of generations and over its range.

Consequently, Sewall Wright's evolutionary optimal structure may in fact be genetically stable, markedly more stable in comparison with a panmictic population of the same size and gene pool. The situation must be more stable, if population gene pools influenced by the balance of genetic drift and migration evolve under the pressure of balancing selection. This problem demands special consideration.

Chapter 5

The Role of Balancing Selection in the Maintenance of Biochemical Polymorphisms in Salmonid Populations

Although the discussion about the adaptive significance of protein polymorphism began more than 30 years ago, it continues today. The discovery of previously unknown large amounts of hidden genetic variation in populations of *Drosophila* and humans (Harris, 1966; Lewontin & Hubby, 1966) raised an acute problem of the mechanisms responsible for its maintenance. Some researchers consider that protein polymorphism is maintained by different forms of balancing (stabilizing) selection (e.g. Ayala *et al.*, 1974a,b). According to another viewpoint, this variation is selectively neutral (Kimura, 1983), i.e. irrespective of the genotype for any protein variant, its carrier gains no adaptive advantage. The objective resolution of the controversy between 'selectionists' and 'neutralists' is not only of academic interest, but also of particular significance for practical applications. Indeed, if protein polymorphism is selectively neutral and, according to Kimura (1983), represents a transitional phase of molecular evolution, it may be used only for solving the problems of phylogeny and microsystematics. Conversely, if protein polymorphism is functionally significant and represents a stable (rather than transitory) stage of evolution, it offers a unique opportunity of using it in breeding programmes, with emphasis on its conservation.

In recent years, an approach has been developed to assess the adaptive significance of protein polymorphism, based on the use of five methods:

(1) The analysis of stationary distributions of gene frequencies in natural population systems that can be reliably characterized with respect to parameters of their breeding structure, such as effective size N_e and the coefficient of gene migration m (Rychkov *et al.*, 1973; Altukhov, 1974; Altukhov *et al*, 1975a,b; Rychkov, 1975; Rychkov & Sheremetyeva, 1976).

(2) The analysis of genotypic distributions and correlations of gene frequencies in consecutive generations of the same subpopulation.

(3) The analysis of distributions of genotypes for allozyme loci at early and late ontogenetic stages (Altukhov, 1983a, 1989b; Altukhov *et al.*, 1991).
(4) The analysis of joint variability of adaptively significant polygenic morpho-physiological traits and Mendelian biochemical loci (e.g. allozyme loci) (Altukhov *et al.*, 1979; Altukhov, 1983a,b, 1985a,b; Altukhov & Kurbatova, 1990).
(5) Comparison of observed and demographically predicted levels of spatial genetic differentiation among populations (Rychkov & Sheremetyeva, 1976; Rychkov & Balanovskaya, 1990).

Let us consider the corresponding data.

5.1 Analysis of stationary distributions

Let us approximate the observed distributions of subpopulations of the sockeye salmon population system of Lake Azabachye in corresponding ranges of allele frequencies of *LDH-B2** and *PGM-2** loci by using Wright's stationary Equations 1.33 and 1.34 (Chapter 1). The values N_e, $\bar{q}$ and m must serve to construct the theoretical curves. Our material enables us to estimate the first two parameters, whereas the migration coefficient value is assessed by the average value of 2%, derived from the data cited above (sections 2.3.1 and 4.1.1).

It is evident from the theory of stationary distributions that in the case of unsatisfactory approximations of empirical distributions to those expected, which take into account only migration and random genetic drift, one is justified in making allowances for the effects of selection, measured by $\bar{W}$, the value of the mean population fitness.

By comparing the values of Equation 1.33 (Chapter 1) and the empirical distribution using the same values of q, we can estimate the $\bar{W}$ value and then calculate the W_i values for each separate genotype. This can be done by several methods. Here, with some small changes, we have applied the method used earlier (Altukhov *et al.*, 1975b).

Dividing the distributions of gene frequencies into three intervals ($\leqslant$0.60, $\leqslant$0.70 and $\geqslant$0.70 for lactate dehydrogenase and $\leqslant$0.75, $\leqslant$0.80 and $\geqslant$0.80 for phosphoglucomutase), we determine the observed and expected probabilities (according to the neutral model) of finding the subpopulation (point) in each of these intervals and the ratio of these probabilities. Taking the $2N$ roots of the resulting values, we obtain w for each interval. Using Equation 1.18 (section 1.5.1), the system of three linear equations in three unknowns is obtained:

$$\bar{w}_1 = w_{AA}\varphi_{11} + w_{AB}\varphi_{12} + w_{BB}\varphi_{13}$$
$$\bar{w}_2 = w_{AA}\varphi_{21} + w_{AB}\varphi_{22} + w_{BB}\varphi_{23}$$
$$\bar{w}_3 = w_{AA}\varphi_{31} + w_{AB}\varphi_{32} + w_{BB}\varphi_{33}$$

where $\bar{w}_1$, $\bar{w}_2$, $\bar{w}_3$ are the values of intralocus fitness for first, second and third intervals relatively; φ_{ij} is the expected proportion of j genotype in i interval, according to the neutral model i.e.

$$\varphi_{11} = \sum_1^n q_i^2 \omega_i, \quad \varphi_{12} = \sum_1^n 2p_i q_i \omega_i, \quad \varphi_{13} = \sum_1^n p_i^2 \omega_i$$

and so on, where n is the number of points (q_i) in the interval; ω_i is weighted within i interval value $\Phi(q)$. By solving the above equation system we will find the estimates for the fitness of genotypes.

Corresponding calculations for the entire community of spawning subpopulations have shown that when $N_e = 200$ and $m = 0.02$ in the case of the lactate dehydrogenase locus, there is agreement between the empirical distribution of subpopulations and the expected stationary distribution in the presupposed selective neutrality of this polymorphism ($\chi^2 = 10.1$, $P > 0.1$; d.f. $= 6$; Fig. 5.1). However, the observed distribution of subpopulations for the phosphoglucomutase locus differs significantly from the expected stationary distribution: $\chi^2 = 35.9$, $P < 0.001$; d.f. $= 3$ (Fig. 5.1).

In dealing with the lack of agreement with expectations as an effect of selection, we have defined the absolute values of the fitness of genotypes as $W_{AA} = 0.987$; $W_{AB} = 1.048$; $W_{BB} = 0.861$.

If the fitness of the heterozygote is set at 1.0, the normalized fitnesses become $W_{AA} = 0.942$; $W_{AB} = 1.000$; $W_{BB} = 0.822$.

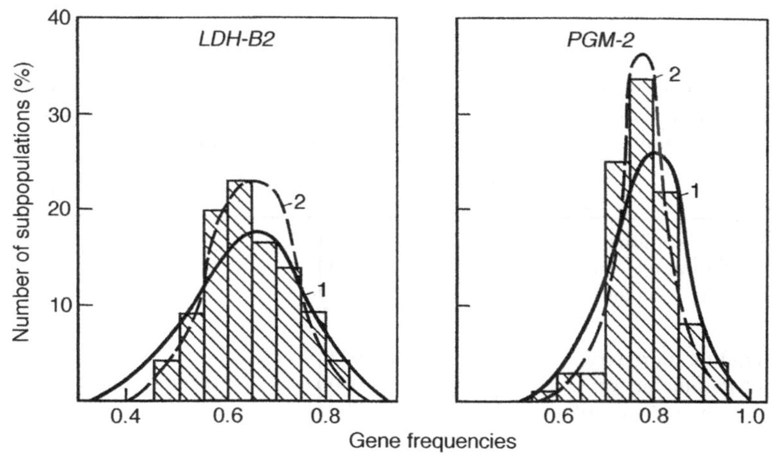

Fig. 5.1 Distributions of spawning subpopulations of sockeye salmon of Lake Azabachye within corresponding ranges of the allelic frequencies of *LDH-B2** (183 samples) and *PGM-2** (171 samples). Empirical distributions: the curves are the expected distributions: 1 – on the basis of genetic drift equilibrated by migration; 2 – under the combined effects of drift, migration and selection. See text for further details.

These estimates attest that the genotypes of the *PGM-2** locus evolve under the strong pressure of balancing selection. Introducing the selection factor to the model also improves the approximation of the *LDH-B2** gene frequencies. In this case $W_{BB}=0.98$, $W_{BB'}=1.00$ and $W_{BB}=0.97$. These estimates give $\chi^2=3.04$, $P>0.5$; d.f. $=6$ (Fig. 5.1).

The precise assessment of gene migration in natural populations of most species (humans are the only exception) is complex and difficult to resolve, and thus the estimates obtained above for the fitness values should be tested by any appropriately independent method. There are two possible methods of analysis. First, stable differences in the interpopulation variances of the allelic frequencies of the two loci studied may be used to hypothesize greater or lesser selective neutrality of genotypes, and accordingly, to evaluate the parameter of structure N_em through the intergroup gene frequency variance (see Equation 1.26, Chapter 1) at the most variable *LDH-B2** locus; this value can be used in Equation 1.34. Second, inasmuch as the population was characterized as a whole by constancy of the distributions of the *PGM-2** gene frequencies in generations with a clearly expressed excess of heterozygotes, by direct comparison of the actual and expected numbers of genotypes by the Hardy–Weinberg equation.

We utilized both of these methods and both revealed virtually complete coincidence of new W_i estimates with what has been indicated above (Table 5.1). At the same time we see also a close similarity of estimates of the $\hat{q}$ equilibrium value with the allele's average frequency in the subdivided population. Facts of this kind are indicative of the effect of overdominance in maintaining polymorphism at the *PGM-2** locus.

Table 5.1 The W_i values of genotypes at the locus *PGM-2** determined by different methods.

Method of determining *W*	Genotype	W_i *values*	Equilibrium (stationary) gene frequency $\hat{q}$	Observed average gene frequency $\bar{q}$
Using Wright's stationary functions. Parameter *Nm* found from the ecological data	AA AB BB	0.942 1.000 0.822	0.756	0.782
Using Wright's stationary functions. Parameter *Nm* found from the variance of allelic frequencies at *LDH-B2** locus	AA AB BB	0.937 1.000 0.799	0.766	0.782
Using the ratio of observed and expected numbers of genotypes	AA AB BB	0.943 1.000 0.796	0.782	0.782

See note under table 4.2 for genotype designation.

It is clear that we discovered this only because the system for collecting material had been planned ahead, envisaging the lengthy observation of the genetic structure of a well-defined natural population. But the same approach allows one to proceed further and to conduct a more detailed analysis of the distribution of genotypes in successive generations of separate subpopulations, taking into account the micro-heterogeneity of an environment when different types of spawning grounds are available (Ryabova *et al.*, 1978; Novosel'skaya *et al.*, 1982).

5.2 Analysis of the genotype distributions and the correlations of gene frequencies in successive generations of the same subpopulations

Tables 5.2 and 5.3 present the genetic parameters of subpopulations from different spawning sites. It follows from this that whereas there are, for the *PGM-2** allele frequency, no significant differences between groups of spawning grounds, the picture is different for the *LDH-B2** locus. When the populations of ponds and streams are compared with lake populations, local differentiation is evident.

In comparing the actual and expected distributions of *LDH-B2** genotypes, the significant deficit of heterozygotes is seen in the population as a whole. The distribution of *PGM-2** genotypes is different: an excess of heterozygotes is typical for the system as a whole and for the subpopulations of ponds, whereas the distribution of genotypes in the lake subpopulations is in accordance with Hardy–Weinberg equilibrium. These findings justify an assumption that *PGM-2** genotypes are

Table 5.2 Observed and expected distribution of genotypes and gene frequencies of the *LDH-B2** locus in sockeye salmon subpopulations preferring different types of spawning grounds.

Types of spawning grounds	Genotype			n	qB	χ^2
	BB	$B'B$	$B'B'$			
Ponds	827 805.40	1080 1123.08	413 391.52	2320	0.5892	3.41
Streams	654 651.51	778 783.15	238 235.34	1670	0.6246	0.07
Totals ('spring race')	1481 1455.62	1858 1908.68	651 625.70	3990	0.6040	2.81
Lake ('summer race')	1253 1244.14	957 974.88	200 190.98	2410	0.7185	0.82
Population system as a whole	2734 2723.2	2815 2903.1	851 773.7	6400	0.6523	10.43*

* $P < 0.01$.

Table 5.3 Observed and expected distribution of genotypes and gene frequencies of the *PGM-2** locus in sockeye salmon subpopulations preferring different types of spawning grounds.

Types of spawning ground	Genotype			*n*	*qA*	χ^2
	AA	*AB*	*BB*			
Ponds	1350 1388.18	886 809.74	80 118.08	2316	0.7742	20.51*
Streams	934 946.62	618 592.63	80 92.75	1632	0.7616	3.01
Totals ('spring race')	2284 2334.69	1504 1402.64	160 210.67	3948	0.7690	20.61*
Lake ('summer race')	1535 1535.61	767 765.89	95 95.50	2397	0.8004	0.00
Population system as a whole	3819 3869.2	2271 2171.2	255 304.6	6345	0.7809	13.32*

* $P < 0.001$.

subject to considerably greater selection pressure than *LDH-B2**, at least in ponds. The effects of selection should, however, not be so great for the distribution of *PGM-2** genotypes in lake subpopulations. This kind of deduction also accords with estimates of the genetic variance at the *PGM-2** locus (Table 5.4), which is minimal for ponds and streams, and maximal for lake subpopulations, although the differences are not large. Comparison of the allele frequency variances for the *LDH-B2** locus reveals a more salient differentiation between the types of spawning grounds: maximal variability is typical in the lake populations, the streams are intermediate, and the ponds have least variability. The differences of variance between these three types of spawning grounds are statistically significant (Table 5.4).

The genetic differentiation of populations thus revealed permits one to specify the previous conclusion about the intensity of selection at the *LDH-B2** locus and the assumption that in ponds, at least, the effects may be more considerable. The same argument can be employed for the genotypes of the phosphoglucomutase locus. There are two possible ways of checking this hypothesis:

(1) By analysing the correlations of gene frequencies in the same spawning grounds in successive generations.
(2) By analysing stationary distributions.

Taking into account the 5-year generation time of Azabachye sockeye salmon (Konovalov, 1980), we correlated the allele frequencies of *LDH-B2** and *PGM-2** loci for the subpopulations of the same spawning sites in successive generations (1972–7; 1973–8; 1974–9).

The significant correlation of gene frequencies in successive generations of the

Table 5.4 The main genetic parameters of a subdivided population of sockeye salmon in Lake Azabachye characteristic of different types of spawning grounds.

Locus	Type of spawning ground	K	n	$\bar{q}$	$F\varphi$	V_q	χ^2 test for homogeneity of variances (d.f.)
*LDH-B2**	Pond	44	2320	0.5892	Pond–stream, 4.98*	0.0026	
	Stream	32	1670	0.6246	Pond–lake, 87.58***	0.0043	$8.62 > \chi^2_{0.05}$ (2) = = 5.99
	Lake	48	2410	0.7185	Stream–lake, 39.69***	0.0058	
	Pond	44	2316	0.7742	Pond–stream, 0.78	0.0026	
*PGM-2**	Stream	31	1632	0.7616	Pond–lake, 4.75*	0.0029	$1.46 < \chi^2_{0.05}$ (2) = = 5.99
	Lake	48	2397	0.8004	Stream–lake, 8.22**	0.0037	

* $P < 0.05$.
** $P < 0.01$.
*** $P < 0.001$.
F_φ – estimated differences of gene frequencies using Fisher's criteria. The remaining designations are as for Table 4.1.

same subpopulation provides an indirect evaluation of the vector of selection, because it is obvious that the more considerable the effect of stabilizing selection, the greater should be the similarity in time of gene frequencies estimated repetitively in the same subpopulations. Indeed, the results obtained do not contradict this model (Fig. 5.2): the pair correlation coefficients at the *PGM-2** locus proved to be adequately high and reliable for the ponds and stream spawning grounds $r = 0.58$; $P < 0.05$ and $r = -0.81$; $P < 0.01$, respectively) and non-significant for the lake ones ($r = -0.13$); at the *LDH-B2** locus a significant positive correlation was found only for subpopulations reproducing in ponds ($r = 0.48$; $P < 0.05$).

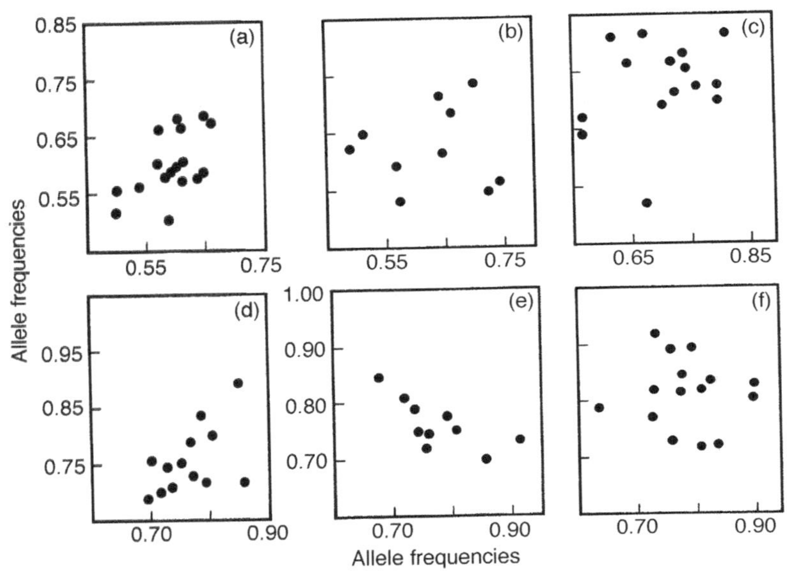

Fig. 5.2 The relationship between gene frequencies in consecutive generations of individual subpopulations of sockeye salmon in the same spawning site. (a, b, c) *LDH-B2** allelic frequencies for subpopulations of ponds, streams and the lake, respectively. (d, e, f) The same for the *PGM-2** allelic frequencies. See text for further details.

It is curious that the correlation for the *PGM-2** locus is positive for the pond subpopulations, but negative for the stream subpopulations. This may be explained as follows: evidently *selection here is of a universal nature and shifts the allelic frequencies in each subpopulation and in every generation to the equilibrium point* (see Chapter 1), which is the same for the total population or even for such an integral component of its structure as represented, for example, by a community of early-spawning populations (the spring race). If this model is correct then one would expect the differences between allelic frequencies in any subpopulation in two successive generations to be the more marked the further the subpopulation was from the equilibrium point in the preceding generation. *In this case, if an allele's* *frequency was lower than the equilibrium frequency in the preceding generation, it should increase in the following; if higher than the equilibrium, it should decrease.*

Indeed, the data represented in Fig. 5.3 are in accordance with this expectation, also satisfying the hypotheses about potential differences in the effects of selection at both loci for the different types of spawning grounds. We see clearly that a correlation can be traced, for both races, for the *PGM-2** locus, which is more subject to selection. The strength of correlation for both loci is maximal in the case of the spring-spawning subpopulations that undergo the most significant selective pressures.

Let us now define again the appropriate coefficients of genotype fitness through approximation of empirical distributions of allelic frequencies to the expected stationary distributions. We can estimate the *Nm* parameter from the F_{ST} statistics of the distribution of *LDH-B2** gene frequencies in lake subpopulations of sockeye salmon (Table 5.5). The estimates of the W_i values obtained are practically indistinguishable from those above.

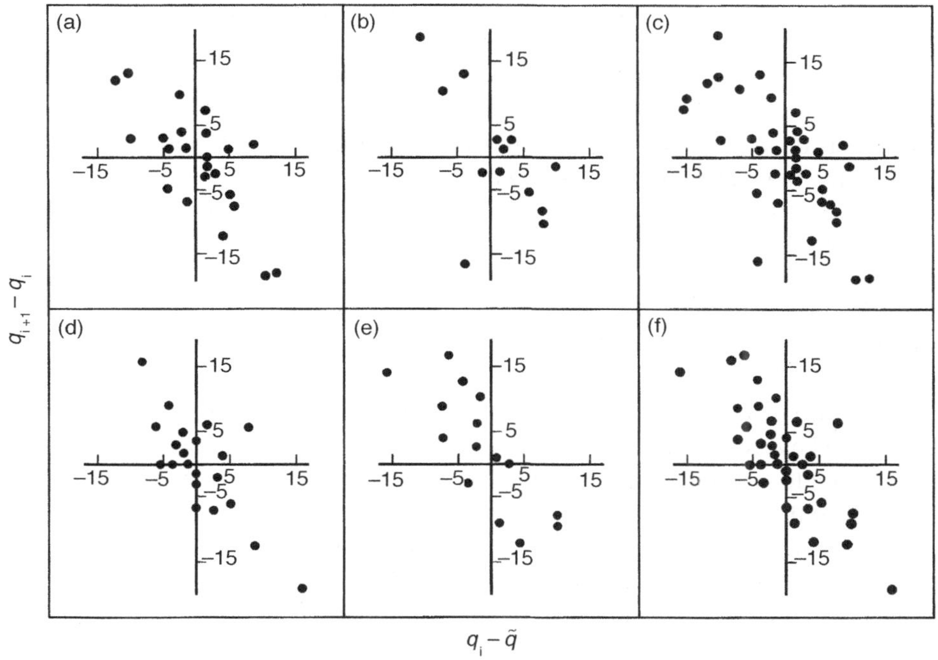

Fig. 5.3 The relationship between the deviation of gene frequency q_i from equilibrium $\hat{q}$ in a given generation (q_i $\hat{q}$; x-coordinate) and the difference between gene frequencies in preceding and subsequent generations of the same subpopulation ($q_{i+1} - q_i$ y-coordinate). (a, b, c) Spring and summer populations and the population system as a whole, locus *LDH-B2**. (d, e, f) The same for locus *PGM-2**. (a) r = −0.70***; (b) r = −0.50; (c) r = −0.62**; (d) r = −0.87***; (e) r = −0.77***; (f) r = −0.75***. ** $P < 0.01$; *** $P < 0.001$.

Table 5.5 W_i values of sockeye salmon genotypes in different kinds of spawning grounds.

Group of subpopulations studied (type of spawning grounds)	W_i values of genotypes at a locus					
	*LDH-B2**			*PGM-2**		
	BB	*B′B*	*B′B′*	*AA*	*AB*	*BB*
Ponds	0.90	1.00	0.86	0.94	1.00	0.80
Streams	0.98	1.00	0.97	0.91	1.00	0.72
Spring race total	0.94	1.00	0.91	0.92	1.00	0.74
Summer race (lake)	—	—	—	0.99	1.00	0.97
Population system as a whole*	0.98*	1.00*	0.97*	0.94	1.00	0.79

The parameter *Nm* is determined from the ecological observations.

Thus, the methods we have used have enabled us to obtain very similar estimates of fitness coefficients of the genotypes at both loci, and moreover reveal finer genetic differentiation of populations by types of spawning grounds. This differentiation (and the corresponding differences in W_i) also appears to harmonize well with the state of the environment.

It is clear, at least at a general level, that ecological conditions in a lake – a 'lake environment' – are more stable than those in streams and ponds, with their varying water regime and accompanying seasonal fluctuations in temperature and other physicochemical parameters, especially during the period which coincides with early ontogenesis of sockeye salmon (the embryonic period and early postnatal stages).

Naturally, the question arises as to which parameters of the environment can act as selective agents on genotypes of the two loci studied. There are several ways of approaching this problem, one of which is to explore links between environmental parameters and frequencies of alleles at different stages of ontogenesis. Although the number of such characteristics may be large and their influence indirect (see, for example, Nevo *et al.*, 1984; Philipp *et al.*, 1985), let us examine the connection between the allelic frequencies at both loci and physicochemical characteristics of spawning sites such as concentration of oxygen, pH and temperature in spawning nests of sockeye salmon.

Appropriate estimates were obtained in one of the field seasons (Table 5.6). It appears that the allelic frequencies for both loci did not relate at all to concentration of oxygen but that there was a reliable correlation with temperature: for *PGM-2**, $r = 0.60$ ($P < 0.05$), and for *LDH-B2**, $r = -0.85$ ($P < 0.001$). This association was similar to a clinal one: the allelic frequencies both at the *PGM-2** locus and (particularly) the *LDH-B2** locus gradually changed, in accordance with a decreasing water temperature in the spawning nests, from between the lake spawning grounds and the ponds. A significant correlation between the frequency of *PGM-2** and pH of the spawning grounds has not been found, but for *LDH-B2**, the correlation coefficient was 0.54, and at the given value of sample size ($n = 14$) the probability approached 0.05. Clearly, additional study is merited here. It would also be inter-

Table 5.6 Physicochemical characteristics of sockeye salmon spawning grounds of the Lake Azabachye basin.*

Type of spawning ground	Number of spawning ground	Date of test sampling	Temperature (°C)	pH	Oxygen content (mg/l)
Ponds	2	20 August	8.2	7.9	9.3
		24 August	8.2	7.9	10.3
		1 September	8.1	7.8	9.8
	3	20 August	9.7	7.3	8.3
		1 September	—	7.3	7.7
	4	20 August	8.8	7.2	8.9
	5	20 August	8.6	7.0	8.4
		24 August	8.6	7.1	8.6
		1 September	8.7	7.1	8.5
	15	20 August	8.2	6.9	8.6
	23	20 August	8.5	7.6	9.7
		1 September	8.4	7.6	10.1
	24	20 August	8.5	7.0	8.8
Average for ponds			8.54 ± 0.12	7.36 ± 0.10	9.0 ± 0.22
Streams	1	20 August	8.3	8.1	11.3
		24 August	8.2	8.2	13.3
	6	20 August	9.1	8.1	10.4
		1 September	9.1	8.1	10.9
	8	20 August	10.0	8.0	10.0
		24 August	9.8	8.1	9.1
		1 September	9.9	8.0	9.3
	18	20 August	9.0	8.1	10.6
	20	20 August	10.1	8.1	8.69
Average for streams			9.17 ± 0.24	8.10 ± 0.06	10.43 ± 0.39
Average for the spring race			8.83 ± 0.14	7.68 ± 0.09	9.65 ± 0.25
Lake	11	20 August	13.3	7.8	9.1
		24 August	13.1	7.9	9.1
		1 September	13.2	7.9	9.4
	13	24 August	12.3	8.6	9.2
		1 September	—	—	10.0
	14	20 August	10.8	8.1	9.0
		24 August	10.7	8.1	9.4
		1 September	10.7	8.0	10.1
	22	20 August	13.0	7.8	8.6
		24 August	13.1	7.8	—
		1 September	13.0	7.8	—
	25	20 August	13.2	8.4	8.2
		1 September	13.2	8.6	8.0
Average for the summer race			12.47 ± 0.31	8.08 ± 0.09	9.10 ± 0.20

* Numeration of spawning grounds is the same as for Fig. 4.1.

esting to partition the time over which selection may be acting. Evidently, it is most effective in the very earliest stages of development, because during the catadromous migration of the young, the genotype distributions within the young cohorts are virtually indistinguishable from the corresponding distributions in the spawning fishes (Novosel'skaya *et al.*, 1982). A similar picture applies for another species of Pacific salmon, namely pink salmon (Altukhov *et al.*, 1987).

We do not exclude the possibility that the heterozygote excess observed in the spring races of sockeye salmon may result from the effects of selective fishing (Chapter 7). None the less, in this case also, if the selection vector is directed in favour of heterozygotes, the comparison of the late ontogenetic stages with the early ones may be an appropriate method of revealing the effect of selection.

5.3 Analysis of genotype distributions at early and late stages of ontogenesis

A study was carried out on several successive generations of a pink salmon population reproducing in the Naiba River basin (south-east coast of Sakhalin Island; see Fig. 6.8 later). It consisted of a comparative analysis of the distributions of allozyme genotypes (at *MDHB-1, 2**, *PGM-2**, *G3PDH-1**, *PGDH** and *MEP-2** loci) in adult fish (spawners) and alevins. For all polymorphic loci studied the proportion of heterozygotes was higher for the spawners than it was for the alevins (Fig. 5.4).

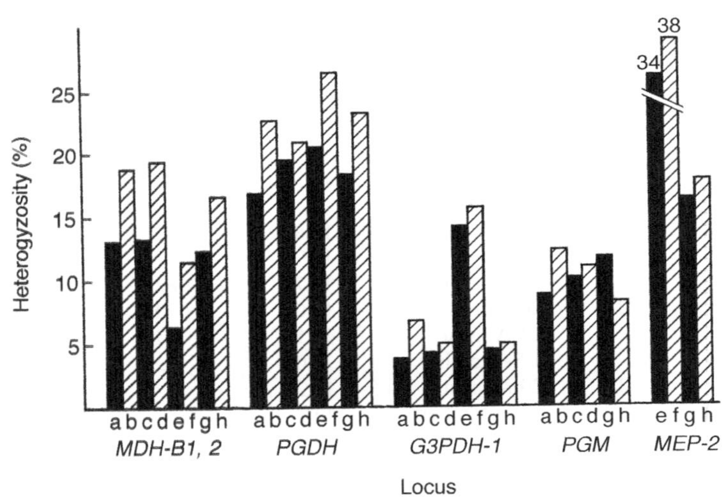

Fig. 5.4 Heterozygosity (%) observed at several allozyme loci in alevins (black columns) and spawners (hatched columns) in successive generations within the same population of pink salmon. (a, c, e, g) Alevins of 1979, 1983, 1984 and 1985 spawning years, respectively. (b, d, f, h) Spawners of 1979, 1981, 1984 and 1985, respectively. The differences between the spawners and alevins are highly significant by the sign-test. The total sample sizes are 1614 spawners and 2316 alevins.

Although the differences between spawners and alevins are small, they are repeated consistently from generation to generation and are observed in different populations, both artificially propagated in hatcheries, and in nature (for details, see Altukhov *et al.*, 1987). At later stages of ontogenesis, when the pink salmon fry begin to migrate, the difference in the levels of heterozygosity between them and the parent generation is no longer marked. These findings are of particular interest in that they limit the intervals of the life cycle during which the selective elimination of the genotypes at the loci studied occurs. At even earlier stages of development the elimination may be directed against heterozygotes (or rare genotypes) (Altukhov *et al.*, 1987; Zhivotovsky *et al.*, 1987) so that, as a whole, the selection discovered is typically balancing and of a cyclical character.

In order to estimate the selection intensity one can, as a first approximation, calculate the fitness coefficients (viability) of the genotypes at individual loci (W_i) or their paired combinations (W_{ij}) by dividing the frequencies of one- or two-locus genotypes in the spawners into corresponding frequencies in the alevins. Here we shall use the data for the generations of odd years, for which sample sizes are larger (Table 5.7). It can easily be seen that the fitness of the heterozygotes is higher than that of the homozygotes at nearly all loci, exept the *PGM-2** locus. However, in the latter case, the deviation has simply a statistical character because of the extreme rareness of one homozygote genotype.

The fitness of genotypes in spawners in relation to alevins was calculated by comparing actual distributions of individual heterozygosity; by taking into account the frequencies of fish that were heterozygous at none, one, two and more loci (Table 5.8). The estimate for this fitness also increases as the class heterozygosity grows. This means that one cannot exclude the possibility of overdominance as the mechanism maintaining variability at the loci investigated in pink salmon.

As follows from the data in Table 5.9, the fitness of the interlocus-paired genotypic combinations W_{ij} is extremely close to the products of the fitness of the genotypes of individual loci, that is $W_{ij} \cong W_i \times W_j$. This corresponds to the multiplicative model, although additional research is required to reach a final conclusion.

Thus, comparing genotype distribution at various stages of ontogenesis for several generations within a population indicates, at least in pink salmon, an important role of stabilizing selection in the maintainance of protein polymorphism.

We now turn to a method of detecting the adaptive nature of protein polymorphism through the analysis of interrelated variability of monogenic and polygenic traits.

5.4 Interrelated variability of monogenic and polygenic traits

This approach is characterized by the analysis of the distribution of individual heterozygosity of various genotypic combinations along the variability curve of a

Table 5.7 Fitness values of genotypes of allozyme loci (W_i) in pink salmon and their normalized values (W'_i) in adult fish in relation to alevins.

Genotypes	*MDH-B1,2**		*G3PDH**		*PGDH**		*PGM-2**		*MEP-2**	
	$W_i \pm$ SE	W'_i	$W_i \pm$ SE	W'_i	$W_i \pm$ SE	W'_i	$W_i \pm$ SE	W'_i	$W_i \pm$ SE	W'_i
AA	0.92 ± 0.02	0.63	0.98 ± 0.01	0.58	0.94 ± 0.03	0.71	0.98 ± 0.02	0.65	0.98 ± 0.03	0.92
AB	1.45 ± 0.16	1.00	0.69 ± 0.39	1.00	1.32 ± 0.14	1.00	1.20 ± 0.18	0.78	1.07 ± 0.21	1.00
BB	—	—	—	—	0.57 ± 0.26	0.43	1.53 ± 2.00	1.00	0.93 ± 0.32	0.93

Designations for genotypes are standard: *AA* – predominating homozygote; *AB* – total heterozygotes at all alleles; *BB* – alternative homozygote or following a predominating one. Average sample size per locus: adult fish – 542; alevins – 910.

Table 5.8 Distributions of individual heterozygosity at allozyme loci in pink salmon spawners and alevins.

Fish groups	Frequencies of classes with a different number of heterozygous loci per individual			Sample size (n)	χ^2 test for differences between distributions in spawners and alevins (d.f. = 3)
	0	1	≥2		
Spawners	0.495 ± 0.016	0.391 ± 0.016	0.144 ± 0.015	606	38.6**
Alevins	0.620 ± 0.020	0.335 ± 0.020	0.044 ± 0.007	901	

The 'fitness' of frequency classes in spawners in relation to alevins equals 0.80 ± 0.04; 1.17 ± 0.08 and 2.57 ± 0.51 tending towards increased individual heterozygosity.
** $P < 0.001$.

quantitative character, or of a set of such characters which appear to be adaptively significant. Here a very important fact is taken into account: those individuals which approximate the population mean value are shown to have the highest fitness, particularly at early stages of ontogenesis (Bumpus, 1899; McAtee, 1937; Dubinin, 1948; Karn & Penrose, 1951).

It is assumed that if allozyme loci are involved to some extent in the adaptive process, the levels of individual heterozygosity must be distributed non-randomly in the field of variability of adaptively significant quantitative traits. We can state at present that in the case of heterotic selection (overdominance), the maximum heterozygosity must be concentrated in the tails of the distribution of adaptively significant polygenic characters or in the centre of such a distribution (under simple additive effects of genes). This approach turned out to be especially effective in the study of sockeye salmon populations (Altukhov, 1983a; Altukhov & Varnavskaya, 1983). We have already given examples that provide evidence for strong heterotic selection at the *PGM-2** locus of sockeye which spawn in Lake Azabachye. The differences between the fitness coefficients of the different genotypes were highly significant in creating the question of how to explain this differentiation in relation to the characteristics which normally mark the effect of heterosis (e.g. the increased rates of growth and maturation). It is possible to reveal clear relationships between intrapopulation differentiation of the species for these characters and allozyme heterozygosity at the *LDH-B2** and especially the *PGM-2** loci.

As noted in Chapter 2, sockeye salmon are characterized by sexual dimorphism for body size and complex reproductive behaviour, including a system of assortative matings. On average, males are larger than females. As a rule, these large males have spent more than 2 years in the sea and are 5 to 7 years old by spawning time. In addition to the large males, a second group of males exists which grow rapidly and

Table 5.9 The fitness of the interlocus-paired genotypic combinations in spawners in relation to alevins.

Interlocus genotypic combinations	*MDH-B1,2* × *G3PDH*		*MDH-B1,2* × *PGDH*		*MDH-B1,2* × *PGM-2*		*G3PDH* × *PGDH*		*G3PDH* × *PGM-2*		*PGM-2* × *PGDH*	
	W_{ij}	$W_i \cdot W_j$	W_{ij}	$W_i \cdot W_j$	W_{ij}	$W_i \cdot W_j$	W_{ij}	$W_i \cdot W_j$	W_{ij}	$W_i \cdot W_j$	W_{ij}	$W_i \cdot W_j$
AA × *AA*	0.89	0.90	0.86	0.86	0.89	0.90	0.93	0.93	0.96	0.95	0.95	0.94
AB × *AA*	1.43	1.43	1.38	1.35	1.44	1.39	1.57	1.61	1.53	1.81	1.19	1.26
AA × *AB*	1.53	1.55	1.26	1.28	1.16	1.10	1.27	1.28	1.13	1.17	1.00	1.13
AB × *AB*	5.44	2.47	2.02	2.01	1.29	1.69	2.71	2.23	4.90	2.24	2.49	1.54
AA × *BB*	—	—	0.69	0.55	—	—	0.58	0.54	1.45	1.42	0.41	0.46
χ^2 (d.f.)	2.0 (3)		0.4 (4)		1.6 (3)		3.5 (4)		3.7 (3)		5.7 (4)	

mature as small, 3-year-old individuals that have spent only 1 year in the sea (they are called 'kayurki' in Kamchatka, and 'jacks' in America and Canada). The mean age of females at spawning varies between 4 and 5 years.

As shown by Hanson & Smith (1967) and McCart (1969), large males possess a selective advantage during the formation of mating pairs by holding a territory on the spawning grounds. However, in years with a water level lower than normal, and in the shallow areas of the spawning grounds, small males will have a reproductive advantage over large ones. Accordingly, with respect to body length at spawning, sockeye populations exhibit a unimodal frequency distribution for females and a pronounced bimodality for males. The best example for such a clear male dimorphism was found for sockeye that spawn in Lake Azabachye (Fig. 2.6). These three groups of fish which can be easily distinguished during spawning were compared with regard to their level of heterozygosity for both the *LDH-B2** and *PGM-2** loci. The estimates were performed with the sockeye salmon of Lake Azabachye. The females show an intermediate level of heterozygosity, while small and large males are characterized by maximal and minimal levels, respectively (Table 5.10). Thus, sockeye males and females represent two adaptive systems with the maximal variance of fitness in males. The high degree of heterozygosity of small males correlates (Altukhov & Varnavskaya, 1983) with their increased rate of growth and early sexual maturation (Krogius, 1960, 1975).

The relationship between increased levels of heterozygosity for the two allozyme genes and the morphophysiological traits, reflecting heterosis for growth and maturation rates, is uncomplicated. The heterozygosity level is especially high in the small males which do not migrate to sea, but remain in fresh water throughout their life (Kirpichnikov, 1981; Altukhov & Varnavskaya, 1983).

Table 5.10 Total levels of heterozygosity for *LDH-B2** and *PGM-2** loci in males and females of sockeye salmon populations (in accordance with distributions in Fig. 2.6).

Research years	Males				Females	
	'Small'		'Large'		Sample size	Heterozygosity
	Sample size	Heterozygosity	Sample size	Heterozygosity		
1971	62	0.74	205	0.62	275	0.63
1972	209	0.66	121	0.63	324	0.66
1973	71	0.69	331	0.65	391	0.67
1978	202	0.61	317	0.60	571	0.65
1979	90	0.65	226	0.63	318	0.65
Totals	705	0.67	1455	0.61	2353	0.65

Statistical analysis (χ^2 test) gives the following results: small males–large males, 8.45 ($P<0.01$); large males–females, 6.86 ($P<0.01$); small males–females, 1.00 ($P>0.05$).

Similar relationships, although not interpreted with certainty, between multi-locus individual allozyme heterozygosity and growth rate and sexual maturation have been found for *Salvelinus fontinalis* (Liskauskas & Ferguson, 1991) and *Oncorhynchus mykiss* (Danzmann *et al.*, 1988; Ferguson 1992). At least a negative correlation between multi-locus heterozygosity and body length for sexually mature males in *S. fontinalis* is evident, as well as the fact that mature males of *O. mykiss* turn out to be more heterozygous compared with young males, simultaneously showing a slow-down in growth rate (Ferguson, 1992). Clearer data completely conforming to the observations for sockeye salmon were obtained for *Salmo trutta* (Makhrov *et al.*, 1997).

The results of different authors concerning the elimination of homo- and heterozygous allozyme genotypes at the fry stage also agree. It is evident from data that a considerably higher mortality among multi-locus homozygotes at early stages of ontogenesis, first discovered in pink salmon (Altukhov, 1983a; Altukhov *et al.*, 1987), is also typical for some other species (Mitton & Grant, 1984; Allendorf & Leary, 1986; Zouros & Foltz, 1987; Quatro & Vrijenhoek, 1989).

Hence, on the basis of all the above considered data we cannot exclude overdominance as one of the mechanisms for maintaining polymorphism of allozyme loci in the wild salmonid populations studied. As the chromosome number in pink salmon is large ($2N = 52$; $NF = 104$), and the genetic differences between alevins and spawners were detected for all the non-linked genes studied, we assume that selection acts on the genotypes of these loci and not on any other genes (or supergenes) whose markers they represent. Arguments in favour of the opposite point of view can be found in Lynch & Walsh (1998).

During the 1990s many papers have been published attempting to analyse the effects of heterosis by investigating allozyme polymorphism in a wide variety of plant and animal species. The results have been reviewed by Zouros & Foltz (1987) and Lynch & Walsh (1998). As these authors note, the study of correlations between individual allozyme heterozygosity and adaptively significant characters has led both to positive and negative results, making it reasonable to summarize these somewhat contradictory findings. One should, however, address the following questions: why are the connections between the variability of quantitative characters and allozyme polymorphism not always discovered? Why are these links, when discovered, associated with a small number of allozyme loci?

When answering both questions it is very important to consider 'genotype × environment' interactions. As noted earlier (Altukhov, 1983a, 1991), under the same environmental conditions various gene loci are subject to different forms of selection, or are sometimes selectively neutral. Correlations can be estimated only when the conditions are strict enough, i.e. when the uniformly vectorized selection simultaneously influences a great number of genes. The early stages of ontogenesis are the most appropriate for revealing such correlations.

The results of our studies show that the *optimal* (intermediate) *level of gene diversity* in a population is especially important for its viability: high heterozygosity at multiple allozyme loci, combined with the extremely high viability of an individual, may be unfavourable for its progeny and, hence, for the fitness of the population as a whole, as some segregated genotypes will be maladaptive (Altukhov, 1989a). High levels of polymorphism and heterozygosity can lead to a sharp increase in recombination frequency, which prevents harmonious interactions between loci. The probability of disturbing the cooperativeness of gene interactions should be especially high at early ontogenetic stages, when many systems of the developing organism are functionally immature and its viability depends primarily on the most vulnerable links in the chain of metabolic processes (Livshits & Kobyliansky, 1985).

If our assumption is valid, individual heterozygosity of parents and viability of their progeny (or its variance) should be interrelated in a certain way. This hypothesis was tested in several crossings between pink salmon spawners with different levels of individual heterozygosity.

We studied eight combinations of parent pairs characterized with respect to the loci *MDH-3,4**, *GPDH**, *PGM-2** and *MEP-2**, which allowed us to rank the families by the average level of heterozygosity. Artificially fertilized eggs were incubated at a hatchery in southern Sakhalin, and mortality was registered throughout the period of incubation and rearing to the smolt stage. The level of individual heterozygosity at the above allozyme genes and body length in alevins of the same age demonstrated a significant positive correlation with one another, i.e. heterosis in growth rate was observed. As expected, a positive correlation was also revealed between the variance of alevin body length and the level of allozyme heterozygosity in the family (Fig. 5.5). Concerning mortality in the progeny (total over the period of incubation and rearing), it was minimal among alevins from intermediately heterozygous parents, and maximal in the progeny of pairs with both high and low levels of heterozygosity (Altukhov *et al.*, 1991). One can see from Fig.

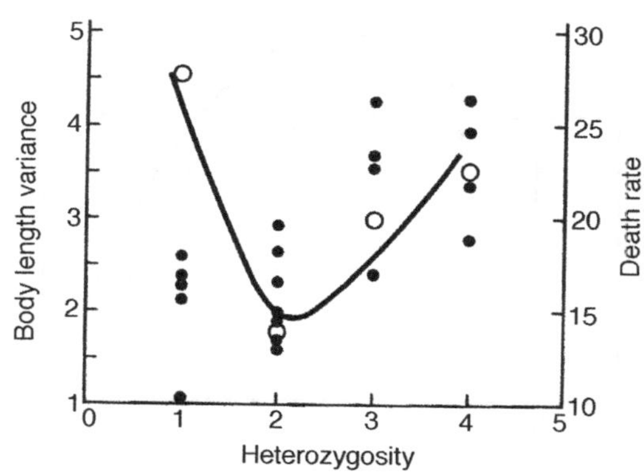

Fig. 5.5 The relationship between the level of individual allozyme heterozygosity of parental pairs (x-axis), variance of alevin body length (y-axis, dark circles; $r = 0.72$; $P < 0.001$; $n = 843$) and death rate (ordinate to the right, light circles; $n = 5184$) throughout incubation of fish eggs and development to the premigration stage in the pink salmon (*Oncorhynchus gorbuscha*).

5.5 that the mortality of offspring was rather high in these experimental conditions. These results agree with all previous data confirming that the genotypes of enzyme loci studied are under strong stabilizing selection. In recent years several works have presented new data on the important role of balancing selection for maintaining allozyme polymorphism in natural populations of various animal species (Karl & Avise, 1992; Pogson *et al.*, 1995; Wilding *et al.*, 1997).

Our next experiment was performed with 20 pink salmon families differing in the level of heterozygosity at eight allozyme loci: *G3PDH**, *MDH-B1**, *MDH-B2**, *MEP-2**, *PGDH*, *PGM-2**, *GPI-A** and *FDHG**. The results of this experiment, in which the rate of progeny mortality was less than in the previous one, showed the dependence between heterozygosity and body length (at the same pre-smolt stage). More detailed analysis showed that this relationship is influenced by the effect of heterozygosity on the interfamilial component of variation. In other words, the average body length in the progeny of a family largely depends on the level of family heterozygosity. Moreover, if the families are ranked with respect to average heterozygosity, the largest average body size is observed in the progeny of intermediately heterozygous families, and the smaller progeny is characteristic of families with the highest and the lowest levels of heterozygosity (Dubrova *et al.*, 1995) (Fig. 5.6).

Especially significant differences are observed between progenies of families with intermediate and high levels of heterozygosity: average body length in the latter is

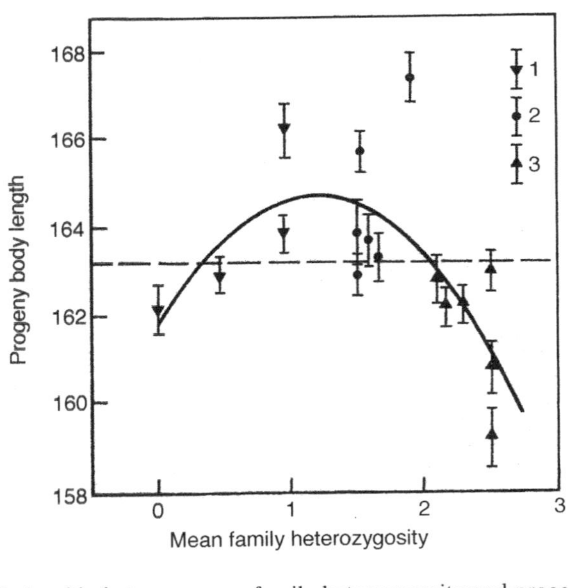

Fig. 5.6 The relationship between mean family heterozygosity and progeny body length of pink salmon. 1 – Families with low heterozygosity; 2 – families with intermediate heterozygosity; 3 – families with high heterozygosity. Broken line indicates the mean body length value for all the progeny examined. Quadratic fit $y = 161.7 + 5.51x \;\; 2.27x^2$; $F_{2,2562} = 189.23$, $P < 0.0001$, $r^2 = 0.1290$.

markedly smaller, and fry from the families with the intermediate level of heterozygosity apparently have a higher growth rate. This result contradicts the predictions based on the heterotic model and does not exclude the possibility that we are dealing with a case of epistatic gene interactions. The latter is characteristic of hybrid and rare genotypic combinations and can have an effect on the type of dependence between individual heterozygosity and variability of quantitative traits (Dubrova & Gavrilets, 1989).

As follows from our data, parent pairs similar to those that gave rise to families with a high heterozygosity rank occur in Sakhalin pink salmon populations with a frequency of approximately 0.05. In other words, the highly heterozygous pink salmon families artificially created in our experiments are rare in nature. A unique combination of alleles of the studied polymorphic loci can provide the basis for the epistatic interaction affecting variation in quantitative traits, including body length in young fish. However, the problem is that mortality at early embryonic stages in this experiment was relatively low, whereas the effects of intrapopulation heterosis (identified by allozyme genes) manifest themselves under suboptimal environmental conditions (Altukhov, 1990, 1991; Altukhov *et al.*, 1991); hence, the heterotic model and the model involving epistasis require further investigation.

Nevertheless, the data reviewed corroborate experimental observations, providing evidence for the hypothesis that the medium (rather than maximum) level of gene diversity, which is maintained by stabilizing selection, is optimal for the population (Altukhov *et al.*, 1987; Altukhov, 1990, 1991; Dubrova *et al.*, 1990, 1991). Body length in young fish has adaptive significance, and individuals with low values of this quantitative trait are characterized by decreased viability (Beacham, 1988; Danzmann *et al.*, 1988). According to our results, a decrease in body length of pink salmon fry is associated with a high heterozygosity of the progeny. Therefore, natural selection, eliminating young fish with a small body length from the population, simultaneously has a strong effect on its genotypic diversity and eventually leads to a decrease in heterozygosity. This agrees with the results of our previous experiments in which minimal mortality was observed in families with intermediate levels of genetic diversity (Altukhov, 1989b; Kartavtsev *et al.*, 1990; Altukhov *et al.*, 1991). From this point of view, one may try to explain a group of interesting facts noticed by Lynch & Walsh (1998). These authors found, in analysing data in the literature on the correlation of multiple-locus heterozygosity with fitness-related characters, that increases in heterozygosity with cohort ageing only appear at loci that initially have heterozygote deficiences. If we are to interpret this as a result of selection directed against multiple heterozygotes at the earliest stages of ontogenesis, but in favour of heterozygotes at the later stages, then it is not necessary to resort to the hypothesis of associative overdominance, and the intrinsic overdominance mechanism appears to be true in the framework of a heterozygosity adaptive optimum.

Generally, if our concept of adaptation as a compromise between its individual and population components is valid, there should be at least several natural mechanisms aimed at maintaining the established optimal levels of gene diversity, which are specific for each population. With respect to salmonids, especially interesting among such mechanisms are *selective matings*, *subdivision into subpopulations*, and *different types of selection varying in the course of ontogeny or in fish of different sex*.

These mechanisms have already been discussed above. It should be noted, however, that the potential for assessing the adaptive value of polymorphism at many biochemical loci simultaneously is also offered by the approach based on the comparison of their standardized genetic variances with variance expected in the case of neutral evolution. Such an approach does depend, however, on the accurate estimation of the parameter of demographic structure $N_e m$ (Rychkov & Sheremetyeva, 1976; Balanovskaya & Rychkov, 1990; Rychkov & Balanovskaya, 1990).

5.5 Comparison of observed and demographically predicted levels of spatial genetic differentiation

If genetic drift and gene migration are in equilibrium, the corresponding coefficient of local genetic differentiation expected for a selectively neutral process and described by Equation 1.30 (Chapter 1) may be designated as $\hat{F}_e$ and compared with the value $\bar{F}_0$ empirically determined from allele frequencies of different allozyme or DNA loci (Equation 1.26):

$$F_0 = 1/K \sum F_i = \frac{1/K \sum V_p}{\bar{p}_i(1 - \bar{p}_i)}$$

where K is number of loci, $\bar{p}_i$ and $(1-\bar{p}_i)$ are mean allele frequencies of the i-th locus in the total subdivided population consisting of n subpopulations, and $V_p = 1/n \sum (p_i - \bar{p}_i)^2$ is Wahlund's variance (Equation 1.24).

If a locus is selectively neutral at a given time and at a certain site of the species range, $\bar{F}_0 \approx \hat{F}_e$; if a polymorphic gene evolves under the effect of stabilizing selection, $\bar{F}_0 < \hat{F}_e$; and in conditions of disruptive or oppositely directed local selection, $\bar{F}_0 > \hat{F}_e$.

Studies on the indigenous Mongoloid populations of the circumpolar zone of Russia showed that, as estimated according to the set of polymorphic loci, the genetic process occurring in them actually is selectively neutral, and, hence, $\bar{F}_0 \approx \hat{F}_e$ (Rychkov & Sheremetyeva, 1976).

On the other hand, the species *Homo sapiens* at different phases of its genetic differentiation was exposed to different types of environmental selection pressure, and the ratio of different gene groups in the gene pool changed with time. Some genes that behaved as adaptively important or, conversely, selectively neutral at the early stages of historical development of the human population system changed their rank at later stages, whereas the rank of other genes remained unchanged

(Balanovskaya & Rychkov, 1990). It is noteworthy that no polymorphic gene loci proved to remain selectively neutral throughout the period of time considered in this study, i.e. since the Upper Palaeolithic period (Balanovskaya & Rychkov, 1990; Rychkov & Balanovskaya, 1990).

The analysis performed in our studies provides evidence that local genetic differentiation (average per locus) is selectively neutral not only in the case of human populations, but also in population systems of biological species in general (Altukhov, 1995), including salmonid fishes (Altukhov & Salmenkova, 1994).

Estimations of N_e and coefficients of gene migration m were obtained from publications and on the basis of our own field studies (Altukhov, 1974; Altukhov *et al.*, 1975a; Altukhov & Salmenkova, 1994). They may be considered approximate, but correspondence between the expected values ($\hat{F}_e$) and those observed from allele frequencies of allozyme loci ($\bar{F}_0$) proved to be so close that it cannot be accidental.

In Fig. 5.7, the correlations between values $\hat{F}_e$ and $\bar{F}_0$ are shown for several biological species, including salmonid fishes, to demonstrate the universality of this relationship.

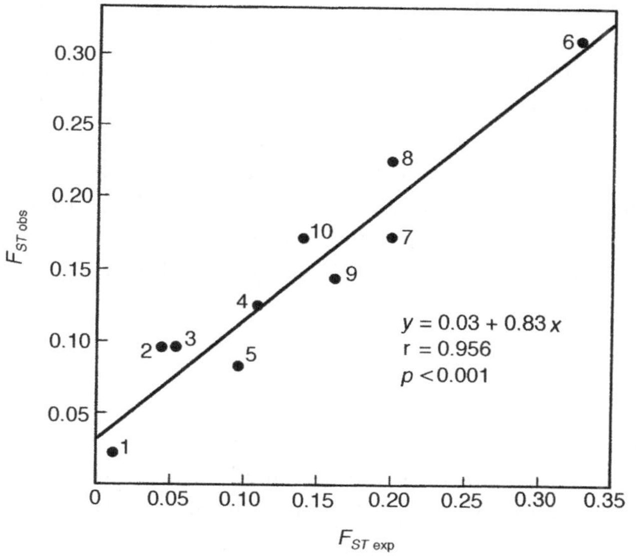

Fig. 5.7 Relationship between observed $F_{ST\text{obs}}$ and expected $F_{ST\text{exp}}$ values of spatial genetic differentiation in different biological species. 1 – pink salmon, *Oncorhynchus gorbuscha*; 2 – sockeye salmon, *Oncorhynchus nerka*; 3 – chinook salmon, *Oncorhynchus tshawytscha*; 4 – Atlantic salmon, *Salmo salar* (data for these four fish species from Altukhov & Salmenkova, 1994); 5 – *Homo sapiens*, Siberian isolate (Rychkov & Yashchuk, 1985); 6 – black-tailed prairie dog, *Thomomys bottae* (Lidicker & Patton, 1987); 7 – mouse, *Mus musculus* (Petras, 1967; Lidicker & Patton, 1987); 8, 9 – red howler monkey, *Alouata seniculus* (Pope, 1992); 10 – green turtle, *Chelonia mydas* (Carr *et al.*, 1978; Bowen *et al.*, 1992; Karl *et al.*, 1992). Reproduced from Altukhov, 1995.

This method obviously needs further improvement because it may be applied only if:

(1) The number of loci is sufficiently large
(2) The limit of estimating the polymorphic state of the gene should not be lower than 5%
(3) The asymmetrical distribution of F_{ST} values for i-th loci should be normalized
(4) Estimating F_0 values for the i-th locus, and comparison of them with the mult-locus $\bar{F}_0$, should be made by using the bootstrap analysis.

Nevertheless, when gene pool reproduction in a population system is in a steady state or at least quasi-stationary process (i.e. when the processes of its differentiation and integration are in equilibrium), the selective status of a gene may be determined reliably enough even if the number of loci is small, but the number of subpopulations is large. To illustrate this conclusion, let us consider the data on the sockeye population system of Lake Azabachye. Although the necessary data are available for only two polymorphic genes, this obstacle is possible to overcome owing to the mathematically equivalent increase in the number of populations characterized genetically (over the entire period of studies, a total of 183 and 171 populations for loci *LDH-B2** and *PGM-2**, respectively). At the corresponding sample size, estimations of F_0 by individual loci are fairly stable. The same applies to N_e determined on the basis of direct ecological observations (see section 4.1.1), which proved to be almost the same as the value obtained by analysing the data found in the world scientific literature (Table 2.4).

Comparing empirical estimates of F_0 with the expected level of selectively neutral local genetic differentiation F_e calculated from demographic parameters (Equation 1.30, Chapter 1), we can determine the approximate intensity of environmental selection pressure on the loci under study using the following equation:

$$S = \frac{\hat{F}_e \quad F_0}{4N\hat{F}_e F_0}$$

where $\hat{F}_e$ is the expected genetic variance under conditions of selectively neutral polymorphism, F_0 is the empirically determined variance of the gene frequency (Equations 1.26, 1.30 and 1.37, Chapter 1), and N is the effective population size (see Kimura & Weiss, 1964; Cavalli-Sforza & Bodmer, 1971).

In the sockeye salmon population system of Lake Azabachye, the $\hat{F}_e$ value, expected for a neutral stationary process and calculated on the basis of demographic parameters ($N_e = 200$, $m = 0.02$), is 0.059. In the case of *PGM-2**, $F_0 = 0.008$. As shown above, polymorphism in this gene is maintained by a strong pressure of stabilizing selection.

Using the above equation to estimate selective environmental pressure on the *PGM-2** locus, we obtain $S = 0.135$. At gene frequencies at the point of equilibrium, this value is approximately similar to the standard deviation σ in genotype fitness w_i. Calculating this deviation on the basis of different estimates of w_i values (from the comparison of observed and expected genotype distribution for the *PGM-2** locus in the spring race (Table 5.3) and from the solution of stationary distributions) we obtain $\sigma_W = 0.118 \pm 0.048$, which does not differ from $S = 0.135$ estimated above.

For the *LDH-B2** locus, which is selectively more neutral, $F_0 = 0.027$, and the estimated $S = 0.025$ and $\sigma_w = 0.032$ are virtually identical.

Thus, the method of comparing standardized variances of gene frequencies for individual loci (F_{0i}) with that expected for a selectively neutral process (F_e), with demographic parameters of population structure N_e and m taken into account, produced the results reliably characterizing the selective status of corresponding allozyme loci and fully confirming the initial conclusion that the *PGM-2** locus has evolved under strong stabilizing selection, whereas polymorphism at the *LDH-B2** locus is nearly selectively neutral. Remember that this conclusion was drawn also after solving S. Wright's stationary functions, which allowed us to obtain accurate estimates of genotype fitness for these allozyme loci (section 5.1).

Conclusion

We have demonstrated above the role of natural selection in maintaining allozyme polymorphism in two salmonid species, sockeye and pink salmon. Although the methods of analysis involved were different, they revealed the important role of only one form of natural selection, i.e. stabilizing selection. We have succeeded in revealing the selective advantage of heterozygotes not only for genotypes of single loci (*PGM-2** or *LDH-B2**) in sockeye salmon, but for a set of unlinked genes when genotypic distributions were compared in alevins and spawners of pink salmon. Evidently such data are interesting in the context of theoretical discussions on the biological significance of protein polymorphism and they also are important in wild population management.

None the less, at least three questions remain unsolved.

(1) How important is the role of balancing selection on a macrogeographical scale?
(2) To what degree can the effects of diversifying selection be detected?
(3) What is the proportion of different gene groups in respect of their selective status throughout a species' range?

To answer these questions it is necessary to exploit the opportunities provided by the method, discussed in section 5.5. For this purpose, the observed values of spatial genetic differentiation should be compared locus by locus with theoretically expected ones for selectively neutral evolution.

In the next chapter, we will compare F_{0i} values for individual loci with the parameter $\bar{F}_0$ as the average reflecting the selectively neutral process. This should allow us to reveal the spatial dynamics of the adaptive status of allozyme genes (in conditions when their large samples are analysed simultaneously), and to show how previously insoluble problems in population genetics and biology of salmonid fishes can be addressed effectively.

Chapter 6
Genetic Variability in Different Levels of Intraspecific Population Hierarchy

Genogeographic analysis applied to many local populations of the main commercial salmonid species reveals a variety of patterns of spatial genetic differentiation – from a mosaic of gene frequencies in brown trout to their significant uniformity in pink salmon.

The question arises as to how to interpret data of this kind, taking into account that the local populations were characterized with different degrees of thoroughness and with respect to differences in the numbers of genes sampled. Samples were taken from populations differing in their degree of exposure to human pressures such as commercial fishing and artificial propagation.

Apparently, to interpret objectively the diversity of patterns of interpopulation hereditary variation and processes determining it, one should take into account patterns discovered in studies on intrapopulation genetic differentiation, described in the previous chapters.

First, allozyme polymorphism is selectively neutral only on average, at the large set of corresponding genes, whereas some of the genes are subject to different forms of natural selection.

Second, the sample of loci under analysis may be characterized by the prevalence of allozyme genes which, at a given moment or place, are evolving under the pressure of stabilizing or disruptive selection. Stabilizing selection leads to uniformity of allele frequencies, and levelling of spatial genetic differentiation; disruptive selection to the enhancement of diversity and hence to the distortion of various phylogenetic schemes based on the theory of random genetic drift.

The corresponding errors can be avoided only by working with native populations or at least with those not yet destroyed by anthropogenic impact. It is apparent that populations meeting this requirement are found mainly in Pacific salmon species, particularly in sockeye, which reproduces naturally in numerous freshwater bodies of the American and Asian continents. Chum salmon is also propagated in hatcheries on a small scale (if at all) on the continental coast of the Sea of Okhotsk, in Kamchatka, western Alaska, and the Anadyr and Yukon river basins. In other areas of the range, especially in Japan and the Sakhalin–Kuril region, its populations have been involved in intensive artificial propagation for decades.

Chinook salmon reproduce naturally in the Asian part of their range, and also in Alaska, the Yukon River basin, and many rivers of British Columbia, whereas in regions located further south (California, Oregon) this species is largely reproduced in hatcheries. This provides an opportunity to compare the values of gene diversity and spatial genetic differentiation in native and hatchery populations (see Chapter 7). Species of the Atlantic basin – Atlantic salmon and brown trout – suffer from especially strong anthropogenic pressure, and the published data on the gene diversity of their populations should be interpreted cautiously.

Taking into account these facts, we will discuss patterns of intraspecific gene diversity in different salmon species and, after that, the role of microevolutionary factors in its maintenance.

6.1 Patterns of intraspecific gene diversity

6.1.1 Chum salmon

To illustrate the principles formulated above, let us consider spatial variation of gene frequencies in chum salmon populations that spawn in river systems of the American and Asian continents. To date, these populations have been studied in more than 300 localities of the regions shown in the maps (Figs 6.1 and 6.2).

Table 6.1 shows parameters of gene diversity that include estimations of total gene diversity H_T, intrapopulation (H_S) and relative interpopulation (G_{ST}) gene diversity in different regions. On average, total regional heterozygosity is 0.071.

In the complex of chum salmon populations from the region of Alaska and the Russian north-east, which was studied with respect to 62 loci, the proportion of polymorphic genes is approximately one-third at $P_{0.99}$ (i.e. if the frequency of a polymorphic locus $P \leqslant 0.99$) or 24% at $P_{0.95}$ (Winans *et al.*, 1994). In more southern populations, these values estimated for 75 loci are 41.5 and 26%, respectively (Phelps *et al.*, 1994b).

The levels of expected intrapopulation heterozygosity (H_S) vary significantly from population to population. Among Far Eastern localities included in our studies (Fig. 6.2), the lowest heterozygosity was revealed in populations from the Amur River, the most abundant in the Far East ($H_S = 0.057$), while small populations spawning in small rivers of Iturup (the Kuril Islands) – Kurilka, Reidovaya and Rybatskaya – proved to be the most heterozygous ($H_S = 0.106$) (Salmenkova *et al.*, 1992).

Beacham *et al.* (1985a, 1987), who studied chum salmon in British Columbia, also found that the highest values of average heterozygosity for allozyme loci are typical for small populations, which inhabit this region. In the large Fraser River, where the size of chum salmon populations is much greater, their heterozygosity is lower. In

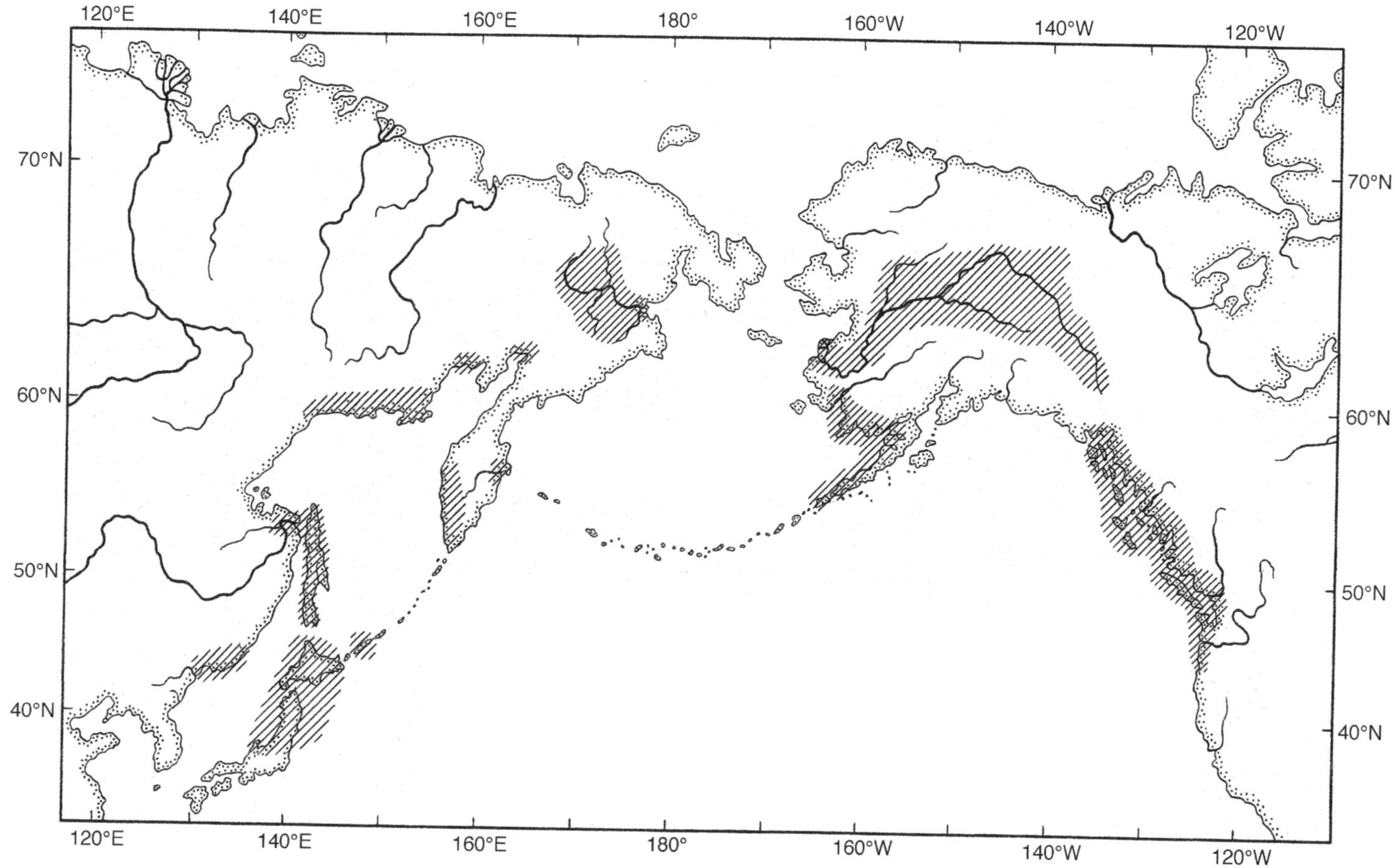

Fig. 6.1 Location of chum salmon populations (hatched areas) studied genetically throughout the range. (From data by Kijima & Fujio, 1979; Okazaki, 1982b, 1983; Beacham *et al.*, 1985a, 1987; Salmenkova *et al.*, 1986, 1992, 1994; Viktorovsky *et al.*, 1986; Omelchenko *et al.*, 1992; Kondzela *et al.*, 1994; Phelps *et al.*, 1994b; Wilmot *et al.*, 1994; Winans *et al.*, 1994.)

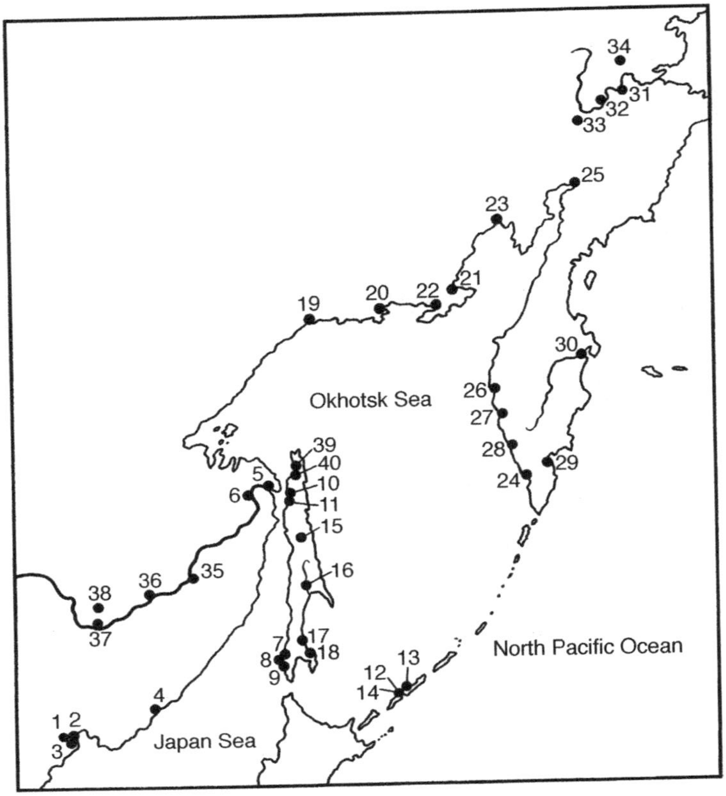

Fig. 6.2 Detailed location of chum salmon populations studied in the Far East of Russia. Rivers: 1 – Narva, 2 – Barabashevka, 3 – Riazanovka, 4 – Avvakumovka, 5, 6 – Amur (summer and autumn races), 7 – Kalininka, 8 – Zavetinka, 9 – Yasnomorka, 10 – Langry, 11 – Chingai, 12 – Kurilka, 13 – Reidovaya, 14 – Rybatskaya, 15 – Tym, 16 – Poronai, 17 – Naiba, 18 – Udarnitsa, 19 – Okhota, 20 – Ola, 21 – Yama, 22 – Taui, 23 – Garmanda, 24 – Bolshaya, 25 – Penzhina, 26 – Khairuzova, 27 – Icha, 28 – Vorovskaya, 29 – Avacha, 30 – Kamchatka, 31 – Vayegi, 32 – Anadyr, 33 – Eropol, 34 – Belaya, 35–38 – tributaries of Amur.

the Yukon River basin, heterozygosity of individual populations proved to be lower on average than in small rivers of Bristol Bay (Wilmot *et al.*, 1994) or British Columbia (Beacham *et al.*, 1987).

To date, chum salmon populations have been studied with respect to many allozyme loci from virtually throughout the species range, and this provides the basis for comparing variability of different genes. According to the data obtained by the authors cited above, the highest variation is characteristic of the loci *ESTD** (the frequency of the main allele varies over the range from 0.16 to 1.0, Fig. 6.3), *sIDHP-2** (0.23–0.74), *mAH-3** (0.21–0.79), *LDH-A1** (0.55–1.0), *sAAT-3** (0.49–0.82), *sAAT-1,2** and *mAAT-1** (both from 0.5 to 1.0). Each of the maximally variable loci *ESTD**, *sIDHP-2**, *mAH-3** and *mAAT-1** can be represented in different localities by alternative alleles occurring with the highest frequency. As to most other

Table 6.1 Gene diversity analysis at allozyme loci in chum salmon (*Oncorhynchus keta*).

Regions	Number of populations (localities)	Number of loci (polymorphic)	H_T	H_S	G_{ST} (%)	Reference
Yukon River basin	17	54 (24)	0.0644*	0.0628*	2.4*	Wilmot *et al.*, 1994
Bristol Bay coast	12	54 (24)	0.0670*	0.0648*	3.2*	Wilmot *et al.*, 1994
South-eastern Alaska	37	46 (25)	0.0871*	0.0847*	2.8*	Kondzela *et al.*, 1994
Northern and central British Columbia	24	46 (25)	0.0891*	0.0859*	3.6*	Kondzela *et al.*, 1994
Southern British Columbia	15	39 (22)	0.0868*	0.0858*	1.1*	Phelps *et al.*, 1994b
Eastern Kamchatka and Anadyr River basin	7	61 (37)	0.0777*	0.0749*	3.6*	Winans *et al.*, 1994
	6	15 (6)	0.0568	0.0550	3.3	Authors' own data
Western Kamchatka and northern Okhotsk Sea coast	9	61 (37)	0.0779*	0.0757*	2.8*	Winans *et al.*, 1994
	10	15 (6)	0.0582	0.0566	2.7	Authors' own data
Amur River basin	10	16 (9)	0.0579	0.0566	2.2	Salmenkova *et al.*, 1994
Japan	17	61 (37)	0.0773*	0.0756*	2.2*	Winans *et al.*, 1994
Japan	43	22 (13)	0.0418	0.0406	2.9	Okazaki, 1982b, 1983
Hokkaido	14	22 (13)	0.0420	0.0415	1.2	Okazaki, 1982b, 1983
Honshu	29	22 (13)	0.0412	0.0395	4.0	Okazaki, 1982b, 1983

* Our calculation from original data.

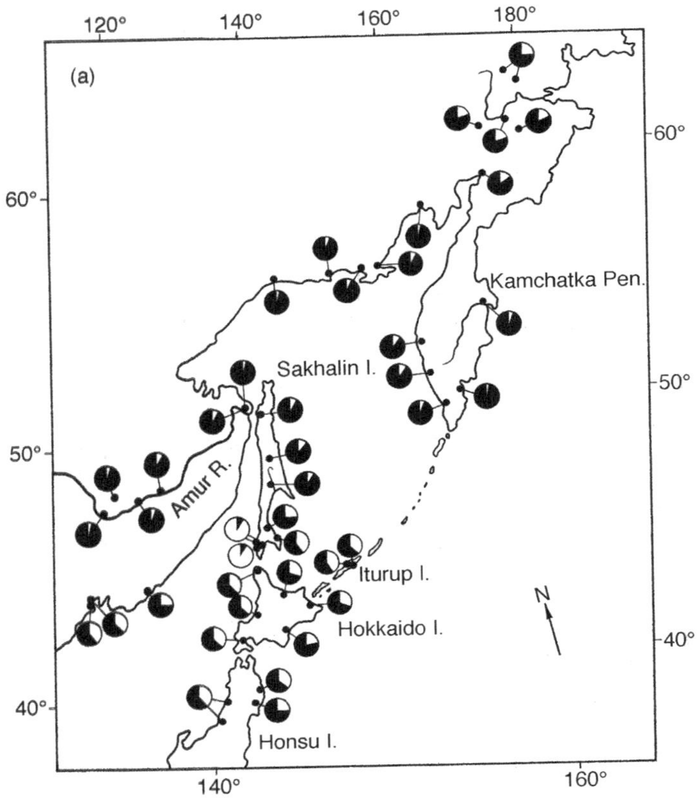

Fig. 6.3 Distribution of alleles of the *ESTD** locus in chum salmon populations throughout (a) Asian and (b) American ranges. Black sector corresponds to the frequency of allele **100*, white sector to allele **82* (or **90*). For references see Fig. 6.1.

loci, the same allele is usually the most frequent throughout the range. Thus, in the majority of loci (except for those mentioned above), the most frequent alleles are universal in their prevalence in different parts of the range.

Some populations demonstrate the presence of private (rare) alleles. Thus, chum salmon from the Amur River carry the allele *ESTD* 118*, which is rare and has not been found in individuals from other localities. Populations from Primorye are characterized by a considerable frequency of variants at the locus *sMEP** encoding a 'faster' isozyme (former designation *MEP-3**), and one of these populations (from the Narva River) contains a rare allele of the monomorphic locus *LDH-A2**. A rare allele of the monomorphic locus *LDH-B2** occurs in the population of the Tym' River (Omelchenko *et al.*, 1992; Salmenkova *et al.*, 1992). Rare private alleles of different loci were also described in other chum salmon populations (Phelps *et al.* 1994b).

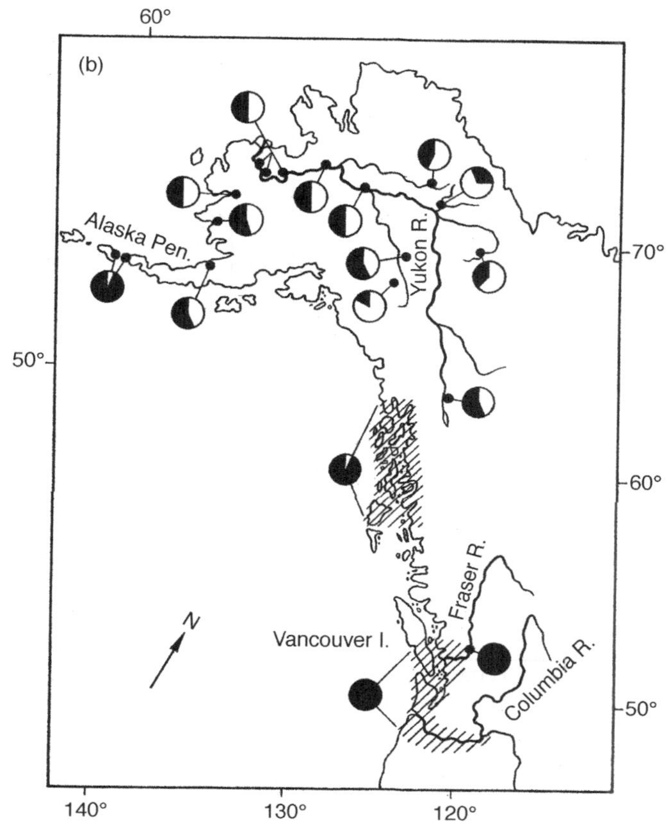

Fig. 6.3 Cont.

As a rule, researchers note that allele frequencies of allozyme loci in chum salmon populations remain stable during several years of observations (Okazaki, 1982b; Salmenkova *et al.*, 1986, 1992; Altukhov & Salmenkova, 1987a; Phelps *et al.*, 1994b).

Dendrograms based on genetic distances between populations (Nei, 1972) and the principal components analysis of variation in allele frequencies demonstrate mainly the geographical or regional pattern of cluster formation. For example, Fig. 6.4 shows that chum salmon populations of the Russian north-east and Japan are distinctly segregated by the degree of genetic similarity. On a smaller geographic scale, however, exceptions to this rule are often observed: some populations fall into 'foreign' clusters, as also observed in Fig. 6.4. This issue will be specially considered in the next section.

The dendrogram and the principal components analysis of Far Eastern populations (Figs 6.5 and 6.6) demonstrated that three chum salmon populations from south-western Sakhalin (the Kalininka, Zavetinka and Yasnomorka rivers) are

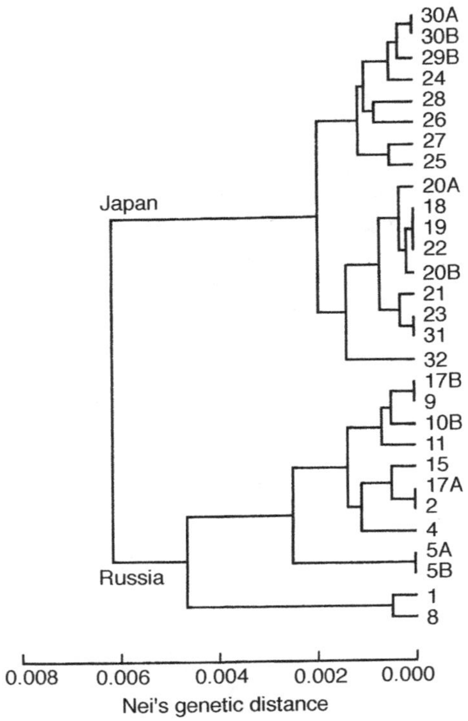

Fig. 6.4 Dendogram based on genetic distances between chum salmon populations in the north-east of Russia (1 – Anadyr River, 2, 4, 5A, 5B – east Kamchatka, 8, 9, 10B, 11 – west Kamchatka, 15, 17A, 17B – mainland coast of the Okhotsk Sea) and islands of the Japanese archipelago (18, 19, 20A, 20B, 21, 24–29, 30A, 30B – Hokkaido, 22, 23, 31, 32 – Honshu). A and B indicate two different years' samples from the same river. (From Winans *et al.*, 1994; 62 loci; reproduced with permission from the National Research Council of Canada.)

similar to one another but differ markedly from all other groups. There are also three additional discrete groups comprising populations of (1) Primorye, the Kuril Islands and south-eastern Sakhalin; (2) the Anadyr and Penzhina rivers, north-western and eastern Sakhalin, and the Amur River; and (3) the northern Sea of Okhotsk and the eastern and western Kamchatka Peninsula (Salmenkova *et al.*, 1992).

Regional genetic differentiation has also been observed in other parts of the chum salmon range (Beacham *et al.*, 1987; Kondzela *et al.*, 1994; Phelps *et al.*, 1994b; Wilmot *et al.*, 1994). The degree of interpopulation differentiation within regions varies in different parts of the range, being especially marked in the Russian group of chum salmon stocks from the Sea of Japan basin, e.g. from Primorye and western Sakhalin (Salmenkova *et al.*, 1992). In Japan, variation between Honshu populations is far greater than between Hokkaido populations because the latter have a greater size and are mainly propagated at hatcheries (Okazaki, 1982b).

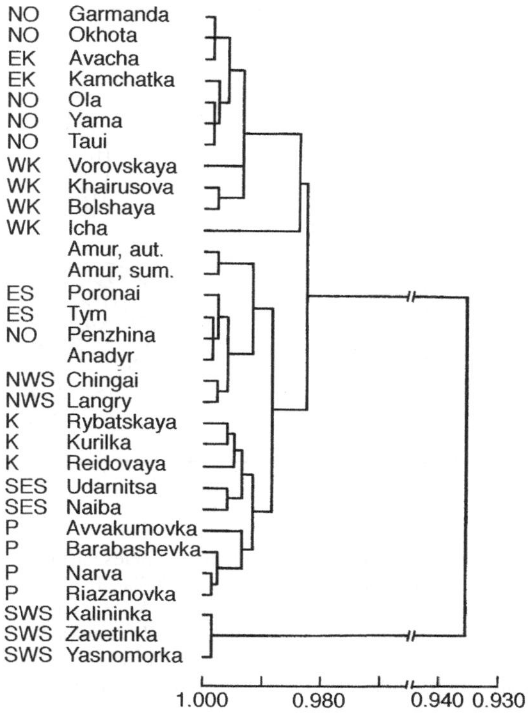

Fig. 6.5 Dendrogram based on genetic similarity between chum salmon populations spawning in Sakhalin rivers (ES, NWS, SES, SWS – east, north-west, south-east and south-west Sakhalin, respectively), Iturup Island (K), Primorye (P), Amur River basin, Anadyr River, east and west Kamchatka (EK and WK), northern Okhotsk Sea coast (NO). Zhivotovsky's (1979) similarity indices were estimated at the allelic frequencies of *sMDH-B1,2**, *sAAT-1,2**, *LDH-A1**, *PGDH**, *mMEP-2** and *ESTD** loci (from Salmenkova *et al.*, 1992).

On the basis of data obtained by Wilmot *et al.* (1994) and Winans *et al.* (1994), we performed a hierarchical analysis of gene diversity in the group of populations from the Russian north-east, the Yukon River basin and rivers of the Alaskan peninsula, i.e. the regions where chum salmon largely retain the natural genetic structure (Table 6.2) with minimal hatchery releases. The results show that the degrees of divergence between regional groups of populations and between populations of the same region are similar.

One of the levels of genetic differentiation in chum salmon is associated with temporal subdivision of its populations. In large rivers, such as the Amur and the Yukon, this species is represented by summer and autumn races. We determined allele frequencies of some allozyme loci in populations of summer and autumn chum salmon from the Amur River and revealed genetic differences between these races, which remained stable during 7 years of observation (Salmenkova *et al.*, 1994).

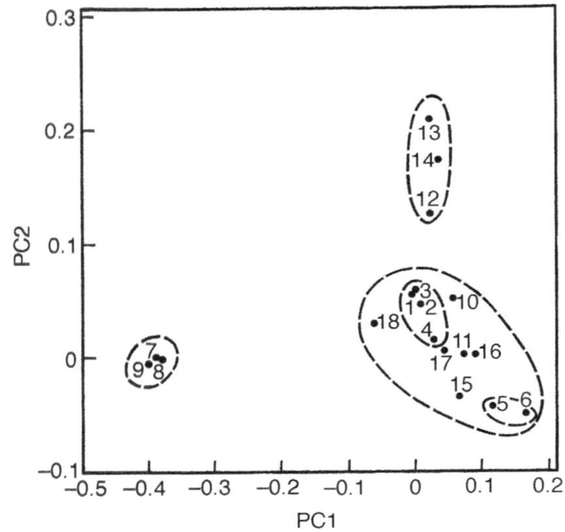

Fig. 6.6 Principal component analysis of allelic frequencies variation in chum salmon populations of Sakhalin, Iturup Island and Amur basin (PC1 – 67.8%, PC2 – 13.7% of variation) (from Salmenkova *et al.*, 1992). Rivers: 1 – Narva, 2 – Barabashevka, 3 – Riazanovka, 4 – Avvakumovka, 5, 6 – Amur (summer and autumn races), 7 – Kalininka, 8 – Zavetinka, 9 – Yasnomorka, 10 – Langry, 11 – Chingai, 12 – Kurilka, 13 – Reidovaya, 14 – Rybatskaya, 15 – Tym, 16 – Poronai, Buyuklinka tributary, 17 – Naiba, 18 – Udarnitsa. Loci studied are as in Fig. 6.5 and the *mIDHP-1** locus.

Genetic differences were also found between summer and autumn races from the Yukon River, which was confirmed by the RFLP analysis of mtDNA (Cronin *et al.*, 1993; Wilmot *et al.*, 1994), and between summer, autumn and winter runs into rivers of the southern part of the American range (Phelps *et al.*, 1994b). It is noteworthy that differences in allele frequencies at the loci *ESTD** and *mIDHP-1** between the summer and autumn races were found both in the Amur and in the Yukon (Salmenkova *et al.*, 1994; Wilmot *et al.*, 1994).

Thus, the analysis of allozyme markers in chum salmon revealed a significant level of intraspecific genetic differentiation. This differentiation manifests itself at

Table 6.2 Subdivision of gene diversity in the complex of chum salmon populations of north-eastern Russia, Alaska and the Yukon basin (45 samples, 21 polymorphic loci; our calculation by data from Wilmot *et al.*, 1994; Winans *et al.*, 1994).

Hierarchy level	Absolute diversity	Relative diversity (%)
Totals, H_T	0.1619	100
Within populations, H_S	0.1517	93.70
Between continents	—	2.57
Among regions within continents	—	1.89
Among populations within regions	—	1.84

different levels: macrogeographic (between groups of populations inhabiting the Asian and the American parts of the range), regional (between populations of individual rivers within regions), and local (within large rivers, between populations spawning in different tributaries or in different seasons). A noteworthy fact is the very high level of genetic divergence in the group of populations spawning in rivers of the northern Sea of Japan basin (Primorye and western Sakhalin), among which populations of south-eastern Sakhalin are especially distinct (see Figs 6.5 and 6.6; Salmenkova *et al.*, 1986; 1992; Omelchenko *et al.*, 1992). Genetic individuality of the latter was confirmed by corresponding differences in the frequencies of mtDNA haplotypes (Ginatulina & Mashkin, 1990; Ginatulina, 1992).

Studies on DNA polymorphism in chum salmon populations confirmed their genetic heterogeneity. MtDNA data revealed the greatest extent of differentiation between the Japanese population group on the one hand, and those of British Columbia and the state of Washington on the other, with population groups of the Russian north-west and Alaska occupying an intermediate position (Park *et al.*, 1993). Studies on variation in polymorphic loci of minisatellite DNA in chum salmon populations from different parts of the species range provided the basis for subdividing them into three main regional groups: (1) Japan, (2) the Russian north-east and the Yukon River basin, and (3) south-eastern Alaska and British Columbia (Taylor *et al.*, 1994). This pattern of genetic clusters is closely similar to that revealed by means of mtDNA analysis. In the authors' opinion, this subdivision reflects the history of postglacial species dispersal from three main refugia: Beringia in the north, Cascadia in the south (the lower Columbia River basin), and the area of the Sea of Okhotsk. Note that a similar differentiation of these population complexes is observed for the distribution of allele frequencies of the highly variable locus *ESTD** (Fig. 6.3).

6.1.2 Pink salmon

Figures 6.7 and 6.8 show the regions in which the gene pools in pink salmon populations were characterized according to the set of allozyme loci. The values of total gene diversity H_T, and its intra- and interpopulation components H_S and G_{ST}, are shown in Table 6.3.

The most characteristic feature of intraspecific differentiation in pink salmon is the genetic difference between generations spawning in odd and even years. As pink salmon are semelparous and their life span is 2 years, these lines of spawning groups are reproductively isolated. According to our estimations, genetic differences between generations account for more than one-half of the total interpopulation variation in pink salmon from the Sakhalin–Kuril region (Salmenkova, 1989).

Differences between the two lines of generations were described for the first time by Aspinwall (1974a) in American pink salmon for the loci encoding two enzymes, MDH and G3PDH. Subsequently, corresponding differences in these and other genes were observed throughout the species range, including the Sakhalin–Kuril

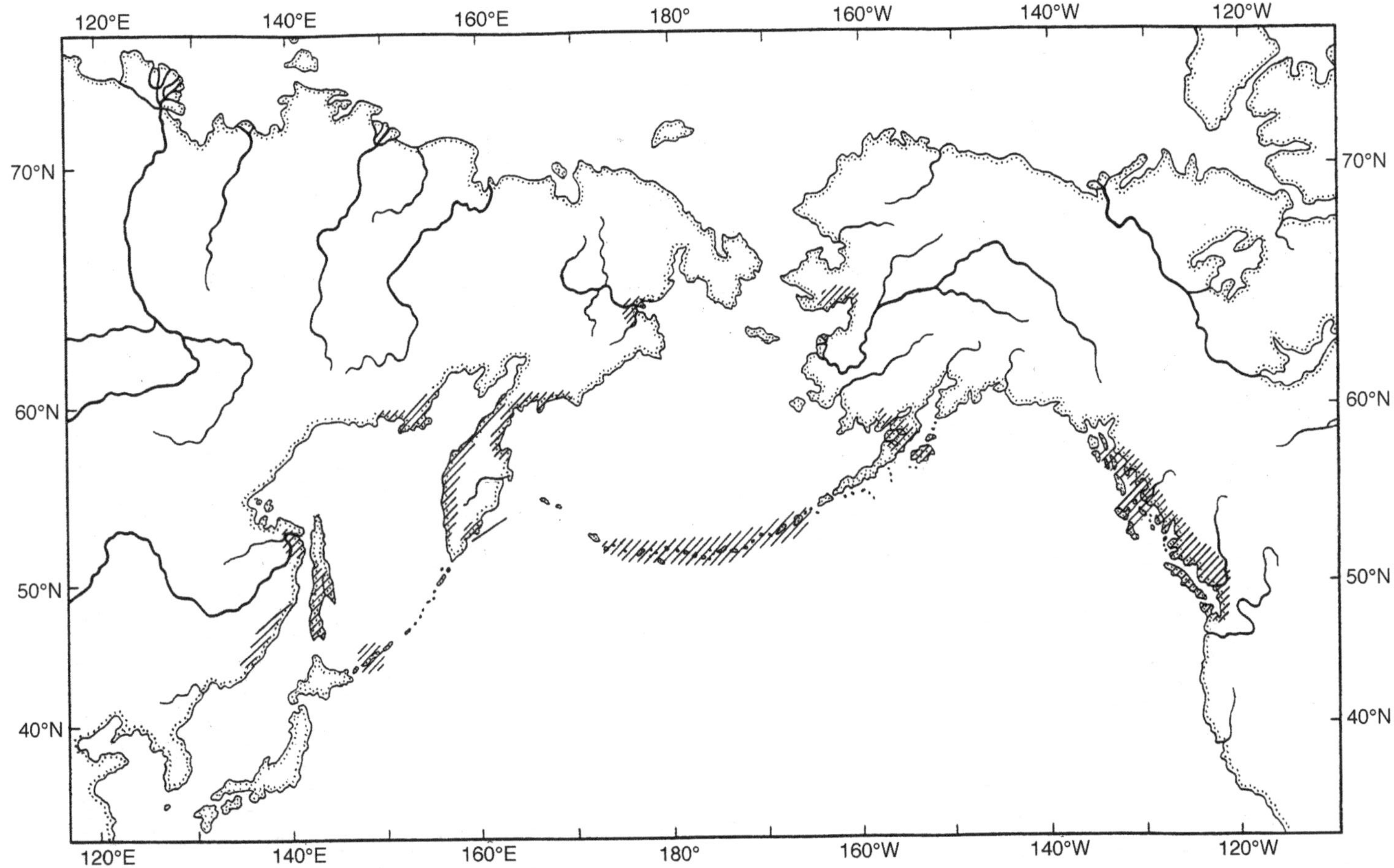

Fig. 6.7 Location of populations of pink salmon (hatched areas) studied genetically throughout the range (from data by Salmenkova *et al.*, 1981; Altukhov *et al.*, 1983; Beacham *et al.*, 1985b, 1988; Gagalchy, 1985, 1986; Gharrett *et al.*, 1988; Zhivotovsky *et al.*, 1989; Kartavtsev, 1991; Shaklee *et al.*, 1991; Varnavskaya & Beacham, 1992; Makoedov *et al.*, 1993; Shaklee & Varnavskaya, 1994; Brykov *et al.*, 1996).

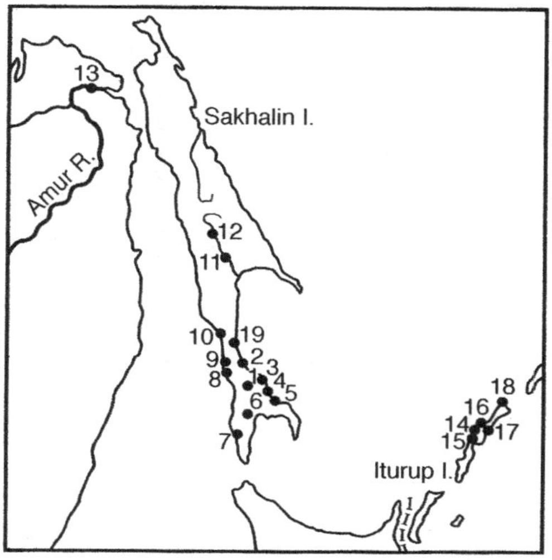

Fig. 6.8 Detailed location of pink salmon populations studied in Sakhalin and Iturup islands. Rivers: 1 – Naiba, 2 – Firsovka, 3 – Bakhura, 4 – Dolinka, 5 – Lesnaya, 6 – Lutoga, 7 – Yasnomorka, 8 – Chernaya, 9 – Pokosnaya, 10 – Ainka, 11 – Buyuklinka, 12 – Kholodnyi, 13 – Amur, 14 – Kurilka, 15 – Rybatskaya, 16 – Reidovaya, 17 – Glush, 18 – Olya, 19 – Pugachevka (from Altukhov *et al.*, 1983; Altukhov, 1989a).

region (Salmenkova *et al.*, 1981), rivers of the Russian north-east (Makoedov *et al.*, 1993), Kamchatka (Gagalchy, 1985, 1986), British Columbia (Beacham *et al.*, 1985b, 1988), and the state of Washington (Shaklee & Varnavskaya, 1994). In Asian pink salmon from the Sakhalin–Kuril region, these differences were revealed with respect to all the loci studied, being most pronounced in the cases of *PGM-2**, *mMEP-2** and *PEPB-1** (Salmenkova *et al.*, 1981; Salmenkova, 1989 and unpublished data). In British Columbia (Beacham *et al.*, 1985b, 1988), the lines of generations differed significantly in 14 out of 15 loci studied, and the greatest differences concerned the genes *mMEP-2**, *PGDH**, *PEPB-1** and *PGM-2**.

It is noteworthy that differences in allele frequencies between the lines of generations have a 'unidirectional' pattern virtually throughout the species range. For example, the *PGM-2** locus is always monomorphic (any variants are extremely rare) in even-year generations, but polymorphic in odd-year generations; frequencies of the common allele at the loci *mMEP-2** (up to its fixation in some American populations), *PGDH** and *G3PDH-1** are higher in odd years than in even years, whereas those at the *MDH-B1,2** loci are lower, and this is observed in any locality of the species range. In the regions where corresponding investigations were performed, such differences were revealed with respect to other genes as well. Cluster and the principal component analysis of variation in allele frequencies confirm the genetic differences between pink salmon generations, segregating them into two distinct lines (Beacham *et al.*, 1985b, 1988; Zhivotovsky *et al.*, 1989) (Fig. 6.9). On the other hand, the uniformity of genetic differences between generations

Table 6.3 Genetic diversity analysis at allozyme loci in pink salmon (*Oncorhynchus gorbuscha*).

Regions, generations of odd and even years	Number of populations (localities)	Number of loci (polymorphic loci)	H_T	H_S	G_{ST} (F_{ST}) (%)	References
British Columbia–Puget Sound						
Odd years	21	14 (13)	0.1244*	0.1222*	1.7*	Beacham *et al.*, 1985b
Even years	4	12 (9)	—	0.132	—	Beacham *et al.*, 1985b
British Columbia–Washington State						
Odd years	26	39 (21)	0.1036*	0.1015*	2.0*	Shaklee *et al.*, 1991
Aleutian Islands–Alaska						
Even years	19	17 (17)	0.114*	0.110*	3.6	Gharrett *et al.*, 1988
Eastern Kamchatka						
Odd year	8	16 (12)	0.1038*	0.1030*	0.8*	Varnavskaya & Beacham, 1992
Odd year	8	32 (26)	0.1012*	0.1003*	(0.9)*	Shaklee & Varnavskaya, 1994
Sakhalin–Kuril Islands						
Both generations	14	17 (5)	0.046	0.045	2.2	Salmenkova, 1989
Far East of Russia						
Both generations	33	4 (4)	0.122	0.119	2.2	Zhivotovksy *et al.*, 1989

* Our calculation from original data.

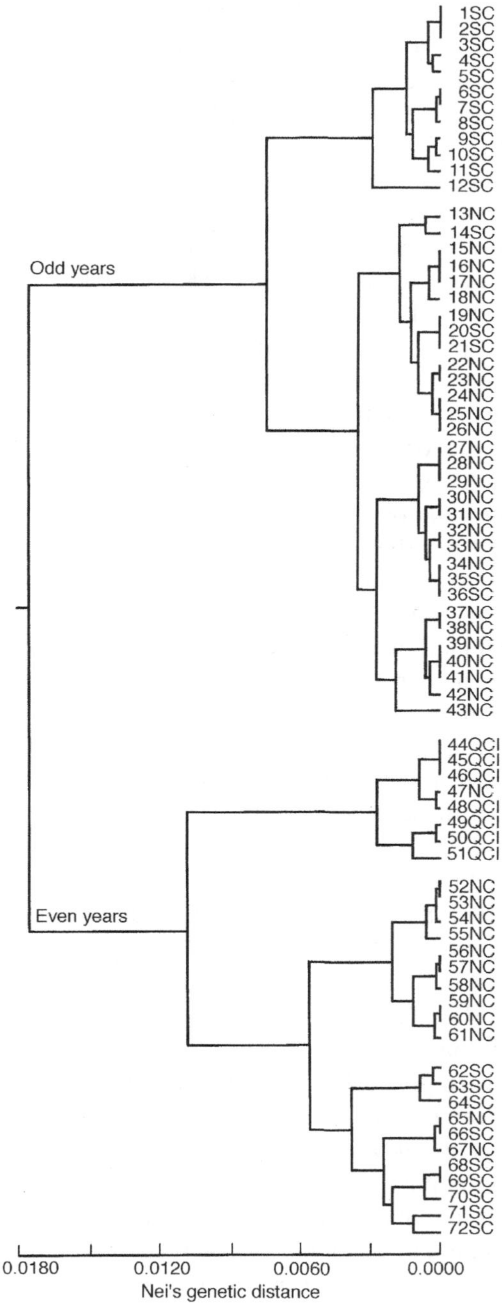

Fig. 6.9 Dendrogram based on Nei's genetic distances between pink salmon populations (1–72) in British Columbia. SC – south coast, NC – north and central coast, QCI – Queen Charlotta Island. (From Beacham *et al.*, 1988; 14 loci. Reproduced with permission of the American Fisheries Society.)

in the entire species area provides evidence that the even-year or odd-year lines of the Asian and American regions have common origins. Genetic differences of these lines were confirmed in studies on mtDNA (Polyakova *et al.*, 1992, 1996; Brykov *et al.*, 1996) and chromosome polymorphisms (Gorshkova & Gorshkov, 1983; Gorshkova *et al.*, 1988).

The level of interpopulation genetic divergence within the lines of generations may be different, e.g in pink salmon from the Sakhalin–Kuril region, this divergence in odd-year generations proved to be higher than in generations of even years, as shown by the analysis of both allozyme genes (Kalabushkin & Salmenkova, 1988) and mtDNA (Brykov *et al.*, 1999a, b).

Asian populations of pink salmon differ from American populations in the frequencies of common alleles and sets of rare alleles (Salmenkova *et al.*, 1981; Varnavskaya, 1992; Shaklee & Varnavskaya, 1994). We used the data of allele frequencies at 26 allozyme loci in odd-year generations of 15 American and 8 Asian populations (from Shaklee & Varnavskaya, 1994) for the estimation of their genetic divergence. The differentiation between continental groups of populations was estimated at 1.8%, and that between populations within these groups (weighted mean) at 2.3%. In the American group, however, the G_{ST} value was significantly higher than in the Asian group: 3.06 vs. 0.90%, which may be explained by both the specific pattern of population divergence and the difference in the number of populations studied in each group.

The distribution of allele frequencies of allozyme loci over the species range provides evidence for regional genetic differentiation, as in the case of chum salmon. In Asian pink salmon, for example, populations of the Sakhalin–Kuril region differ significantly from those of eastern and western Kamchatka (Salmenkova *et al.*, 1981; Altukhov *et al.*, 1983; Gagalchy, 1985, 1986). Groups of populations from British Columbia, the Fraser River basin and the Puget Sound differ from one another (Beacham *et al.*, 1985b). Within British Columbia, population groups of the northern and central regions differ from the southern group, and populations from rivers of the Queen Charlotte Islands also demonstrate genetic individuality (Beacham *et al.*, 1988; Fig. 6.9). Populations of the Aleutian Islands differ from Alaskan populations (Gharrett *et al.*, 1988).

Differentiation between pink salmon populations from rivers of the same region is usually less pronounced than interregional differences in some areas (Gharrett *et al.*, 1988; Zhivotovsky *et al.*, 1989; Kartavtsev, 1991). Analysing allele frequencies of several allozyme loci in populations of the Sakhalin–Kuril region for several years, we revealed only weak spatial differentiation (Altukhov *et al.*, 1983, 1987). Nevertheless, although the number of loci analysed is relatively small, estimates of allele frequencies in each generation line for individual rivers, obtained by averaging the data on numerous samples, provide the possibility of demonstrating that such 'pooled' samples form clusters corresponding to geographic origin. It is important that the patterns of clusters obtained on the basis of data on spawners and fry from the same river prove to be similar and are reproduced in each generation line. This approach also reveals significant divergence of populations from neighbouring rivers (Kalabushkin *et al.*, 1998).

Slight but appreciable genetic differentiation was revealed between pink salmon runs entering the rivers of eastern Sakhalin in different time seasons (Altukhov *et al.*, 1983). These stocks differ in the areas of feeding migration: early-run fish migrate from the Sea of Japan, whereas late-run fish migrate from the Pacific waters off the Kuril Islands (Efanov, 1989). Subsequently, temporal genetic heterogeneity of pink salmon stocks during spawning migration to some Sakhalin rivers was confirmed by means of mtDNA analysis (Brykov *et al.*, 1999a,b). Genetic differences between early and late runs were also observed in different pink salmon populations from Alaska (McGregor *et al.*, 1998; Seeb *et al.*, 1998).

Compared with sockeye and chum salmon, in which variation in some allele frequencies over the range and even within a certain region (in G_{ST} values) may exceed 0.20, the highest G_{ST} values for individual loci in pink salmon are low and usually do not exceed 0.072 (e.g. for the *PEPLT** locus in the group of populations from British Columbia and the state of Washington, as we calculated from the data by Shaklee *et al.*, 1991). Among many allozyme genes studied by different authors, only at the *AAT-4** locus (expressed in liver tissue) does the frequency of one allele vary over the species range from 0.3 to 0.7, i.e. any of the two common alleles may prevail in different populations (Shaklee & Varnavskaya, 1994); at other loci, the same allele **100* prevails throughout the range. On the whole, pink salmon is characterized by the lowest level of spatial genetic differentiation among the Pacific salmon species. This fact will be discussed further in section 6.3 below.

6.1.3 Sockeye salmon

The geographic distribution of genetically studied populations of this species is shown in Fig. 6.10. The level of total gene diversity H_T in sockeye populations, which has been determined by different authors for no less than 17 loci, averages 0.043 (see Tables 6.4 and 6.12 later), i.e. is lower than in pink and chum salmon. The proportion of polymorphic loci is also relatively small. Thus, according to Varnavskaya *et al.* (1994a), only 12 out of 87 loci studied (14%) proved to be polymorphic at the $P_{0.99}$ level. Interpopulation genetic divergence in sockeye is significantly higher than in chum and pink salmon. Polymorphic loci fixed at different alleles in different geographic localities have not been revealed in sockeye as well as in other Pacific salmon species. In populations of Kamchatka, Alaska, British Columbia, the state of Washington and the Columbia River basin, the loci *LDH-B2**, *ALAT**, *PGM-1** and *PGM-2** are the most polymorphic (Withler, 1985; Varnavskaya *et al.*, 1994a,b; Wood *et al.*, 1994).

We analysed the intraspecific hierarchy of gene diversity in sockeye salmon at three main polymorphic loci (*LDH-B2**, *PGM-2** and *ALAT**) using the data on populations studied in the Asian (Kamchatka Peninsula) and American parts of the species range and classifying large rivers (e.g. Kamchatka, Skeena and Fraser) as

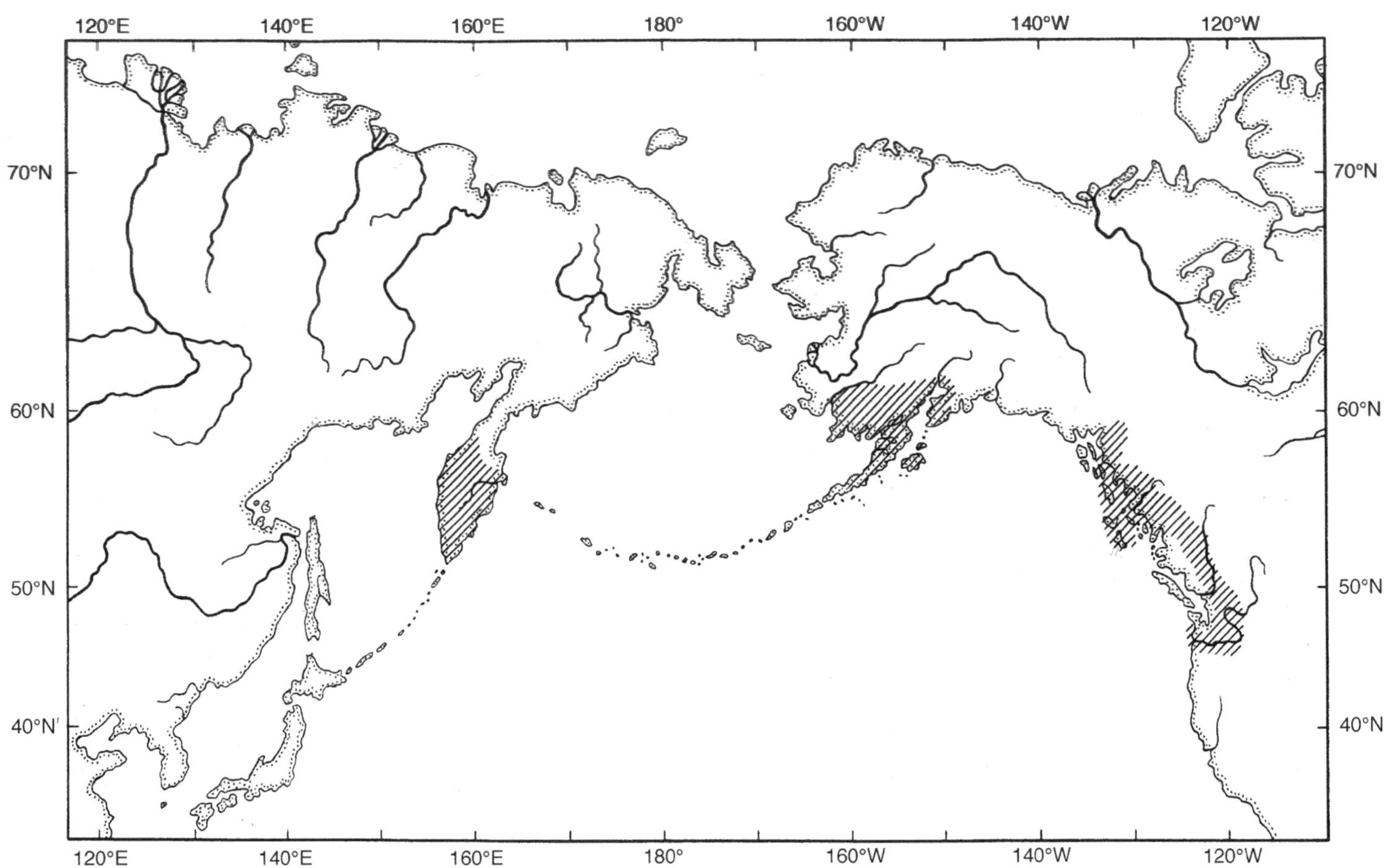

Fig. 6.10 Location of populations of sockeye salmon (hatched areas) studied genetically throughout the range (from data by Altukhov, 1974; Altukhov *et al.*, 1975a,b; Grant *et al.*, 1980; Kirpichnikov & Muske, 1981; Muske, 1983; Utter *et al.*, 1984; Wilmot & Burger, 1985; Pustovoit, 1993; Varnavskaya *et al.*, 1994a; Wood *et al.*, 1994; Bickham *et al.*, 1995; Taylor *et al.*, 1996).

Table 6.4 Gene diversity analysis at allozyme loci in sockeye salmon (*Oncorhynchus nerka*).

Regions	Number of populations (localities)	Number of loci (polymorphic loci)	H_T	H_S	G_{ST} (%)	Reference
Alaska (Cook Inlet)	13	26 (6)	0.046	0.043	5.8	Ryman, 1983
West coast of North America	16	50 (12)	0.0347	0.0313	9.6	Utter *et al.*, 1984
Bristol Bay	22	39 (5)	—	—	5.9*	Varnavskaya *et al.*, 1994a
East and west Kamchatka	7	39 (9)	0.0483*	0.0468*	3.0*	Varnavskaya *et al.*, 1994a
Fraser River basin	15	33 (8)	0.0431*	0.0404*	6.4*	Wood *et al.*, 1994
Skeena River basin	15	33 (9)	0.0459*	0.0434*	5.4*	Wood *et al.*, 1994
British Columbia						
North coastal	11	33 (10)	0.0413*	0.0357*	13.5*	Wood *et al.*, 1994
Vancouver Island	6	33 (10)	0.0374*	0.0350*	6.4*	Wood *et al.*, 1994

* Our calculation from original data.

individual regions (for references and the structure of intraspecific subdivision, see Fig. 2.7). The results showed that relative values of gene diversity (G_{ST}) at different levels of this subdivision are very similar, e.g. 2.49% for rivers within the regions, 2.37% for regions within the continents, and 2.54% for continents (for details, see section 2.4; Table 6.5).

Table 6.5 Subdivision of interpopulation gene diversity in sockeye salmon (*Oncorhynchus nerka*) (according to Fig. 2.7; at *LDH-B2**, *PGM-2** and *ALAT** loci).

Hierarchy level	Value of local genetic differentiation	
	G_{ST_i}	Proportion of total $G_{ST_{i-t}}$ (%)
Between continents, $G_{ST_{3-4}}$ (Asia and America)	0.0254	34
Among regions within continents, $G_{ST_{2-3}}$	0.0237	32
Among rivers within regions, $G_{ST_{1-2}}$	0.0249	34

* $G_{ST_{i-t}} = 0.074$.

The hierarchical analysis of gene diversity among sockeye populations from Canadian rivers and lakes, performed using the data obtained by Wood *et al.* (1994), also revealed similar G_{ST} values at different levels of population structure (Table 6.6).

Table 6.6 Subdivision of gene diversity in Canadian sockeye salmon populations (78 samples, 33 loci; our calculation by data from Wood *et al.*, 1994).

Hierarchy level	Absolute diversity	Relative diversity (%)
Total, H_T	0.0470	100
Within samples, H_S	0.0389	82.75
Among all samples	—	17.25
Among regions	—	6.10
Among population within regions	—	4.83
Among samples within populations	—	6.32

According to the results of cluster analysis of data on the seven most polymorphic loci, performed by Varnavskaya *et al.* (1994a), all sockeye populations that have been studied within the range form three regional groups: (1) western Alaska, (2) the Fraser and Columbia rivers, and (3) the Kamchatka Peninsula and northern British Columbia. The pattern of clusters formed by Canadian populations studied by 33 loci (Wood *et al.*, 1994) differs in some details but generally agrees with this subdivision. Within each cluster, significant interpopulation heterogeneity can be observed.

The microspatial genetic differentiation in sockeye salmon was discussed in detail using an example of the system of subpopulations of Lake Azabachye (see Chapter 4). Genetic subdivision into subpopulations within sockeye populations of individual lakes and rivers has also been observed by other researchers (Varnavskaya *et al.*, 1988, 1994b; Pustovoit, 1993; Wood *et al.*, 1994; Burger *et al.*, 1997; Taylor *et al.*, 1997).

The degree of interpopulation gene diversity in this species does not depend on the geographic distance between populations. Indeed, genetic heterogeneity within the same large river basin may exceed that between different regions, as demonstrated by analysing spatial variation of allele frequencies at the *LDH-B2** locus (Withler, 1985; Varnavskaya *et al.*, 1994a,b).

The analysis of minisatellite and mitochondrial DNA polymorphism in samples of sockeye from several lakes of Kamchatka and different water bodies of Alaska and British Columbia showed that corresponding populations form two major complexes: north-western and southern (Bickham *et al.*, 1995; Taylor *et al.*, 1996). The former comprises populations of Kamchatka, western Alaska and northern British Columbia; the latter, populations of the Fraser and Columbia river basins. The analysis of DNA markers applied to the totality of sockeye populations revealed virtually the same pattern of genetic subdivision as the analysis of allozymes. Apparently, notwithstanding the selective effects at several allozyme loci of this species (Chapter 5), the averaged spatial distribution of their alleles largely reflects the natural history of the formation of recent populations in the postglacial period.

It is noteworthy that resident and anadromous forms of sockeye analysed for the same DNA loci do not form individual clusters but are distributed according to the geographic location between the two main clusters, north-western and southern (Taylor *et al.*, 1996). This fact confirms the concept that the non-anadromous form is polyphyletic and originated from ancestors that inhabited different parts of the species range (Foote *et al.*, 1989).

As in some other salmonid species, genetic differentiation in sockeye is associated not only with spatial but also with temporal subdivision into the early (spring) and late (summer) races (Chapters 4 and 5). The values of genetic differences between the races in G_{ST} values are relatively low and vary from 0.002 to 0.04 in basins of different lakes (Varnavskaya *et al.*, 1994a).

6.1.4 Chinook salmon

To date, chinook populations have been studied only in the North American part of the species range; there are no data on Asian populations (Fig. 6.11). The value of total gene diversity of regional population systems averages approximately 0.07, with strong fluctuations in intrapopulation heterozygosity, from 0.027 to 0.143. The average G_{ST} value is 0.065 for native populations and 0.090 for populations of all regions studied (see Tables 6.7 and 6.12 later). According to Bartley & Gall (1990), the proportion of polymorphic loci in Californian chinook populations varies in a

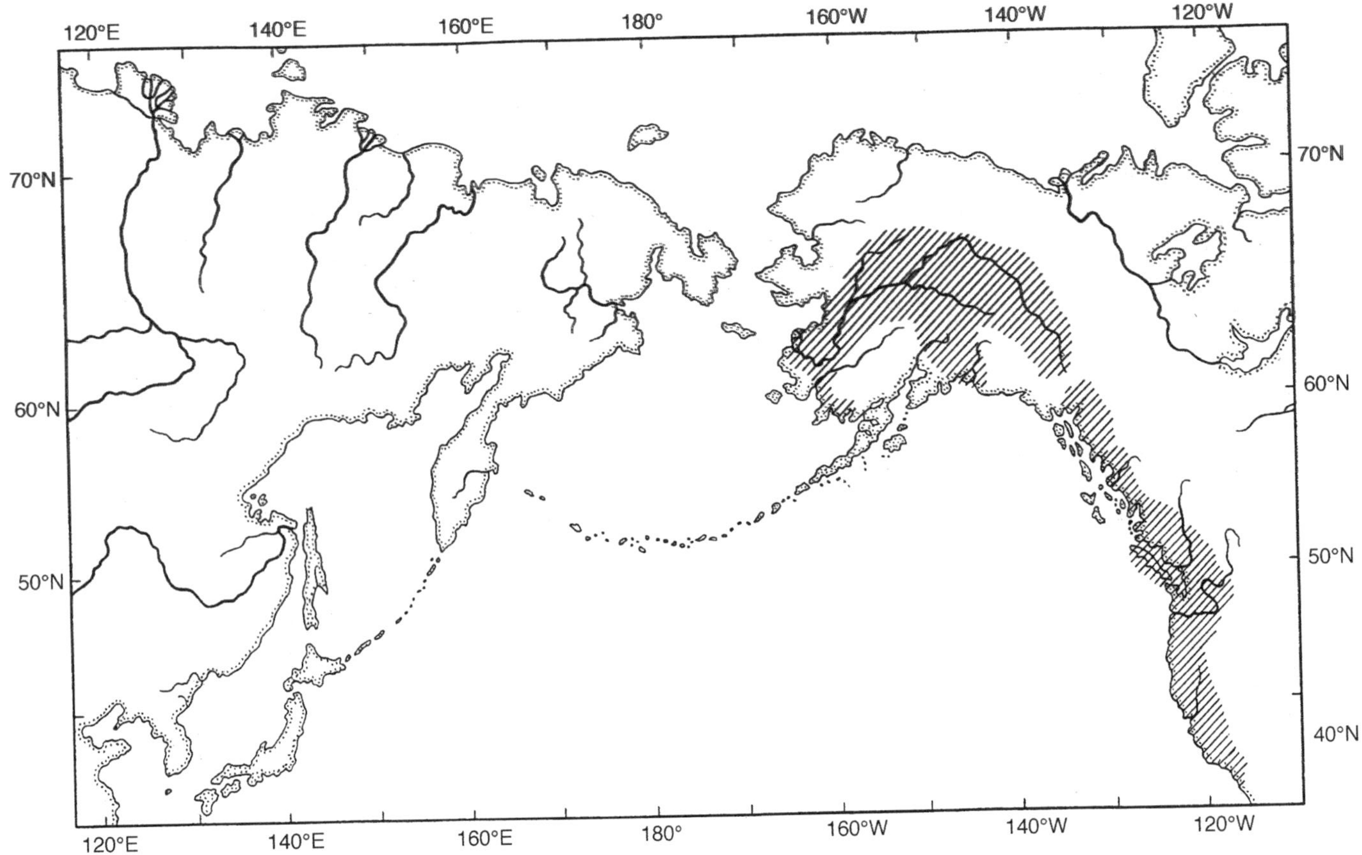

Fig. 6.11 Location of populations of chinook salmon (hatched areas) studied genetically throughout the range (from data by Gharrett *et al.*, 1987; Beacham *et al.*, 1989; Utter *et al.*, 1989; Bartley & Gall, 1990; Bartley *et al.*, 1992a; Wilmot *et al.*, 1992).

Table 6.7 Gene diversity analysis at allozyme loci in chinook salmon (*Oncorhynchus tshawytscha*).

Regions	Number of populations (localities)	Number of loci (polymorphic)	H_T	H_S	G_{ST} (%)	References
Alaska	23	28 (13)	0.0347*	0.0327*	5.9*	Gharrett *et al.*, 1987
Yukon River basin	17	20 (17)	0.0524*	0.0485*	7.4*	Beacham *et al.*, 1989
	16	22 (21)	0.0963*	0.0911*	5.4*	Wilmot *et al.*, 1992
Northern America (from British Columbia to California)	65	25 (25)	0.1161	0.1018	12.3	Utter *et al.*, 1989
British Columbia	7	25 (25)	0.1110*	0.1030*	7.4*	Utter *et al.*, 1989
California	35	53 (22)	0.0431	0.0370	14.4	Bartley & Gall, 1990
California–Oregon	37	78 (47)	0.0620	0.0554	10.6	Bartley *et al.*, 1992b

* Our calculations from original data.

broad range (5.8–32.1%, $P_{0.99}$) because many loci are polymorphic only in a small part of populations, whereas in the rest of them the common allele is fixed. A similar situation is observed in chinook populations from other regions (Gharrett *et al.*, 1987; Utter *et al.*, 1989; Bartley *et al.*, 1992b).

The most heterozygous loci in populations from northern regions are *AAT-3**, *ADA-1**, *sAH-1**, *MPI** and *SOD-1**; in more southern populations, these are the same loci (except *AAT-3**) and, in addition, *PGK-2**, *PEPB-1**, *PGM-4** and *GPI-2** (Utter *et al.*, 1989).

The distribution of common and private allozyme alleles over the range, the hierarchical analysis of gene diversity, cluster and the principal components analyses of allozyme variation in populations demonstrate a clear geographical (regional) pattern. In this sense, chinook does not differ in principle from other Pacific salmonid species. Thus, populations from the North American Pacific coast in the section from northern British Columbia to the Sacramento River in California are subdivided into nine major genetically individual groups, which are confined to the following regions (Utter *et al.*, 1989):

(1) Fraser River basin
(2) Rivers flowing into the Strait of Georgia
(3) Rivers flowing into the Puget Sound
(4) Rivers of the coastal region from Vancouver Island to northern California
(5) Lower Columbia River basin
(6) Middle Columbia River Basin
(7) Snake River
(8) Klamath River
(9) Sacramento River.

Populations within each regional group also differ genetically.

A similar picture is revealed by the analysis of minisatellite DNA, when estimating divergence at the number of restriction fragments and frequencies of allelic variants. Firstly, the northern and the southern groups of populations are distinguished between the Yukon and Fraser rivers; secondly, populations of individual regions within these groups are also well differentiated by the same traits. In the southern group, these are populations of the Fraser River basin, the southern continental coast of British Columbia and the eastern and western coasts of Vancouver Island; in the northern group, populations of the northern continental part of British Columbia, the Skeena River and the Yukon River (Beacham *et al.*, 1996).

Alaska is the region of reproduction of natural chinook salmon populations where anthropogenic pressure (in particular, that of commercial fishing) is less intense than in southern territories. Analysing the totality of Alaskan populations, it becomes apparent that the values of relative gene diversity between population systems of geographic regions and between populations of individual rivers in these regions are similar. Thus, G_{ST} values determined at the levels of three regions of

Alaska and of river systems within these regions are 2.6 and 3.3%, respectively (Gharrett *et al.*, 1987).

In chinook salmon, two ecological types – the ocean and the stream type – are distinguished on the basis of differences in the freshwater period of young fish life: in the former, fish migrate to the sea after living in fresh water for 1–2 months; in the latter, they remain in fresh water for 1–2 years (Healey, 1991). Moreover, chinook salmon spawning in many water bodies have races isolated temporally: spring, summer and autumn races. Genetic divergence between young fish of different types from the same water body was revealed at some allozymes (Carl & Healey, 1984); however, these are only limited data, requiring further investigation. Anadromous salmonid fishes are characterized by a strong genetic component determining the time of the spawning run, which is marked by allozyme differences between seasonal races and temporal intrapopulation groups in chum, sockeye and pink salmon (Altukhov, 1974; Altukhov *et al.*, 1975a, 1983; Kirpichnikov & Muske, 1981; Okazaki, 1982b, 1983; Salmenkova *et al.*, 1989; see also Helle, 1981). In chinook salmon, temporal or interracial genetic differentiation is relatively weak and accounts for only 7.3% of the total interpopulation component of gene diversity at allozyme loci (Utter *et al.*, 1989). Nevertheless, this differentiation is also seen at the level of mtDNA (Wilson *et al.*, 1987; Adams *et al.*, 1994; Nielsen *et al.*, 1994). The observed pattern of clusters formed by temporal components of chinook populations is usually regional and provides evidence that these components diverged on the basis of already existing spatial genetic differentiation. It cannot be excluded, however, that the early and late seasonal races of chinook salmon in some localities could have originated from different ancestral populations in the course of post-glacial dispersal of the species from the northern and the southern refugia (Adams *et al.*, 1994).

6.1.5 Coho salmon

The data on investigated regions, the number of populations studied and their genetic parameters are shown in Table 6.8. As in the case of chinook salmon, there is no corresponding information on Asian populations of coho salmon. A noteworthy fact is that the level of intrapopulation heterozygosity in coho salmon is lower than in other Pacific salmonid species, as was also observed in earlier studies (Utter *et al.*, 1973). According to Wehrhahn & Powell (1987), it is especially low in populations from southern British Columbia. This finding is explained firstly by the fact that heterozygosity was estimated without taking into account the transferrin and esterase loci, which are the most polymorphic loci in coho salmon. Secondly, populations of this species were apparently exposed to a ‘bottleneck effect’ and underwent especially strong reduction in size, which could have resulted in the reduced gene diversity, when the territory of British Columbia was covered with ice during the last, Wisconsin glaciation (Wehrhahn & Powell, 1987).

The interpopulation component of gene diversity, G_{ST}, in coho salmon popula-

Table 6.8 Gene diversity analysis at allozyme data in coho salmon (*Oncorhynchus kisutch*).

Regions	Number of populations (localities)	Number of loci (polymorphic)	H_T	H_S	G_{ST} (F_{ST}) (%)	Reference
Washington State (USA)	12	55 (21)	0.0210	0.0190	9.5	Reisenbichler & Phelps, 1987
British Columbia	43	26 (15)	—	0.0025*	(8.6)	Wehrhahn & Powell, 1987
California	27	45 (23)	0.0380	0.0320	15.8	Bartley *et al.*, 1992c

* Our calculation from original data.

tions from different regions is significant. Its values are greater than in chum and pink salmon but comparable to those in chinook salmon (see Table 6.12 later). Researchers note that an important contribution to both intra- and interpopulation components of gene diversity is made by the highly polymorphic transferrin locus *TFN**, as noted previously (Utter *et al.*, 1970a,b). Remarkably significant genetic differentiation between coho populations persists irrespective of the intensive transfer of these fish from one river into another in many regions, especially in California, for artificial reproduction and restoration of populations (Bartley *et al.*, 1992c). In particular, there is still a north–south cline in the *TFN* 103* allele frequency, revealed by Utter *et al.* (1970a,b). It was shown that carriers of certain *TFN** genotypes are resistant to bacterial kidney disease (Suzumoto *et al.*, 1977; Winter *et al.*, 1980), and fish differing in the *TFN** genotype also differ in growth and survival rates (McIntyre & Johnson, 1977). This is suggestive that certain transferrin genotypes have a selective advantage (Bartley *et al.*, 1992c) and, hence, the genetic structure of natural coho populations at the *TFN** locus has possibly been influenced by selection. Variation in selectively neutral allozyme loci is largely affected by genetic drift, as most coho populations have a relatively small size (see Chapter 2, Table 2.3). Interpopulation gene exchange is apparently limited, which is clearly illustrated by the fact that coho stocks of two neighbouring rivers in California differ sharply genetically: fish from one river are monomorphic, whereas those from the other river are highly polymorphic at the same allozyme loci under study (Bartley *et al.*, 1992c). According to the data on allele frequencies, gene flow was estimated at five to six individuals per generation within the limits of river systems (Wehrhahn & Powell, 1987).

There are only a few population studies with DNA markers in coho salmon. The analysis of mtDNA applied to this species by Carney *et al.* (1997) proved to be promising. Parameters of interpopulation diversity estimated in this way for five populations from different regions of Alaska exceed corresponding values obtained from allozyme markers in the same population group and in populations from other regions. The analysis of mtDNA differentiates efficiently regional groups of populations (Carney *et al.*, 1997), whereas minisatellite DNA markers are more effective in distinguishing individual populations within regions (Miller *et al.*, 1996). Analysing microsatellite DNA loci, Small *et al.* (1998) revealed divergence between 16 coho salmon populations inhabiting the basin of the Fraser River in Canada.

6.1.6 Atlantic salmon

To date, allozyme electrophoresis has been employed in the genetic analysis of more than 120 Atlantic salmon populations throughout the species range, including territories of northern Russia, Scandinavia, British Isles, Iceland, France, Spain and the Atlantic coast of North America (Ståhl, 1987; Elo, 1993; reviews of Kazakov & Titov, 1995; Kazakov, 1998).

The estimates of gene diversity in Atlantic salmon are shown in Table 6.9. The number of highly polymorphic allozyme loci is relatively small in this species. According to Ståhl (1987), among a total of 38 loci studied, the proportion of polymorphic loci is 23%, and many of them have rare alternative alleles only in single populations. The loci *AAT-3**, *MEP-2** and *IDDH-1** are the most polymorphic.

On the macrogeographic scale, the most pronounced differences exist between North American and European population complexes, and in the European complex, between North Atlantic and Baltic groups (Fig. 6.12). Each of these regions is characterized by significant genetic differentiation between population groups belonging to individual hydrographic basins, between populations of individual rivers, and between subpopulations inhabiting different tributaries of the same river (Ryman, 1983; Ståhl, 1987; Elo *et al.*, 1994; Verspoor, 1994; Kazakov & Titov, 1995). Microgeographic heterogeneity in *S. salar* is revealed even more distinctly by means of mini- and microsatellite DNA markers, which have been studied actively in this species in recent years (Galvin *et al.*, 1996; Tessier *et al.*, 1997; Beacham *et al.*, 1998).

We used the sum of data on allozyme genes of Atlantic salmon from northeastern Europe (the territory of the former Soviet Union), including populations of different sea and lake basins and rivers (Kazakov & Titov, 1992), for comparing parameters of gene diversity at different hierarchical levels (Table 6.10). Estimates of gene diversity between population complexes of different sea basins and between river populations within a basin are fairly similar: 10% and 7.8%, respectively. As in the case of generalized data on the entire species range (Ståhl, 1987), the higher genetic divergence is observed between regional species subdivisions belonging to different sea basins.

The American and European population complexes differ markedly in allele frequencies at the loci *TFN**, *MDH-3**, *mMEP-2** and *ESTD** (Verspoor, 1986, 1994; Ståhl, 1987; Blanco *et al.*, 1992; Kazakov & Titov, 1992, 1995; Makhrov *et al.*, 1998), in sets and frequencies of mtDNA haplotypes (Bermingham *et al.*, 1991), and virtually fixed and unique alleles of microsatellite DNA loci (McConnell *et al.*, 1995; Taggart *et al.*, 1995b).

As mentioned above, Baltic reproductive localities in the European population complex can be distinguished reliably by a typical spectrum of allele frequencies, especially at the *MEP-2** and *IDDH-1** loci (Ståhl, 1987; Koljonen, 1989; Kazakov & Titov, 1992, 1995). In populations of the Barents and White sea basins, a distinctive feature is polymorphism of the *ESTD** locus. The latter is monomorphic not only in all other regions, but also in Atlantic salmon populations from the Pechora, Northern Dvina, and Onega rivers, which makes them similar to Baltic populations (Semenova & Slynko, 1988; Kazakov & Titov, 1995).

In more western and southern European populations, researchers revealed genetic divergence with respect to both allele frequencies of allozyme loci (Davidson *et al.*, 1989b; McElligot & Cross, 1991; Blanco *et al.*, 1992; Jordan *et al.*, 1992; Moran *et al.*, 1994; Apostolidis *et al.*, 1996a and various DNA loci (O'Connell *et al.*, 1995; Nielsen *et al.*, 1996; Sanchez *et al.*, 1996; Nilsson, 1997; Stone *et al.*, 1997).

Table 6.9 Gene diversity analysis at allozyme loci in Atlantic salmon (*Salmo salar*).

Regions	Number of populations (localities)	Number of loci (polymorphic)	H_T	H_S	G_{ST} (%)	Reference
Basins of the Baltic Sea, Eastern and Western Atlantic, average (natural populations)	29	38 (9)	0.029	0.027	6.4	Ståhl, 1987
Finland, sea basins, average	20	25 (6)	0.047*	0.042*	11.2	Koljonen, 1989
Finland, river basins	27	36 (8)	0.038	0.035	6.4	Elo *et al.*, 1994
Russia, the White and Barents sea basins, average (natural populations)	9	30 (6)	0.074*	0.068*	7.8*	Kazakov & Titov, 1995
Newfoundland, anadromous populations	12	21 (9)	0.0655*	0.063*	3.9*	Verspoor, 1994

* Our calculation from original data.

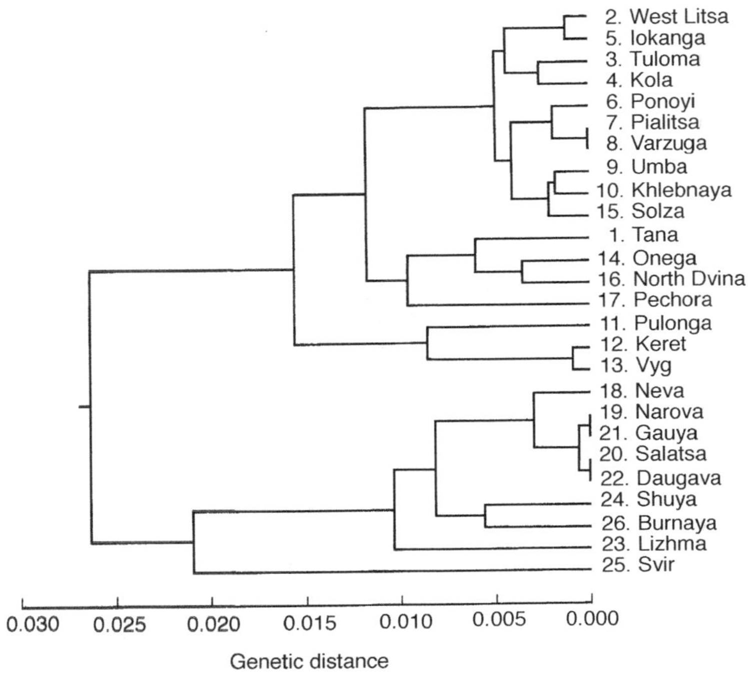

Fig. 6.12 Dendrogram based on genetic distances between Atlantic salmon populations in north-east part of the range (30 protein loci; from Kazakov & Titov, 1995).

Populations from rivers of the North American Atlantic coast also differ significantly by allele frequencies of microsatellite loci (McConnell *et al.*, 1995, 1997; Fontaine *et al.*, 1997)

Freshwater populations of Atlantic salmon from European lakes, as with anadromous populations, are genetically subdivided into the Baltic and East Atlantic clusters corresponding to the location of their home ranges in a certain sea basin (Ståhl, 1987; Kazakov & Titov, 1995). The degree of genetic differentiation among

Table 6.10 Subdivision of gene diversity in *Salmo salar* populations of north-eastern Europe (our calculation for *sAAT-3**, *IDDH-1**, *IDHP-3**, *mMEP-2**, *ESTD**, *PEX-2** loci by data from Kazakov & Titov, 1992).

Hierarchy level	Absolute diversity	Relative diversity (%)
Total, H_T	0.3722	100
Within river populations	0.3059	82.2
Among sea basins (the White, Barents, Baltic seas)	—	10.0
Among river populations within sea basins	—	7.8

isolated freshwater populations is much greater than among anadromous populations of the same region (Verspoor, 1994). Sympatry of freshwater and anadromous forms did not prevent their significant genetic divergence for both allozyme and mtDNA markers (Vuorinen & Berg, 1989; Birt *et al.*, 1991). These data suggest reproductive isolation among forms although no ethological factors preventing their hybridization were revealed under experimental conditions, and the eggs obtained in experiments were viable (Hutchings & Myers, 1985). It is assumed that reproductive isolation of freshwater and anadromous salmon in nature is maintained by selection for preference for different sites and periods of spawning and by assortative mating (Birt *et al.*, 1991). Similarity between these forms in the pattern of their regional differentiation provides evidence for the polyphyletic origin of freshwater populations (Behnke, 1972; Ståhl, 1987; Kazakov, 1993; Verspoor, 1994).

The analysis of the relationship between genetic and geographic distances in all Atlantic salmon populations studied throughout the species range showed that genetic distances do not depend on the latter parameter in populations within the same basin or geographic region but correlate with it in remote populations located more than 1000 km apart (Elo, 1993). This may be accounted for by the structure of migrations or supposed selective pressure on some allozyme loci used for determining genetic distances. Thus, selective value is ascribed to the locus *mMEP-2**: its allele frequencies in populations proved to correlate with geographic latitude and thermal parameters of spawning grounds (Verspoor & Jordan, 1989; Verspoor *et al.*, 1991), and carriers of different genotypes at this locus differed in growth rate at the stages of parr and smolt (Jordan & Youngson, 1991) and in the age of maturity (Jordan *et al.*, 1990). There is some evidence for selective effects at the transferrin and trypsin loci (Verspoor, 1986; Torrisen *et al.*, 1993) and some other allozyme genes of Atlantic salmon, e.g. on *PGM-1** (Pollard *et al.*, 1994; Verspoor, 1994). None the less, the question of selective pressure on allozyme loci in *S. salar* calls for more thorough research involving additional methods, as was argued in Chapter 5. Admittedly at present only the first steps have been taken in this direction (see also Jordan *et al.*, 1997).

The pattern of hierarchical distribution of intraspecific gene diversity in Atlantic salmon suggests that the divergence between the European and North American complexes was the most ancient in the evolution of this species, and subsequently each complex evolved independently (Kazakov & Titov, 1995). Nevertheless, the possibility that Atlantic salmon migrated from Europe and colonized some rivers on the North American coast cannot be excluded. Such a hypothesis proposed by Behnke (1980) in relation to the colonization of Newfoundland is supported (but not confirmed) by genetic evidence obtained from allozyme (Verspoor, 1994) and mtDNA markers (Bermingham *et al.*, 1991).

Virtually the entire recent range of Atlantic salmon was covered with ice during the last glaciation, which ended 10 000–15 000 years ago (Behnke, 1972; Ståhl, 1987; Kazakov, 1993). Water bodies that became free from ice were colonized by fish that survived in isolated refugia located in the West and East Atlantic. The geological history of the Baltic Sea, which was a lake during the last glaciation, provided for isolation of local Atlantic salmon from all other populations of the East Atlantic, which lasted for almost 9000 years (Ståhl, 1987). During such a long period of isolation ($\approx$ 2000 generations), Atlantic salmon from the Baltic Sea acquired a distinctive genetic structure, which differs from that in other East Atlantic populations. Genetic divergence between recent *S. salar* populations, which developed in the postglacial period within geographic regions, was influenced by colonization of fish from different sources, the founder effect and subsequent genetic drift. Further effects of restricted migration because of strict homing, and selection pressure on some allozyme loci, may have also been operative.

6.1.7 Brown trout

The population genetics of this species have been studied in detail. The data on parameters of gene diversity and coefficient of gene differentiation are shown in Table 6.11.

Ferguson (1989) reviewed the available data on within-population allozyme variation in brown trout until the late 1980s. The observed heterozygosity of populations varied over the species range from 0 to 0.122, and the proportion of polymorphic loci from 0 to 0.398. It is noteworthy that in two closely located lakes in Scotland, Prodöhl *et al.* (1997) discovered brown trout populations that were monomorphic with respect to both protein and various DNA markers.

The value of total gene diversity H_T in brown trout is similar to that in other salmonid species, whereas the value of its interpopulation component G_{ST} within regions is significantly greater, reaching 53–65% in population groups from water bodies of France and Spain (Krieg & Guyomard, 1985; Garcia-Marin *et al.*, 1991; Presa *et al.*, 1994). On the whole, interpopulation gene diversity in the Atlantic part of the brown trout range is 60% (Ferguson, 1989).

Strong genetic differentiation is observed between population complexes of northern and southern territories of the former Soviet Union (Osinov, 1984; Osinov & Bernatchez, 1996), as well as North Atlantic and Mediterranean basins in western Europe (Krieg & Guyomard, 1985; Estoup *et al.*, 1993; Presa *et al.*, 1994). In general, G_{ST} values in northern populations of brown trout are lower than in southern counterparts (Table 6.11). Two factors are responsible for this situation: first, southern populations are more ancient because the corresponding regions were not affected by glaciation; second, they are hydrographically associated with different sea basins: the North Atlantic and the Mediterranean. Indeed, the hierarchical analysis shows that the interpopulation component of gene diversity is largely accounted for by differentiation between complexes of populations belonging to these sea basins (Presa *et al.*, 1994; Garcia-Marin & Pla, 1996). The highest interpopulation variation in these regions is observed at the loci *CK-2**, *GPI-2**, *MEP-1** and *MEP-3**. In populations of the North Atlantic basin, these are *PEP-LT**, *LDH-5** (*LDH-C**), *GPI-2** and *LDH-1**.

Table 6.11 Gene diversity analysis at allozyme loci in brown trout (*Salmo trutta*).

Regions	Number of populations (localities)	Number of loci (polymorphic)	H_T	H_S	G_{ST} (%)	Reference
White Sea basin	9	16 (8)	0.153	0.139	8.9	Makhrov, 1999
Sweden	38	35 (9)	0.040	0.025	36.7	Ryman, 1983
France, native populations	12	46 (26)	0.112	0.050	55.0	Krieg & Guyomard, 1985
France, Atlantic coast, native populations	19	47 (27)	0.077*	0.059*	32.8*	Presa *et al.*, 1994
Spain, native populations						
Atlantic coast	10	49 (25)	0.073*	0.018*	76.7*	Garcia-Marin & Pla, 1996
Mediterranean coast	5	49 (25)	0.047*	0.029*	38.0*	Garcia-Marin & Pla, 1996
Spain, Biscay Bay basin	18	9 (9)	—	—	21.5	Moran *et al.*, 1995

* Our calculation from original data.

Genetic divergence on the microgeographic scale, i.e. between subpopulations inhabiting tributaries of the same river or the basin of a small lake or bay, is even more prominent in this species (Ryman *et al.*, 1979; Ferguson & Mason, 1981; Ryman, 1981, 1983; Ferguson, 1989; Makhrov, 1999). This fact was confirmed in recent studies with the use of mtDNA and microsatellite markers (Hansen & Loeschcke, 1996; Estoup *et al.*, 1998; Hansen & Mensberg, 1998). Brown trout subpopulations demonstrate genetic divergence even in the absence of any physical isolating barriers, which is often accompanied by pronounced morphological and ecological differentiation. This provides evidence for well-developed homing in this species, which ensures reproductive isolation of subpopulations.

The picture of the distribution of alleles of several loci over the brown trout range probably reflects the natural history of recent populations formed during postglacial colonization of northern Europe, which began 10 000–15 000 years ago (Ferguson, 1989). Thus, the geographic distribution of *LDH-5** alleles (**100* and **90*) is likely to be determined in part by the history of this colonization. In brown trout populations inhabiting the Mediterranean, Black and Caspian sea basins, the allele **100* of this locus is fixed, whereas populations of the Baltic and White sea basins carry both alleles, **100* and **90* (Osinov, 1984, 1988, 1990; Krieg & Guyomard, 1985; Ferguson, 1989; Karakousis & Triantaphyllidis, 1990; Garcia-Marin & Pla, 1996). There is a hypothesis that northern Europe was populated by two 'races' of brown trout. The first, 'ancestral' race carried the allele **100* and probably dispersed from the refugium located in the Bay of Biscay basin. The second, 'modern' race, which carried the **90* allele, may have spread from the refugium (refugia) in the area of the Baltic or White seas and displaced the 'ancestral' race to some extent (Ferguson & Fleming, 1983; Hamilton *et al.*, 1989). In this context, the presence of allele **90* in brown trout populations is interpreted as evidence for a more recent origin. Moreover, the carriers of this allele possibly have certain selective advantages in some northern populations (Henry & Ferguson, 1985; Ferguson, 1989; Zelinsky & Gruzdev, 1992; Makhrov *et al.*, 1994).

The quite recent generalization of the information on geographic distribution of alleles of 11 polymorphic allozyme loci throughout the *S. trutta* species range proposed a model of colonization from three preglacial lineages. During glacial retreat, the migration and recolonization could have occurred possibly from (1) Mediterranean–Caspian refugia to the north-west, (2) Atlantic drainages of Iberia and southern France to the north, and (3) the English Channel area to the north and east (Garcia-Marin *et al.*, 1999).

mtDNA studies on brown trout from populations inhabiting different regions of the species' range resulted in identification of five phylogenetic groups: three in the Mediterranean basin, one in the northern sea basins of Europe (the Atlantic group), and one group, named Danubian, in basins of the Black, Caspian and Aral seas (Bernatchez *et al.*, 1992; Osinov & Bernatchez, 1996). The pattern of distribution of mtDNA haplotypes suggests that the postglacial dispersal of brown trout in some regions proceeded in a more complicated way than had been assumed previously, and involved secondary contact and introgressive hybridization between populations of different phylogenetic groups or races (McVeigh *et al.*, 1995; Hansen & Loeschcke, 1996; Hynes *et al.*, 1996; Osinov & Bernatchez, 1996).

As noted above, brown trout is one of the fish species studied in greater detail in terms of population genetics, and currently it is used for testing new molecular approaches to the analysis of differentiation (Ferguson *et al.*, 1995). Thus, estimating spatial genetic variation (F_{ST}) from polymorphism of mtDNA, microsatellite and minisatellite loci was performed for the same groups of brown trout populations from different regions of the range that were previously studied with respect to allozyme genes. The comparison of results showed that F_{ST} values obtained by analysing allozyme and minisatellite loci are similar to one another, but not necessarily similar to those estimated by means of mtDNA analysis. For example, F_{ST} values based on allozymes, single-locus minisatellites and mtDNA for Icelandic populations are 0.38, 0.40 and 0.00, and for Scottish sea trout, 0.07, 0.06 and 0.31, respectively (data from Ferguson *et al.*, 1995; see also Laikre *et al.*, 1995; Hansen & Loeschcke, 1996; Estoup *et al.*, 1998).

Note that F_{ST} values estimated for brown trout populations on the basis of individual minisatellite loci (and minisatellites are regarded as selectively neutral regions of the genome) are fairly similar to one another, as it should be in the case of neutral markers (Ferguson *et al.*, 1995). Hence, F_{ST} values obtained by averaging the data on many allozyme genes, being similar to corresponding values estimated by minisatellites, can be regarded as a parameter reflecting the level of selectively neutral differentiation. However, F_{ST} estimations based on individual allozyme genes are much more variable than those based on the analysis of minisatellites, suggestive of differences in the selective significance of these genes (Chapters 4 and 5).

It is possible to determine the extent to which the interpopulation component of gene diversity for a set of allozyme loci can reflect the level of selectively neutral divergence of populations accounted for by a random microevolutionary factor only, i.e. by genetic drift, by comparing average F_{ST} (G_{ST}) values obtained by means of allozyme analysis with those expected on the basis of demographic parameters of populations. The results of such a comparison performed for different animal species and humans, which were discussed in Chapter 5 (see Fig. 5.7), demonstrated good correspondence between the observed and expected values of local genetic differentiation. Here, we performed this comparison for six salmonid species, using the latest data to make $F_{ST\text{obs}}$ and $F_{ST\text{exp}}$ more accurate (Fig. 6.13). It can be seen that correspondence between these estimates is fairly convincing.

As follows from Table 6.12, the total level of heterozygosity in salmonid species differs approximately by a factor of 3, whereas the coefficient of gene differentiation (G_{ST}) differs by more than an order of magnitude. With respect to the extent of spatial genetic differentiation, G_{ST}, the species shown in this table can be arranged in the following ascending series: pink salmon < chum salmon < sockeye salmon <

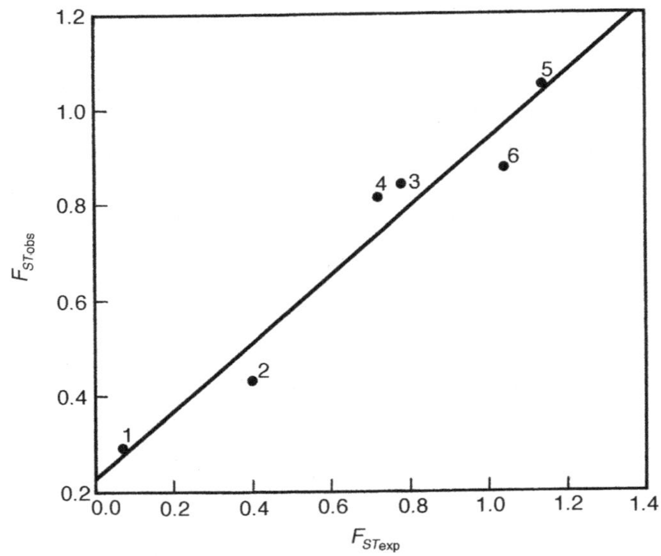

Fig. 6.13 Relationship between observed (F_{STobs}) and expected (F_{STexp}) values (%, log-transformed) of spatial genetic differentiation in six species of salmon. 1 – *Oncorhynchus gorbuscha*, 2 – *O. keta*, 3 – *O. nerka*, 4 – *O. tshawytscha*, 5 – *O. kisutch*, 6 – *Salmo salar*, $y = 0.224 + 0.717x$; $r = 0.972$; $P < 0.01$. (Data from Tables 6.1, 6.3, 6.4, 6.7, 6.8 and 6.9, and Altukhov & Salmenkova, 1994; for estimate of F_{STobs} in *O. tshawytscha*, the data for regions with intensive artificial propagation are not included.)

Atlantic salmon < chinook salmon < coho salmon < brown trout. This raises the questions as to the origin of this species-specificity, and the nature of the evolutionary factors that determine it.

To answer these questions, we will generalize the data discussed above in order to examine the roles of different microevolutionary factors on intraspecific genetic structure.

Table 6.12 Average regional values of the components of gene diversity in salmonid species (using data from Tables 6.1, 6.3, 6.4, 6.7, 6.8, 6.9 and 6.11).

Species	H_T	H_S	G_{ST}
Oncorhynchus gorbuscha	0.1044	0.1024	1.91
O. keta	0.0709	0.0688	2.90
O. nerka	0.0424	0.0394	7.00
O. tshawytscha	0.0736	0.0688	6.52*
O. kisutch	0.0295	0.0262	11.30
Salmo salar	0.0508	0.0470	7.14
S. trutta	0.0837	0.0526	37.10

* Data for regions with intensive artificial propagation are excluded.

6.2 Effects of random and systematic evolutionary factors on intraspecific genetic differentiation

Firstly, we should recall the data from the previous chapters that are relevant to the approach, namely comparison of empirical G_{ST} (F_{ST}) values with those expected under conditions of neutral evolution and calculated on the basis of the parameter N_em, that is, the main characteristics of the breeding structure of subdivided populations (section 5.5).

As shown above, G_{ST} values for different loci vary strongly, and those for the same locus in different regions are also unequal. Assuming that the average $\bar{G}_{ST}$ ($\bar{F}_{ST}$) values (estimated according to the large sample of allozyme loci) correspond to the level of genetic differentiation in conditions of a selectively neutral process, it may be concluded that genes with low F_{ST} values are most probably exposed to stabilizing selection, and those with high F_{ST} values, to diversifying (disruptive) selection. We have employed this reasoning to analyse the spatial variation of the same set of individual genes in chum salmon populations from different regions (Table 6.13). The values of intrapopulation (H_S) and interpopulation (G_{ST}) gene diversity per locus vary depending on the region. Statistical estimates of difference between G_{STi} and $\bar{G}_{ST}$ values in each region were detected by means of bootstrap analysis. It can be seen that the adaptive status of the gene varies over the species range, i.e. a given gene may undergo stabilizing selection in populations of one region and remain selectively neutral or evolve under the effect of disruptive selection in populations of another region. It is interesting that the proportion of different classes of genes in respect of their selective significance is approximately the same in different regions: 43–50% for genes evolving under stabilizing pressure, 36–38% under disruptive selections, and 14–21% for selectively neutral ones.

Geographically and historically related populations are known to demonstrate the effect of isolation by distance with respect to selectively neutral genes, i.e. the genetic divergence of populations increases proportionally to the geographic distance between them. Slatkin (1993) suggested such effects could be revealed by the method in which the logarithm of gene flow N_em between pairs of populations (calculated from F_{ST} values for these pairs) is regressed on the logarithm of corresponding geographic distance. A significant negative regression means isolation by distance. If the loci evolve under stabilizing selection, the effect of isolation by distance should be absent. We applied this approach to examine the relationship between N_em (estimated according to different groups of loci) and geographic distance in pairs of more or less linearly connected populations of pink and chinook salmon using published data.

In pink salmon populations from coastal rivers of eastern Kamchatka, which were studied for 12 loci (data from Varnavskaya & Beacham, 1992), no isolation by distance was revealed by this method (Fig. 6.14a). This fact provides evidence for the prevalence of genes exposed to stabilizing selection among the 12 genes studied, confirming our data that such selection at several allozyme loci does exist in pink

Table 6.13 Values of mean population heterozygosity (H_S) and interpopulation differentiation (G_{ST}) at shared polymorphic loci in chum salmon of three regions.

Loci	Alaska including Yukon River basin		British Columbia		North-eastern Russia	
	H_S	G_{ST}	H_S	G_{ST}	H_S	G_{ST}
*mAAT-1**	0.176	0.036	0.373	0.062***	0.080	0.087***
*sAAT-1,2**	0.192	0.021**	0.182	0.018**	0.253	0.030**
*ALAT**	0.169	0.023**	0.276	0.036***	0.271	0.018**
*ESTD**	0.452	0.101***	0.109	0.033	0.249	0.067***
*mIDHP-1**	0.063	0.057***	0.124	0.052***	0.237	0.097***
*sIDHP-2**	0.561	0.028**	0.687	0.041***	0.618	0.034
*LDH-A1**	0.332	0.049***	0.062	0.023**	0.218	0.033**
*sMDH-B1,2**	0.034	0.024**	0.092	0.026**	0.031	0.021**
*mMEP-2**	0.264	0.052***	0.222	0.028**	0.317	0.055***
*MPI**	0.211	0.024**	0.150	0.022**	0.180	0.015**
*PGDH**	0.561	0.011**	0.002	0.044***	0.103	0.033**
*PEPB-1**	0.198	0.033	0.438	0.031	0.183	0.034
*PEPLT**	0.051	0.044***	—	—	0.027	0.034
*PEPA**	0.001	0.019**	0.001	0.010**	0.022	0.072***
$\bar{G}_{ST}$, average per locus		0.032		0.032		0.039

* H_S and G_{STi} were calculated from data by Kondzela *et al.*, 1994; Wilmot *et al.*, 1994; Winans *et al.*, 1994. Average $\bar{G}_{ST}$ value and statistical estimates of differences between G_{STi} and average $\bar{G}_{ST}$ in each region are obtained by means of bootstrap analysis (1000 permutations) of log-transformed G_{STi} values. ** G_{STi} is less than $\bar{G}_{ST}$. *** G_{STi} is more than $\bar{G}_{ST}$; G_{STi} without marks does not differ significantly from $\bar{G}_{ST}$. In all cases probability level $P < 0.01$.

salmon populations (Altukhov, 1983a, 1989a; Altukhov *et al.*, 1987). Different results were obtained on the basis of data by Shaklee *et al.* (1991) for pink salmon populations from the North American coast (southern Canada): isolation by distance was traced when the parameter N_em was calculated according to all 20 loci studied (Fig. 6.14b) or to the set of loci used in the cited work on populations of eastern Kamchatka (Fig. 6.14c). Isolation by distance appeared to be absent, however, when the group of genes characterized by minimal F_{ST} values were used for calculating N_em (Fig. 6.14d). According to these data, stabilizing selection actually plays a key role in the spatial variation of some allozyme loci, and their selective rank varies in different parts of the species range.

Similar analysis performed for the group of chinook salmon populations from the Yukon River tributaries demonstrates the existence of a highly significant correlation between the log of geographic distances and the log of N_em calculated from F_{ST} values averaged over 16 loci (Fig. 6.15a). When N_em is estimated using only the loci with minimal F_{ST} values (falling beyond confidence intervals of average values per locus), such a correlation is absent, i.e. isolation by distance is not revealed (Fig. 6.15b). As migration between populations affects all the loci equally, the absence of isolation by distance with respect to only some of them suggests that they are

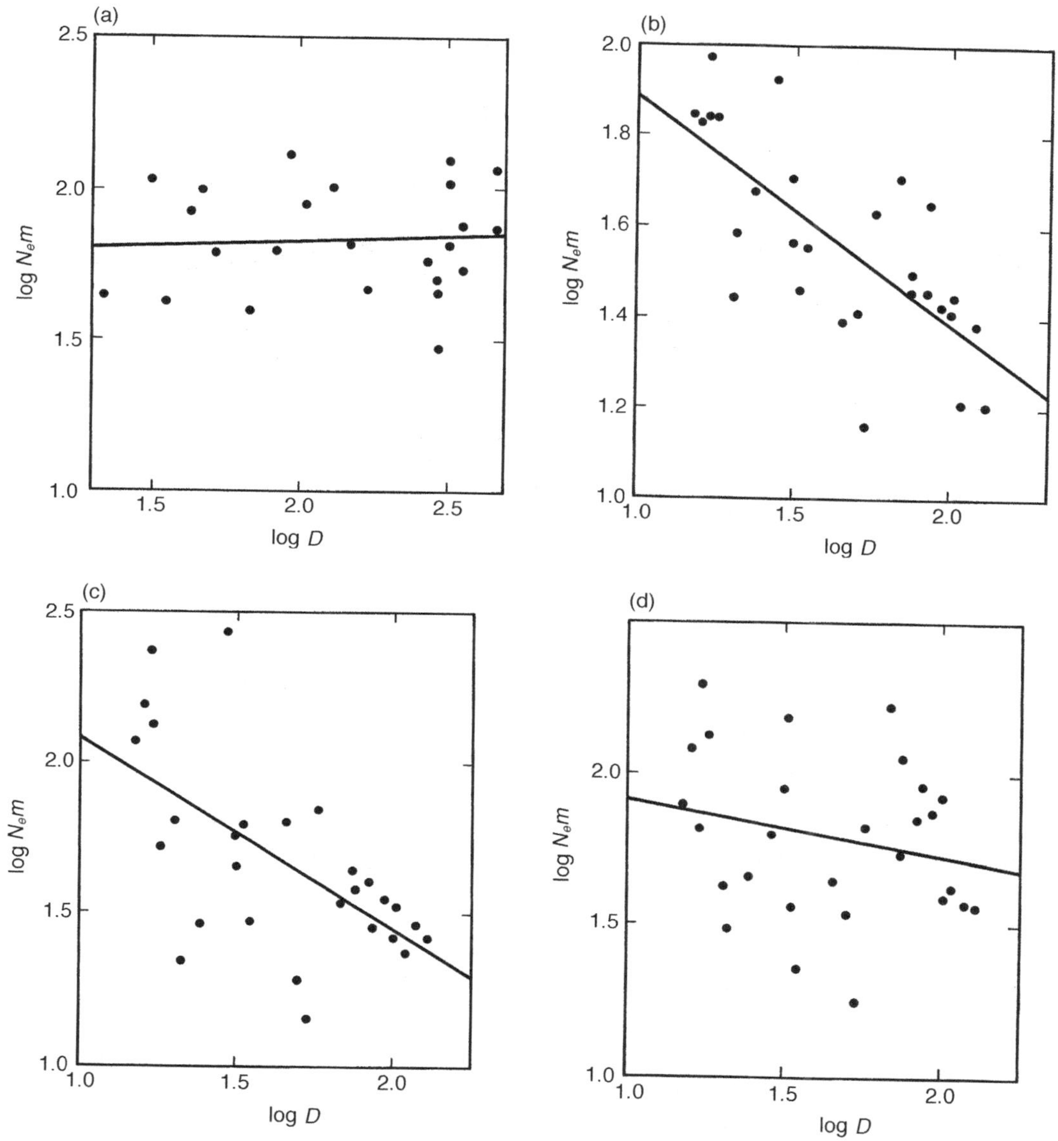

Fig. 6.14 Relationship between geographic and genetic distances for pink salmon populations. Calculations of the authors based on published allele frequency data. (a) Correlation between log D and log N_em for eight populations from east Kamchatka; N_em is estimated using 12 allozyme loci (by data from Varnavskaya & Beacham, 1992); correlation is non-significant, $y = 1.75 + 0.05x$, $r^2 = 0.128$. (b) Correlation between log D and log N_em for eight populations from British Columbia rivers (20 allozyme loci, by data from Shaklee *et al.*, 1991); correlation is significant, $y = 2.40 - 0.51x$, $r^2 = 0.511$. (c) Correlation between log D and log N_em for the same eight populations from British Columbia, but N_em is estimated using the same 12 loci, which were used in (a); correlation is significant, $y = 2.55 - 0.60x$, $r^2 = 0.391$. (d) Correlation between log D and log N_em for the same eight populations, but N_em is estimated using four loci with minimal F_{ST} values; correlation is non-significant, $y = 2.08 - 0.18x$, $r^2 = 0.043$.

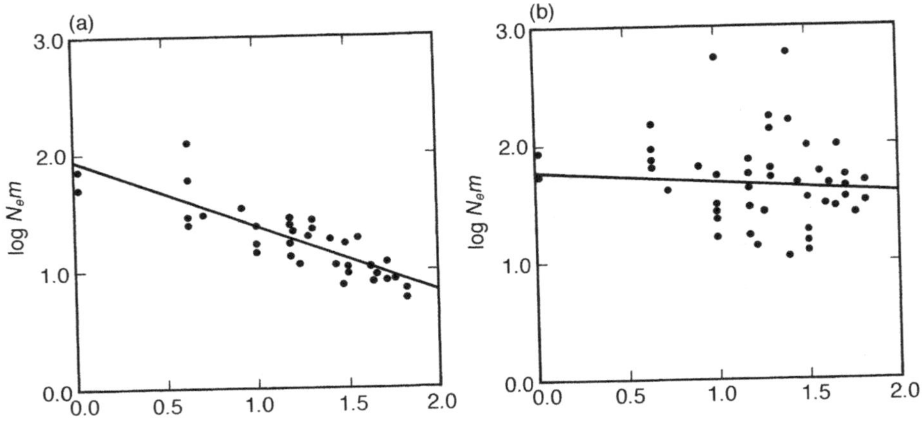

Fig. 6.15 Relationship between log D and log N_em for ten populations of chinook salmon spawning in tributaries of the Yukon River (calculated by the authors on the basis of allele frequency data from Wilmot *et al.*, 1992). (a) N_em is estimated using 16 allozyme loci; correlation is significant, $y = 1.97 - 0.57x$, $r^2 = 0.729$. (b) N_em is estimated using four loci with minimal F_{ST} values, correlation is non-significant; $y = 1.90 - 0.16x$, $r^2 = 0.032$. See text for details.

exposed to stabilizing selection. Therefore, this example also confirms the role of stabilizing selection in the formation of the uniform spatial distribution of allozyme gene frequencies. In the next section, we will discuss whether this reasoning is suitable for explaining uniformity or weak differentiation of allele frequencies observed in large areas of the pink salmon range.

Let us now consider the so-called phylogeographic component (Avise *et al.*, 1987) of spatial genetic differentiation common for different Pacific salmon species. We have already addressed this question with reference to Atlantic salmon and brown trout in section 6.1.

As shown above, spatial genetic differentiation in all the salmonid species (as in many other groups of organisms) has a regional geographical pattern, which is especially clear in Atlantic salmon. On a macrogeographical scale, this differentiation is largely determined by the natural history of recent populations, which is directly associated with the ice ages (McPhail & Lindsey, 1970). During the last Pleistocene glaciation, large areas in the ranges of Pacific and Atlantic salmonid species were covered with ice, and only a few populations survived in ice-free refugia. As inhabitants of different refugia were isolated from each other, their adaptive and neutral evolution led to the development of specific genetic features in each isolate, i.e. to intraspecific genetic divergence. After the retreat of ice (10 000–15 000 years ago), the surviving populations began to colonize appropriate ecological niches, and new populations could be formed by immigrants from one or several refugia.

During the last (Wisconsin) Pleistocene glaciation in the North American continent, the main refugia of salmonid populations adjoined the ice sheet in the north and south. The northern (Beringian) refugium was in Alaska. It included a major part of the Yukon River basin and the land mass that emerged in the Bering Strait and the northern Bering Sea after a regression of sea level, forming the so-called Bering Land Bridge, or Beringia, which connected the Asian and North American continents (Hopkins, 1967; Flint, 1971; Kontrimavichus, 1973). The southern refugium, or Cascadia, was in the lower Columbia River basin (McPhail & Lindsey, 1970). Between them, several small refugia could exist, and all of them were separated by the ice sheet.

In north-eastern Asia, refugia were in the Gulf of Anadyr, the Anadyr River basin, and some small areas on the northern coast of the Sea of Okhotsk. Almost the entire Kamchatka Peninsula and, to a lesser extent, the continental coast of the Sea of Okhotsk were covered with the ice sheet, while Sakhalin, Iturup, the Amur River basin and more southern regions were not affected by the last Pleistocene glaciation. Thus, probable Asian refugia included the Anadyr River basin, separate parts of the Okhotsk Sea coast, and areas located south of the Kamchatka Peninsula. The drainages of large rivers, such as the Yukon and Amur, changed strongly during the Pleistocene because of sea regression and transgression, whereas rivers of the Anadyr Lowland remained hydrologically more stable (Lindberg, 1972; Bespalyi & Glushkova, 1987; Chereshnev, 1996).

The postglacial dispersal of salmonids from refugia provided the basis for spatial genetic differentiation. The history of this dispersal is reflected in specific genetic features of populations, ranging from the geography of individual allozyme genes, such as alleles of the locus *ESTD** in North American chum salmon populations (Fig. 6.3b), *LDH-B2** in sockeye (Withler, 1985), or *LDH-5** alleles in brown trout, to DNA markers in different species (section 6.1).

With respect to both the set of allozyme loci and DNA markers, virtually all Pacific salmon species – chum, sockeye, chinook and, to a lesser extent, pink salmon – demonstrate differentiation between the northern and southern complexes of North American populations. A certain similarity is observed between populations of Alaska and north-eastern Russia, whereas populations spawning in the rivers located south of British Columbia or populations of Japanese rivers (chum salmon) are segregated from them (Varnavskaya & Beacham, 1992; Taylor *et al.*, 1994; Winans *et al.*, 1994; Wood *et al.*, 1994; Bickham *et al.*, 1995).

The general pattern of spatial genetic differentiation is in good agreement with the biogeographic hypothesis (McPhail & Lindsey, 1970, 1986; Lindsey & McPhail, 1986), according to which the recent populations of Yukon, Anadyr and Kamchatka originated from their joint Beringian refugium, whereas more southern regions were colonized by immigrants from southern refugia. Probable repeated immigration to the same water body from different refugia, genetic drift and pressure of stabilizing selection are apparently responsible for the greater similarity between remote but more ancestral populations than between ancient and more recent populations from geographically close regions. As will be shown below, these factors account for aberrations in the pattern of regional population clustering on the basis of genetic distances. Thus, analysis of dendrograms for chum salmon populations from the Russian far east and north-east (based on the data by Salmenkova *et al.*, 1986, 1992; Viktorovsky *et al.*, 1986; Bachevskaya, 1990) shows that genetic and geographic distances between localities correspond mainly to one another, but there are several exceptions to this rule (Fig. 6.5). First, populations of the Anadyr and Penzhina rivers fall into the same cluster as geographically distant Sakhalin populations. Second, populations from the northern coast of the Sea of Okhotsk cluster with populations from eastern Kamchatka instead of those from western Kamchatka to which they are closer geographically. Third, populations of south-western Sakhalin are unique and have no analogues among other chum salmon populations included in the analysis.

Let us consider in more detail this disparity of geographic and genetic distances between populations with reference to the data on the history of their distribution over recent ranges and on the principle of population systems (Altukhov & Rychkov, 1970; Rychkov, 1973; Altukhov, 1974; Chapter 4). As we have stressed, the total gene diversity of recent populations corresponds to a certain ancestral gene pool (pra-population), which can be reconstructed by averaging gene frequencies of all the studied populations of the species. The relation of recent populations to the pra-population was assessed using Nei's genetic distances (Nei, 1975). To make the results more illustrative, they are presented graphically as polygons generalizing the genetic image of the ancestral and several recent populations of chum salmon (Fig. 6.16).

On the basis of palaeogeographic data on Siberia and the Far East (Baranova & Biske, 1964; Khudyakov *et al.*, 1972; Alexandrov, 1973; Melekestsev *et al.*, 1974; Glushkova, 1984; Bespalyi & Glushkova, 1987), it is possible to demonstrate that both specific features of interpopulation genetic differentiation observed today and the divergence of each recent population from the ancestral gene pool generally correlate with the history of late Pleistocene glaciations. As noted above, the latter strongly affected the Kamchatka and the Okhotsk Sea coasts, whereas large areas in the Anadyr and Penzhina river basins remained ice-free, as well as the Amur River and the rivers of Sakhalin, Primorye and Iturup.

In strict agreement with these data, more ancient populations of the corresponding regions proved to be much closer genetically to the reconstructed ancestral population than the historically younger localities of Kamchatka and the northern Okhotsk Sea coast. This situation may be interpreted as a consequence of ancestral gene pool conservation in ice-free refugia that existed in the late Quaternary Period, inhabited by populations as large and structurally complex as the present-day populations, so supporting their internal gene diversity owing to systemic organization. Such an interpretation allows us to understand why the time of divergence for populations related closely to the ancestral gene pool or to one another deviates from the values calculated by the formula expressing this time (t) as function of genetic distance (D): $t \approx 5 \times 10^6 D$ (Nei, 1987). As follows from the palaeogeographic data, these populations are evolutionarily ancient, whereas calculations contradict this fact, estimating the time of their autonomous existence

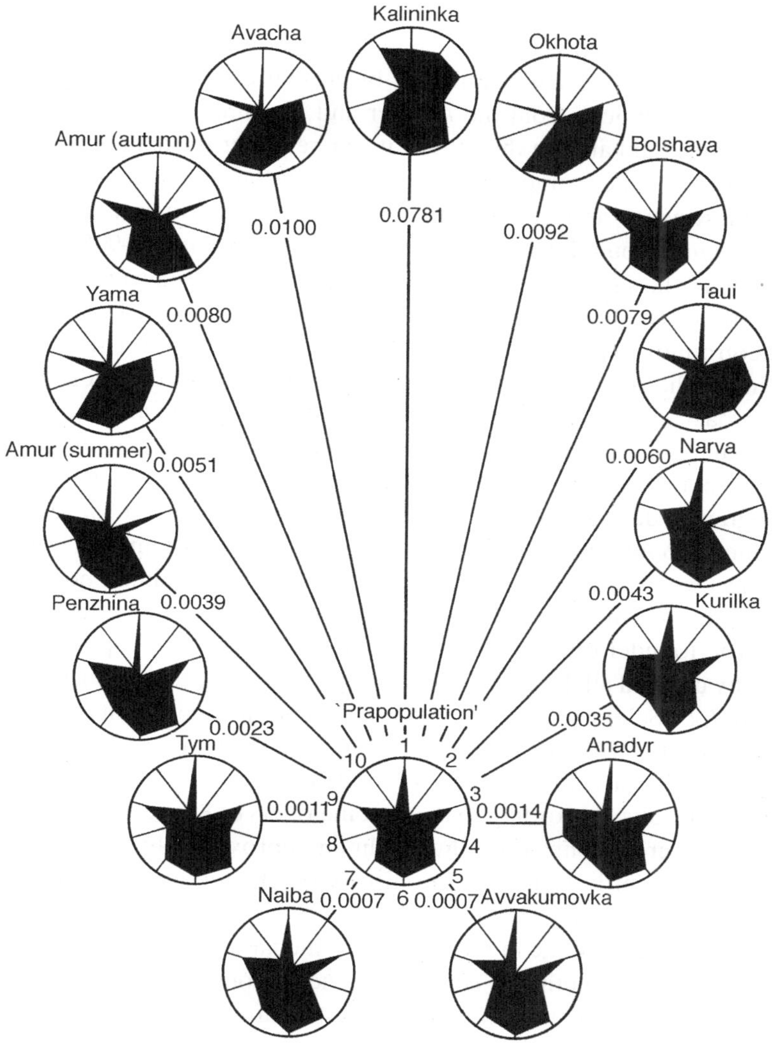

Fig. 6.16 Schematic representation of the reconstructed 'prapopulation' (ancestral population) and several contemporary populations of chum salmon spawning in the Asian part of the range. At radii are the frequencies of following alleles: 1, 2 – *sMDH-B1,2** (100) and (75); 3, 4 – *sAAT-1,2** (100) and (118); 5 – *LDH-A1** (100); 6 – *PGDH** (100); 7, 8 – *mMEP-2** (100) and (124); 9, 10 – *ESTD** (100) and (82). The interval of allele frequencies (from centre of circle): 0.5–1 at radii 1, 3, 5, 6 and 7; 0–0.5 at radii 2, 4 and 8; 0–1 at radii 9 and 10. Values of Nei's genetic distances (Nei, 1975) estimated at these polymorphic loci are on lines connecting studied populations with the 'prapopulation'. See text for details.

approximately as 1300 to 7000 years (for these very rough estimates we used *D* values, calculated at six polymorphic and ten monomorphic loci analysed in all these populations). As to the genetic similarity of present-day remote populations (e.g. chum salmon populations from the Anadyr and Penzhina rivers on the one hand, and from eastern Sakhalin on the other), it can be either incidental, because of the limited number of genes compared, or it can reflect the similarity of adaptations in carriers of these genes, as in chum salmon from the Anadyr River that clustered with *autumn spawning* populations. It is noteworthy, however, that 80–90% of these fish, which belong to the *summer race* according to the dates of spawning migration and spawning proper (Kulikova, 1972; Smirnov, 1975), reproduce in water bodies fed from underground sources (Putivkin, 1989), i.e. those typical for the *autumn race* reproducing in the rivers of Primorye, Sakhalin and the Kuril Islands.

This approach can elucidate anomalies such as the seeming ancientness of chum salmon populations from south-western Sakhalin, which, according to calculations, deviated from the ancestral gene pool 150 000 years ago. As follows from the palaeogeographic data on the development of the coastline in Sakhalin, small rivers in which these populations spawn have changed significantly in size (Alexandrov, 1973). Hence, we cannot exclude that corresponding chum salmon populations are of relatively recent origin associated with a founder effect or the influence of strong selection. The founder effect appears more probable, as these populations are distinguished by unusual allele frequencies (for this region) at several loci (e.g. *ESTD** and *MDH-B1,2**; Salmenkova *et al.*, 1986, 1992), and specific haplotypes of a neutral marker such as mtDNA (Ginatulina & Mashkin, 1990; Ginatulina, 1992).

Meanwhile, reconstruction of the time of divergence for other populations produces credible results. According to them, some rivers of Kamchatka and the Sea of Okhotsk basin were colonized by chum salmon approximately 14 000–18 000 years ago, which agrees fairly well with the history of Quaternary glaciations in these regions (Melekestsev *et al.*, 1974; Glushkova, 1984). The analysis described here is no more than the initial stage in reconstructing the natural history of corresponding populations by genetic methods. To improve such reconstructions, it is necessary to continue detailed investigations of population structure and to broaden the range of allozyme and DNA markers used in the phylogeographic analysis.

6.3 Similarity of allozyme gene frequencies in spatially separated populations as a consequence of stabilizing selection: an example from pink salmon

In the first section of this chapter, we noted a weak genetic differentiation of pink salmon from spatially separated localities, which is observed in both even- and odd-year generations. Regional G_{ST} (F_{ST}) values in pink salmon are only 0.01–0.03, being two to five times as low as in other species of the genus *Oncorhynchus*.

This fact, along with incorrectly interpreted data on fish marking (marked fish caught in the sea were regarded as evidence for the exchange of members between populations), provided the basis for a hypothesis concerning periodic large-scale migrations of pink salmon between regions (Glubokovsky & Zhivotovsky, 1986, 1989). Its authors proposed a system of 'fluctuating stocks' in which population boundaries, number and size are subject to periodic changes determined by fluctuations in the intensity of gene exchange between populations, which, in turn, depend on changes in migration barriers. The authors claimed that, under such circumstances, pink salmon cannot have a stable population structure (Glubokovsky & Zhivotovsky, 1986).

In the framework of this hypothesis, the spatial similarity of allele frequencies is explained by the interaction of two processes: selection, which differentiates populations over the species range, and migration, which largely eliminates this effect (Zhivotovsky *et al.*, 1987, 1989, 1994, 1996; Glubokovsky *et al*, 1989).

The uniformity of allele frequencies, however, is not necessarily the consequence of gene migration: it can also result from stabilizing selection proceeding with similar intensity in different localities (Altukhov, 1983a, 1990; Altukhov *et al.*, 1983, 1987). The existence of such selection was demonstrated by comparing genotypic distributions, allele frequencies and levels of heterozygosity at five unlinked polymorphic loci in young and mature fish of natural and hatchery pink salmon populations from Sakhalin (see Chapter 5).

Selective fishing is likely to be a selective factor, as will be discussed in detail in Chapter 7. The main point here is that the interregional genetic differentiation in some cases can be traced fairly easily, notwithstanding considerable similarity and even uniformity of allele frequencies in representatives of the same generation in individual regions of the species range (section 6.1). The degree of difference (or similarity) varies in space and, hence, selection can retain its vector but differ in intensity, giving rise to numerous points of equilibrium located around the same peak of the adaptive landscape characteristic for a certain generation line.

Here, however, another point arises: how can one explain stable genetic differences at the same loci between odd- and even-year generations of pink salmon?

Without ruling out the possibility of equally significant differences in the environment of these two genetically isolated lines (e.g. density-dependent selection; see Altukhov *et al.*, 1987; Glubokovsky & Zhivotovsky, 1989), we believe, however, that a more probable explanation is the effect of genetic drift during the period of initial differentiation of generations (the founder effect). Genetic differences that turned out to be intrinsic to the initial populations were subsequently retained in reproductively isolated lines of even- and odd-year generations owing to their systemic organization (a balance of genetic drift and gene migration).

Apparently, a crucial point in solving the problem concerning contributions of different microevolutionary factors in the spatial genetic differentiation of pink salmon (or any other species) is the assessment of basic demographic parameters such as effective population size and coefficient of gene migrations.

Our estimates of these parameters were presented in Chapter 2 (section 2.5).

Although they are approximate ($N_e \cong 660$, $m \approx 0.032$), the level of local genetic differentiation expected in conditions of a selectively neutral process, $F_{ST\text{exp}} = 0.012$, does not differ significantly from $F_{ST\text{obs}} = 0.019$, empirically determined according to the aggregate of many polymorphic genes (see also Figs 5.7 & 6.13).

As noted above, when the sample of loci included in the analysis is sufficient, selective neutrality of the genetic process (average per locus) in pink salmon can also be determined by revealing the effects of isolation by distance for population groups linearly connected in geographic space (Fig. 6.14b). The prevalence of overdominant genes with minimal F_{ST} values in the sample makes this effect indistinguishable. For example, when F_{ST} values for the populations shown in Fig. 6.14b are calculated with respect to four loci characterized by the minimal variance of frequencies, the correlation between log N_em and log D is statistically insignificant (Fig. 6.14d), i.e. isolation by distance is not revealed.

Facts and observations contradicting the hypothesis of extensive gene flow in pink salmon are as follows:

(1) Well-manifested morphological and ecological differentiation of many populations studied (Vaughan, 1947; Gritsenko, 1981, 1990; Beacham, 1985; Ivankov, 1986; Beacham *et al.*, 1988; Gorshkov, 1988; Efanov, 1989; Ivankov *et al.*, 1996).
(2) Stable differences (observed in a series of generations) in the dynamics of biological and genetic parameters between artificially propagated and natural populations spawning in neighbouring regions (Altukhov *et al.*, 1989; see also section 7.3).
(3) Differences between geographically close populations with respect to parasites used as markers (Omelchenko & Vyalova, 1990; Omelchenko, 1994).
(4) Strict homing demonstrated in experiments with genetic marking (Lane *et al.*, 1990).
(5) Critical analysis of the data on fish tagging used by the authors of the 'fluctuating stocks' hypothesis, which showed that parameters of straying in pink salmon were strongly overestimated (Omelchenko & Vyalova, 1990); according to published data, average straying intensity in pink salmon is commensurate with that in other salmonid species (Altukhov & Salmenkova, 1994; Table 2.1). Moreover, the intensity of effective gene migration is lower than that of physical migration of individuals.
(6) The existence of significant genetic differentiation between regional populations and between neighbouring river populations (section 6.1).

Recent studies on modelling the dynamics of numerical size of pink salmon populations and their genetic structure, performed by the authors of the 'fluctuating stocks' hypothesis, showed that even a weak migration flow (from tenths of a percent to a few percent) provides for the existing relatively low level of genetic differentiation and alternation of dominance in abundance in consecutive generations (Zhivotovsky *et al.*, 1994, 1996). If the actual intensity of migration in this species is similarly low, the proposal concerning annual changes in boundaries, number and size of its local populations is groundless.

As we think, the concept of 'fluctuating stocks' is not only erroneous in scientific terms, but also may have dangerous practical consequences if accepted in salmon fisheries. Fortunately, practical fisheries are always based on the consideration of salmon stocks as stable and predictable objects of exploitation. Furthermore, the results of artificial propagation of pink salmon in hatcheries also contradict the practical conclusions of the concept of 'fluctuating stocks' (Khorevin, 1986, 1994; McKean, 1992; Mahnken *et al.*, 1998; Chapter 7).

Experience gained over more than a century of studies on fish ecology, demography, morphology and genetics has provided conclusive evidence that non-fluctuating populations do not exist in nature; however, any population has its own biological optimum determined by its natural history, evolution and current status in the ecosystem. The boundaries of such an optimum with respect to population size are determined by the minimal and maximal levels of the latter, which can be estimated reliably only by averaging the results of systematic observations performed for many years. If the steady-state zone of the system is unknown, its rational commercial management (exploitation and reproduction) is extremely difficult. Unfortunately, such systematic and extensive observations on many spawning populations of pink salmon are not performed. In this field, carefully planned investigations based on fundamental principles essential to the concept of population systems are necessary. As follows from the previous chapter, any population system of salmonids is nothing else but the community of *fluctuating stocks* and its basic property – *genetic stability in a series of generations* – is determined by variation (fluctuations) of components (subpopulations or stocks of different rank) comprising the structure of this system.

If corresponding facts and generalizations are ignored, it is indeed possible to arrive at an erroneous conclusion about the relative contributions of selection, migration and random genetic drift to the spatial genetic differentiation of a species. An even more striking example is the discussion between selectionists and neutralists about the possible number of overdominant genes and the related interpretation of uniformity in allele frequencies of allozyme loci, which is often observed within vast species ranges. It is possible to demonstrate many examples of such effects, and the situation with pink salmon is only one of them.

Estimates of effective population numbers from both the neoclassical and balanced schools (amounting to thousands, if not millions of individuals) show that the segregational load is very large even at insignificant selection coefficients (section 1.5.2). The major point here is that there are no such vast populations in nature: by virtue of subdivision they are composed of numerous subpopulations limited in effective and total size. In the course of their natural history these subpopulations may die out, become re-established through recolonization, or experience sharp fluctuations of numbers in the reproductive period. Under these conditions, and even for a species as a whole, the genetically effective number (total) is much lower than direct estimates would lead one to expect (Maruyama & Kimura, 1980).

There is also the problem of estimating heterozygosity levels where the total number of a species is usually taken into account without allowing for its subdivision into isolated subpopulations. This kind of approach may be feasible if the rate of migration between separate subdivisions of a species is known to be higher than the spontaneous rate of mutation. But in the event of these values becoming comparable (see, for example, Larson *et al.*, 1984), then estimates of the heterozygosity or effective number of alleles, anticipated by making allowance for their selective neutrality, should be made taking into account the numbers of isolated populations and not of a species as a whole. In these cases, which are not so rare in nature, the segregational load problem no longer becomes so acute. It is true that the absence of spatial differentiation is often explained away as migration in discussion between selectionists and neutralists who consider that literally one migrant per generation is sufficient to delete the genetic differences at a subpopulation level. The next logical step is taken from here and it is claimed that if migration is so effective, spatial differences in gene frequencies between separate samples only reflect the *heterogeneity of the environment* over an area of large panmictic populations. It should be pointed out, however, that this conclusion is valid only for Moran's model of population structure (Moran, 1959), in which the intensity of gene exchange increases in proportion to the number of subpopulations. The picture is different in a natural population system. In a one- or two-dimensional area having an 'island' or a 'stepping-stone' structure of gene migrations, their intensity is either independent of the number of subpopulations or inversely related to it. Of course, environmental heterogeneity must not be ignored, but nor should one forget that in nature there are not random mechanical accumulations of individuals which stochastically interact with each other, but real subpopulation units characterized by the unity of their behaviour and morphophysiological features. This emerges both from the results of our own research and from other works in which the biology of populations is given due attention (see Ehrlich & Holm, 1963; Lebedev, 1967; Petras, 1967; Madison, 1969; Selander, 1970; Shilov, 1977; Yablokov, 1987).

Relying on these data, one can indicate that in a wide variety of organisms, from invertebrates to humans, the N_e estimates (or 'neighbourhood' size) come within an interval of 10 to 10^4 individuals with clearly expressed left-handed asymmetry of distribution and a mode in the range of 10–200 individuals (Altukhov, 1990; see also Lande, 1979). Even allowing for the legitimately exaggerated figure that N_e comprises 75% of a population's reproductive numbers and approximately 30–40% of the total numbers, the population size that should be used in Equation 1.23 (Chapter 1) for determining the segregational load, at least insofar as many species are concerned, must be measured not in hundreds of thousands of individuals but by a considerably smaller figure. Clearly in this case the segregational load is sharply reduced while the number of polymorphic loci, maintained by heterozygote advantage, increases significantly. But as demonstrated previously (Chapter 2), effective population size, at least in Pacific salmon, is considerably smaller than 30% of the total.

Table 6.14 presents the results of corresponding determinations of the $e^{\bar{L}}$ value based on realistic estimates of N and coefficients of selection in favour of heterozygotes in a different number of diallelic loci whose effects on fitness are presumed to be multiplicative. Estimates have also been made for the equilibrium frequencies of alleles of hypothetical 'symmetrical' ($\hat{p}=0.5$) and 'asymmetrical' ($\hat{p}\neq 0.5$) selection (Altukhov & Dubrova, 1981). We see that when the task is formulated in this way, made possible by many years' field research experience, the best adapted individual's excess fertility is entirely 'meaningful' in the biological sense and, hence, the hypothesis of overdominance's important role as the support mechanism for the hereditary variability of populations cannot be refuted. This conjecture becomes even more valid if one realizes that the segregational load is markedly

Table 6.14 Excess fertility $e^{\bar{L}}$ depending on population size, number of overdominant loci and selection coefficients against homozygotes.

Number of overdominant loci	Population size, number of individuals							
	100	200	500	1000	5000	10 000	50 000	100 000
$s_1 = s_2 = 0.05; \hat{p}^* = 0.5$								
100	1.97	2.10	2.27	2.37	2.65	2.77	3.04	3.16
500	4.56	5.23	6.17	6.92	8.84	9.74	12.03	13.11
1000	8.56	10.38	13.11	15.43	21.81	25.02	33.73	39.07
5000	121.71	187.41	315.61	454.58	984.79	1339.23	2610.90	3422.47
$s_1 = 0.01; s_2 = 0.09; \hat{p} = 0.1$								
100	1.28	1.30	1.34	1.36	1.42	1.44	1.49	1.51
500	1.73	1.81	1.92	2.01	2.19	2.27	2.45	2.52
1000	2.17	2.32	2.52	2.68	3.03	3.19	3.55	3.71
5000	5.63	6.58	7.94	9.05	11.96	13.35	16.98	18.72
$s_1 = s_2 = 0.1; \hat{p} = 0.5$								
100	3.89	4.39	5.09	5.64	7.02	7.66	9.25	9.99
500	20.84	27.38	38.07	48.95	78.19	94.97	144.87	171.81
1000	73.31	107.86	171.91	238.25	475.70	626.24	1137.81	1449.47
5000	14 812	35 124	99 611	206 640	969 821	1 793 529	6 816 819	11 713 309
$s_1 = 0.01; s_2 = 0.19; \hat{p} = 0.05$								
100	1.29	1.32	1.36	1.39	1.45	1.45	1.53	1.55
500	1.78	1.87	2.00	2.09	2.29	2.37	2.57	2.66
1000	2.26	2.43	2.66	2.83	3.23	3.40	3.81	3.99
5000	6.20	7.30	8.90	10.23	13.72	15.42	19.88	22.03

*$\hat{p}$ is the equilibrium frequency of one of the alleles.

reduced due to asymmetry of selection effects, in other words, at equilibrium frequencies significantly differing from 0.5 (Table 6.14).

And so, taken all together, the material that has been examined makes it possible to suggest with reasonable justification that the uniformity of gene frequencies in areas of *natural* populations should be attributed to both balancing selection and the effects of gene migration. Evidently selection of this kind is particularly significant in the early stages of ontogenesis (section 5.3).

Selective pressure in favour of heterozygotes is the factor responsible for the situation where local populations that are reproductively isolated prove to be similar genetically, and they do not exhibit spatial differentiation. This may be illustrated by our data on pink salmon populations from the Sakhalin–Kuril region.

During several years, we studied 19 spawning groups (Fig. 6.8) and found that they were fairly similar in allele frequencies at virtually all allozyme loci studied; note that these frequencies differed markedly from 0.5 (Table 6.15). Similarity in gene frequencies was also observed for some loci while comparing many samples taken from known isolated populations throughout the period of the spawning run.

It is clear that the pattern of variation suggests the simplest conclusion that panmixia is the basic factor responsible for such weakly expressed genetic differentiation. We have shown, however, that this conclusion is erroneous. Indeed, after broadening the spectrum of allozyme loci and increasing the number of populations studied in different parts of the species range, researchers discovered more neutral loci and demonstrated that spatial differentiation with respect to the latter is significantly higher (Shaklee *et al.*, 1991; Varnavskaya & Beacham, 1992; Shaklee & Varnavskaya, 1994).

Moreover, studies on separate samples taken from an individual native population of pink salmon during the spawning run showed that temporal variation in gene frequencies at certain loci is often observed against the background of the absence of such variation at other loci. This was demonstrated using the data on a natural pink salmon population from the Bakhura River, obtained in 1985 (Figs. 6.17a,b; see also Altukhov, 1990, Fig. 5.3). Such temporal genetic variation, which correlated with biological parameters of fish from different samples (Fig. 6.17c) and was subsequently confirmed by analysing mtDNA in the same population (Brykov *et al.*, 1999a,b), provided conclusive evidence that in pink salmon, as in other salmonid species, there are not only regional and local ('river') levels of population hierarchy, but also the elementary level of spawning subpopulations reproducing in the same river basin. The existence of the subpopulation genetic structure in pink salmon of Alaska (although less distinct than in other species, e.g. sockeye), was also confirmed by the analysis of allozymes and DNA polymorphisms (Lane *et al.*, 1990; McGregor *et al.*, 1998; Seeb *et al.*, 1998).

In the next section we will present additional evidence that salmonid populations are not conglomerate stochastically fluctuating communities, but rather represent historically established systems with intrinsic regulation of the ratio of intra- and interpopulation components of gene diversity.

Table 6.15 The frequencies of common alleles of allozyme loci in pink salmon of the Sakhalin–Kuril region (odd-year generations).

Populations (rivers)	Allozyme loci																	
	*sMDH-B1,2**		*G3PDH**		*PGDH**		*PGM-2**		*mMEP-2**		*GPI-A**		*PEPD-1**		*PEPB**		*FDHG**	
	n	100	*n*	100	*n*	100	*n*	100	*n*	100	*n*	100	*n*	100	*n*	100	*n*	100
Bol'shoi Takoi	1474	0.9012	1477	0.9722	1378	0.8763	1477	0.9425	1223	0.8936	1585	0.9401	144	0.7570	144	0.9970	144	0.9760
Lesnaya	521	0.9080	520	0.9705	492	0.9000	521	0.9348	792	0.8580	592	0.9330	200	0.6550	200	0.9850	200	0.9650
Firsovka	1115	0.9147	1213	0.9701	1034	0.8861	1162	0.9329	1265	0.8890	495	0.9390	—	—	—	—	—	—
Poronay with tributaries	1358	0.9246	1329	0.9663	779	0.9025	1381	0.9375	673	0.8940	—	—	—	—	—	—	—	—
Dudinka	196	0.9209	186	0.9680	197	0.9035	200	0.9400	199	0.8945	—	—	—	—	-	—	—	—
Bakhura	447	0.8909	447	0.9575	439	0.8773	449	0.9178	698	0.8730	300	0.9463	20	0.6660	20	0.9800	100	0.9500
Lyutoga with tributaries	329	0.9360	329	0.9507	267	0.9104	329	0.9477	176	0.8542	—	—	—	—	—	—	—	—
Pokosnaya	285	0.8932	296	0.9633	258	0.8801	295	0.9238	100	0.8950	—	—	—	—	—	—	—	—
Chernaya	277	0.9133	274	0.9524	164	0.9061	277	0.9518	99	0.8990	150	0.9800	—	—	—	—	—	—
Ainka	80	0.8750	80	0.9500	79	0.8671	80	0.9063	—	—	—	—	—	—	—	—	—	—
Kurilka with tributaries	1276	0.9317	1275	0.9797	1118	0.8785	1277	0.9265	595	0.8970	500	0.9300	—	—	—	—	—	—
Reidovaya	86	0.9477	86	0.9593	90	0.8555	86	0.9302	197	0.9110	—	—	—	—	—	—	—	—
Olya	99	0.9394	100	0.9750	84	0.9048	99	0.9798	100	0.9400	100	0.9350	—	—	—	—	—	—
Novoselovka	297	0.8970	295	0.9540	296	0.9210	297	0.9390	197	0.8660	199	0.9730	—	—	—	—	—	—

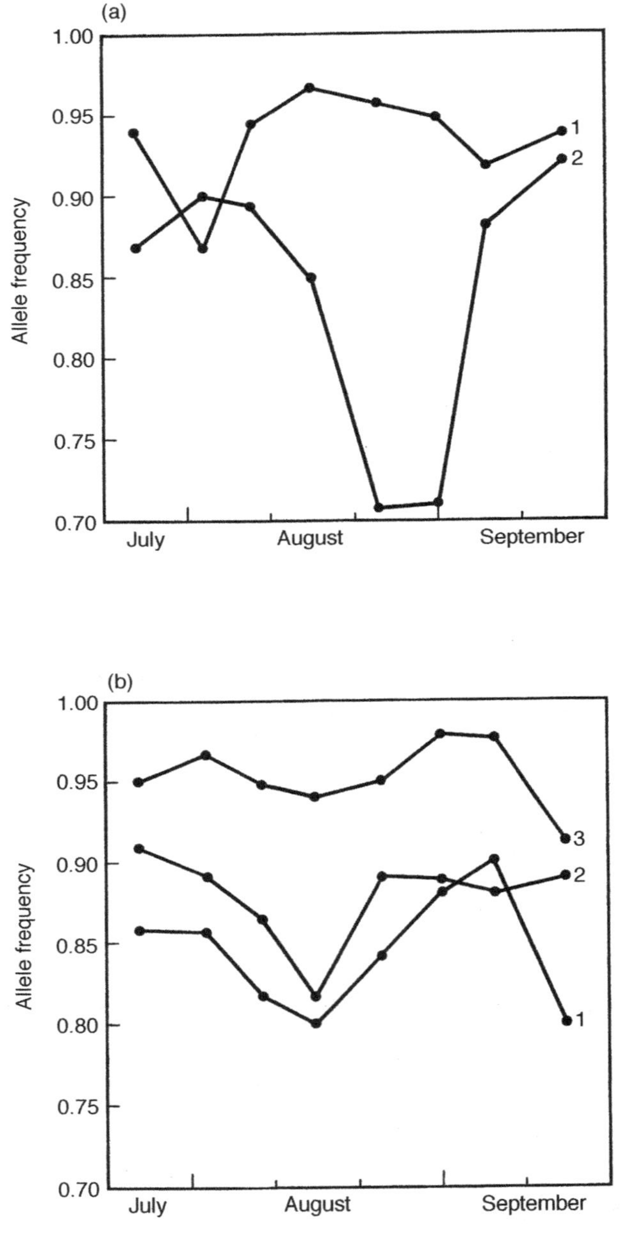

Fig. 6.17 (a, b) Dynamics of allele frequencies of some loci, (c) proportion of males (continuous line) and average body length of females (dotted line) in samples of pink salmon during spawning migration into the Bakhura River (Sakhalin) in 1985. (a) 1 – *PGM-2** (heterogeneity test $\chi^2 = 11.3$; $P > 0.4$), 2 – *PGDH** ($\chi^2 = 63.5$; $P < 0.001$); (b) 1 – *MDH-B1,2** ($\chi^2 = 10.05$; $P > 0.7$), 2 – *mMEP-2** ($\chi^2 = 4.4$; $P > 0.7$), 3 – *G3PDH** ($\chi^2 = 6.04$; $P > 0.5$). (c) Male proportions in samples are heterogeneous, $\chi^2 = 50.0$; d.f. $= 7$; $P < 0.001$. Average body length of females are significantly different among samples: $F = 4.70$, d.f.$_1 = 165$, d.f.$_2 = 7$, $P < 0.05$.

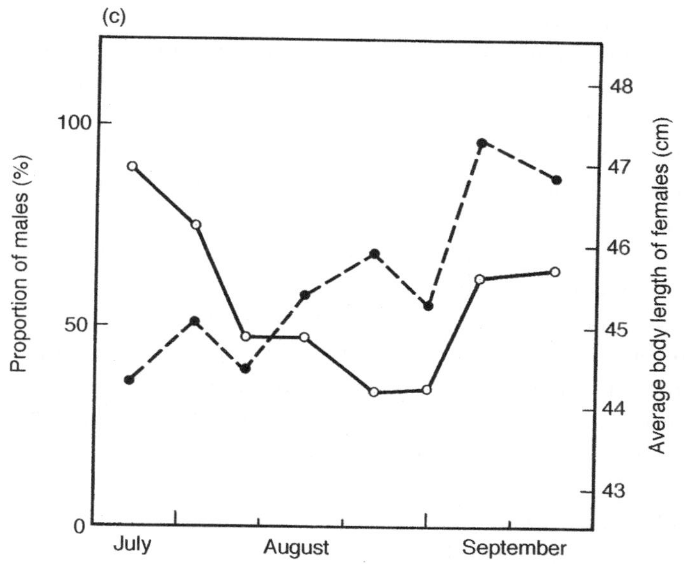

Fig. 6.17 Cont.

6.4 Intraspecific genetic differentiation as a process with 'memory'

In theoretical population genetics, intraspecific differentiation is usually interpreted as a Markov's chain in which the evolution of a population can be predicted for no more than one generation, and protein polymorphism is regarded as a transient phase of molecular evolution (for example, see Kimura, 1983; Lewontin, 1993; Chapter 5).

However, the data discussed in Chapter 5 and section 6.2 contradict this opinion: owing to its systemic organization, the subdivided population maintains the stable level of total gene diversity inherited from the ancestral population. This is accounted for by a reciprocal balance between factors of microevolution. However, stochastic regulation is not the only mechanism responsible for the genetic stability of subdivided populations. There is evidence that gene immigration within a population system of the island type is not random but subject to negative feedback regulation: when local subpopulations comprising this system decrease in size, they begin to receive increasing gene flow from the system core (or from the common gene pool performing the same function). Therefore, the core of the system ('continent') operates as its regulatory component, and the peripheral local populations ('islands') are the component responding to regulation (Altukhov, 1990).

This non-random structure of gene migrations was revealed for the first time in studies on genetic processes in an experimentally subdivided *Drosophila melanogaster* population conforming to Sewall Wright's island model. The results showed that the smaller the size of 'island' subpopulations, the more intense was gene immigration from the 'continent', and vice versa (for details, see Altukhov, 1983a, 1990). The same phenomenon was subsequently observed in natural populations of other species (Altukhov & Salmenkova, 1994).

For example, it was shown that in the circumpolar system of indigenous human populations of northern Eurasia, $N_e = 45$ and $m = 0.034$, whereas in the Siberian isolate as a whole, $N_e = 200$ and $m = 0.01$ (Rychkov & Sheremetyeva, 1976). In the butterfly *Euphydryas editha*, the intensity of dispersal was very high (up to 100%) when their colony comprised about 100 individuals but decreased to only 0–7.3% when colony size increased to approximately 1500 individuals (Gilbert & Singer, 1973; cited from Yablokov, 1987). In chinook salmon, straying intensity and the number of fish returning to the home river for spawning proved to be inversely correlated with one another (Quinn & Fresh, 1984). Likewise, observations on marked pink salmon showed that straying intensity is high in the years of decrease in population size, and decreases when the returning generations of spawners are abundant (Zolotareva, 1980).

Recently new evidence has been obtained concerning statistically significant negative correlations between effective size and immigration rate for rural human populations (Evsyukov *et al.*, 1999). These new findings are in strong support of our conclusion about the autoregulation of gene diversity levels within native population systems.

Such self-regulation consists of *the maintenance of a stable (optimal) ratio between the intra- and interpopulation components of gene diversity in the subdivided population*, i.e. the state of equilibrium *between inbreeding and outbreeding*. Effects resulting from disturbances in this equilibrium will be considered later, when discussing the consequences of anthropogenic impacts on salmon stocks (Chapter 7). Here, the following should be noted: the data cited above and published earlier (Altukhov, 1983a, 1989a, 1990) allow us to interpret the genetic divergence of a species as a branching process involving reorganization of the gene pool of the ancestral population during its differentiation into subordinate groups in a series of generations and over its range. Under normally fluctuating environmental conditions, the average gene frequency and, after approaching the steady state, variance in the system remains unchanged by mutually counterbalancing factors of microevolution. In other words, *an isolated population either disappears in the course of its evolution or expands 'into itself', maintaining a dynamic balance with the environment.*

Therefore, when studying a population system, we deal with a population superorganism capable of *retaining gene diversity as 'memory' of the preceding stages of its historical development* for dozens, hundreds, and thousands of generations (Rychkov, 1969, 1973; Altukhov & Rychkov, 1970; Altukhov, 1974; Altukhov & Kalabushkin, 1974; Altukhov & Pobedonostseva, 1978, 1979a, 1979b).

Thus, intraspecific genetic differentiation in the naturally fluctuating environment is, on average, a selectively neutral process with 'memory', rather than a

Markov's chain. This 'memory', which retains information about the genetic composition of the ancestral population and the optimal ratio of homo- and heterozygous genotypes, is maintained owing to the ordering effect of gene immigration.

As shown in studies on native human populations (Rychkov, 1969, 1984; Rychkov & Movsesyan, 1972; Rychkov & Sheremetyeva, 1977; Sheremetyeva & Rychkov, 1978; Rychkov *et al.*, 1982; Rychkov & Yashchuk, 1986), the depth of genetic 'memory' provides a unique opportunity of reconstructing its *structure* and *volume*.

The *structure of 'memory'* is associated with the constancy of total gene diversity owing to the constancy of gene frequencies in a population system, which can be reconstructed by averaging the corresponding data on all the present-day populations. The *volume of 'memory'* is determined by the accumulation of interpopulation gene diversity during the formation of the population system structure in time and space. It is possible to analyse such characters using Wright's F_{ST} statistics.

The possibility of performing reconstructions of the first type (the structure of genetic memory) has been demonstrated repeatedly for population systems of different species, including humans (Serebrovsky, 1935; Rychkov, 1969; Altukhov & Rychkov, 1970; Rychkov & Movsesyan, 1972; Altukhov & Kalabushkin, 1974; Altukhov & Pobedonostseva, 1978; see also Chapter 4 and section 6.2). As to reconstructions of the second type (the volume of 'memory'), such studies were, until recently, restricted to human populations. This is certainly not accidental, as the history of human populations is the domain shared by many sciences, and its chronology may be reconstructed by methods independent of genetics. It is noteworthy that the genetic chronology of historical events as the method for determining the age of populations consistently gives estimates agreeing with the data on radiocarbon dating of ancient campsites or other evidence provided by historical chronicles (Rychkov, 1984, 1986).

Moreover, the dynamics of accumulation of interpopulation gene diversity demonstrated the previously unknown trend indicating the equivalence of its volumes at different stages of population structure formation. This phenomenon was named the *genetic equidistance of stages of ethnogenesis* (Rychkov & Yashchuk, 1980, 1985). The analysis of data on frequencies of many genes in indigenous population systems of Europe, northern Asia and America showed that the same value of gene differentiation coefficient, expressed in G_{ST} units, is characteristic for every level of population hierarchy, irrespective of the historical time of its establishment. Note that this principle can only be revealed if the total genetic differentiation of the system (G_{STi-t}) is subdivided into regional components (G_{STi}) on the basis of ethnic classifications that are independent of genetics but, at the same time, reveal the actual hierarchical levels of unity, historically developed in the process of microevolution of a subdivided population (e.g. linguistic, ethnoconfessional, etc.). The equivalence of F_{ST} (G_{ST}) values at different levels of the system hierarchy means the existence of a certain limit in the volume of total gene diversity, accumulated by the system in the course of its historical development. Above this limit, a new stage of microevolution of the system begins.

In studies on other species, it is difficult to treat the initial material in the same way because similarly reliable phyletic classifications are usually absent. However, it is sometimes possible to recognize natural–historical features of species range formation and to identify the corresponding 'levels of antiquity' (i.e. stages in the development of population structure). In such cases, the results can be comparable to those obtained for human populations. Studies on the sockeye (*Oncorhynchus nerka*), as a species with a complex subpopulation structure, provide an illustrative example (Table 6.5). G_{ST} estimates are based on our own results and published data on allele frequencies of several allozyme loci in 54 samples from native populations spawning in the basins of 20 rivers of northern Asia and North America (see Fig. 2.7). These data demonstrate clearly that the degree of local genetic differentiation remains the same at any of the three hierarchical levels specified in Table 6.5.

In section 6.1, we already noted that the same regularity was observed for another Pacific salmon, chinook (*O. tschawytscha*) from Alaska, in which estimates of the interpopulation component of gene diversity at the levels of regional population groups and individual river populations within these regions were very similar ($G_{ST}=0.026$ and 0.033, respectively, as estimated from the data obtained by Gharrett *et al.*, 1987). In Atlantic salmon (*S. salar*) from north-eastern Europe, genetic differentiation between population complexes of individual sea basins and between populations within these complexes was characterized by similar values (Table 6.10). In chum salmon (*O. keta*) from the northern regions of the Asian and American ranges, almost equal G_{ST} values were obtained at the levels of regional population groups and river populations within regions (Table 6.2).

The analysis of published data shows that the same trend is also characteristic of other biological species whose spatial differentiation was studied with respect to many allozyme genes simultaneously. Thus, in the barramundi perch (*Lates calcarifer*) inhabiting coastal waters of Queensland (Australia), G_{ST} values for different hierarchical levels were quite similar: 0.008 for localities within subregions, 0.012 for subregions within regions, and 0.009 for regions within Queensland (calculated from the data obtained by Shaklee & Salini, 1993). In the ant (*Myrmica rubra*), the values of F_{ST} estimated between the sites within localities and between localities were 0.205 and 0.199, respectively (Seppa & Pamilo, 1995). The same trend is observed for the rodent prairie dog (*Cynomys ludovicianus*), if we consider differentiation of subpopulations (wards) within two populations studied in greater detail and designated CAPU and PORT by Chesser (1983): in the first case, $F_{ST}=0.0554$; in the second, $F_{ST}=0.0541$. F_{ST} values for the Atlantic–Mediterranean and Indian–Pacific groups of green turtle (*Chelonia mydas*) populations, estimated from haplotype frequencies revealed by RFLP analysis of nuclear DNA, are also identical: $F_{ST}=0.13$ in both cases (Karl *et al.*, 1992).

An analysis of the literature from this point of view shows that similar examples are numerous, but, unfortunately, no known publications contain data allowing us to make a comprehensive and independent assessment of the hierarchical classification of populations without genetic data, and to identify the actual levels of unity established in the process of species microevolution. This problem is likely to attract the attention of researchers, and the required data will probably appear in scientific publications.

In this context, we should emphasize an important property of Wright's F_{ST} coefficient. Being a measure of local genetic differentiation and, simultaneously, a parameter of structural inbreeding in a subdivided population, this parameter has great biological significance: it reflects the ratio between homo- and heterozygous genotypes in a population system and, hence, a balance between inbreeding and outbreeding (Chapter 5). In general, at the level of population systems, the stable ratio between intra- and intergroup components of gene diversity provides evidence that the processes of differentiation and integration of the species gene pool counterbalance one another. This balance can only be revealed in specifically planned studies on natural population systems that have not yet been completely destroyed by anthropogenic impact and still can reproduce their gene pools at a natural rate. In Chapter 7, we will discuss the consequences associated with different types of balance disturbances resulting from commercial fishing and artificial reproduction of salmon populations.

Now it is necessary to consider the cause for the similarity of the F_{ST} (G_{ST}) values at different hierarchical levels in the population systems of different biological species on the one hand, and of human populations on the other. The question is whether this similarity is purely superficial or is based on a certain universal regulatory mechanism?

Rychkov & Yashchuk (1985) studied evolution of the indigenous human population system in Siberia and concluded that such a mechanism actually exists: its role is played by gene migration that decreases with time as each new hierarchical level is formed in the system. They argued that the increasing numerical size of the population system is associated only with an increase in the number of subpopulations, whereas the average effective size of a local population remains unchanged throughout the history of differentiation of the vast Siberian isolate. A comparison of corresponding F_{ST} values – empirical, calculated for a large set of loci ($\bar{F}_0 = 0.0810$), and expected, for a selectively neutral microevolutionary process ($F_e = 0.0974$) – allowed the authors to estimate the time during which the system has approached the existing level of gene diversity (according to the model of its exponential growth):

$$\bar{F}_0 = F_e(1 \quad e^{\quad t/2N_e})$$

The result was 19 500 years (approximately 780 generations) (Rychkov & Yashchuk, 1985), which agrees well with archaeological data on radiocarbon dating and recent estimates based on mtDNA polymorphism (Torroni *et al.*, 1993).

Therefore, the equality in proportions of the total F_{ST} (G_{ST}) at different hierarchical levels in subdivided human populations reflects a continuous process of genetic microdifferentiation directed toward a certain limit, which gradually recedes during the historical development of the human population as a system (Rychkov & Yashchuk, 1985). As for zoological species, they probably achieved an ecological–genetic balance thousands of years ago (Altukhov, 1983a). Since then, the structure of intra- and interpopulation gene diversity has been maintained at an optimal level (reflecting the maximal adaptation of the species to the historically established conditions of its natural reproduction), owing to self-regulation of the parameter N_em (Altukhov, 1989a). It is interesting that such regulation is apparently found even in plant species pollinated by insects or birds. In any case, the term *optimal outcrossing distance*, which reflects a higher fitness of the progeny from a certain combination of parents in the herbaceous plant *Ipomopsis aggregata*, has long been successfully used in botanical literature (for details, see Waser & Price, 1989). There is more direct evidence that in plants, as in animals, the spatial genetic differentiation at comparable levels of natural hierarchy is virtually identical. Thus, studies on the same perennial plant, *I. aggregata*, showed that, at the level of its geographic races (subspecies), N_em values estimated from G_{ST} statistics for 18 allozyme loci are similar (if not identical): *I. aggregata* – 2.62; *I. a. formosissima* – 2.46; *I. a. attenuata* – 2.44; etc. (the mean for eight subspecies is 2.49) (Wolf & Soltis, 1992).

Our assumption that genetic processes in natural populations, in contrast to human populations, are of the steady-state type is supported by one more fact: geochronologically dating via parameters of the subpopulation structure in wild species is ineffective. The age of the population estimated by the above equation is always less than the actual value. Thus, the age of a subdivided sockeye salmon population from Lake Azabachye (Kamchatka) was estimated at approximately 200 generations or 900–1000 years. However, this lake, as a water reservoir with virtually the same characteristics as today, was formed at least 5000–7000 years ago, and fish populated it during approximately the same period (Altukhov, 1983a). The 'age' determined in this case most probably corresponds only to the period during which the population system approaches genetic equilibrium. It also suggests that the natural populations occupied and assimilated their modern ranges in a short period of time. Thus, in addition to a decrease in the extent of gene migration with time, our model postulates that both the number of subpopulations and their effective size increase simultaneously until the system attains a stationary reproductive state. However, we cannot exclude the possibility that, during the exponential growth of the size of a natural population system, the mean N_e value for the constituent local subpopulations remains constant, as it does in human populations. *In any case, when the hierarchy of structural levels in the system is finally established, the subdivided population exists as if it was beyond the time scale of natural history*. In such conditions, the only way to estimate the age of this population (of course, roughly) is to apply the concept of the molecular clock and the theory of genetic distances, as undertaken for chum salmon populations of the Russian far east (section 6.2).

Conclusion

By means of gene diversity analysis, average regional estimates of interpopulation divergence (F_{ST}, G_{ST}) were detected for individual salmon species in the absence of excessive anthropogenic pressure. Similar and distinctive features of many populations on a macrogeographical scale have been revealed, and the role of random (genetic drift) and systematic (natural selection and migration) factors has been shown to influence the spatial distribution of gene frequencies. In several cases, we traced evident relations between the genetic characteristics of populations and the natural history of a species range, and by clustering the populations evaluated the time of their divergence. In spite of large interspecific differences in gene diversity components, it has been possible to identify clear patterns allowing the interpretation of intraspecific genetic differentiation not as a Markov's chain, but as a process with 'memory'. This permitted us, at least qualitatively, to develop a model of this process and to argue it as far as is possible today, taking into account the fact that a natural hierarchical classification of a species population structure, independent of genetic data, is absent in salmonid fishes.

Although this model needs further proof, it still leads to the following important conclusion: because the F_{ST} (G_{ST}) coefficient for each hierarchical level of the population structure is constant, it can be regarded as the criterion of the norm (optimum). Hence, *the normal genetic process can be defined as the type of reproduction of species gene pools when the ratio of intra- and interpopulation components of gene diversity is retained at the evolutionarily established optimal level specific for each individual species* (Altukhov, 1995). It is clear that exactly this ratio should be used as the basic reference point in the *genetic monitoring* of population systems exposed to anthropogenic pressure.

This approach permits us to estimate the degree and the type of deviations from the optimum in the regional gene pool and, on this basis, to predict and prevent negative consequences of such deviations. In other words, when organizing programmes of population monitoring, we should take into account the unique possibility of 'overcoming time through space' by exchanging long-term surveillance over the genetic structure of the same group of populations for an analysis of samples taken uniformly over the whole species range or separate geographical region with due regard to its natural history. Such an approach will be considered in the last chapter of the book.

Chapter 7
Consequences of Anthropogenic Impacts on Salmonid Populations: Genetic and Biological Monitoring

Before considering salmonid populations, we should remember that human activities during the twentieth century have brought to extinction thousands of species of higher plants and animals. Currently, hundreds of unique breeds of domestic animals may share the same fate (IUCNNR, 1980). According to expert estimations, the biosphere may lose up to 10–15% of its constituent species in the coming decades (Wilson, 1988; Sokolov & Shatunovsky, 1995). The rate of extinction under anthropogenic pressure exceeds any such loss known from palaeontological records. The only way of minimizing the effects of (more precisely, of mitigating) such an ecological catastrophe is to revise and radically change the entire strategy of human–nature interactions. Making efforts in this direction, however, it is necessary to realize that adverse factors responsible for the increasing reduction of biological diversity are not limited to environmental pollution and destruction of habitats. One more factor, although less obvious, is the incompetent approach to commercial exploitation of natural resources, disregarding the genetic subdivision of species and the structure of intraspecific hereditary variation (Altukhov, 1994b, 1995; Altukhov *et al.*, 1996).

All these factors have strongly affected salmonid fish resources. Thus, Kazakov (1990) and Andriyasheva (1996) noted that over the period since World War II, registered catches of Atlantic salmon in Russia have decreased fourfold, and populations of the rivers flowing into the Gulf of Finland, the White Sea (Karelian coast), and lakes Onega and Ladoga have lost their commercial significance; catches in the White Sea decreased by a factor of 5–6. Freshwater brown trout from the Caspian and Black sea basins are on the verge of extinction. Atlantic salmon populations inhabiting European water bodies and the Atlantic coast of the United States and Canada have been seriously depleted. Many populations disappeared or are endangered (Hesthagen & Hansen, 1991; for reviews, see Bielak, 1992; Weaver & Stolte, 1992).

The natural populations of Pacific salmon also have been reduced, although less drastically overall. This follows from a decrease in the abundance of spawners in spawning grounds of the Far East and other regions and even a complete disappearance of these fish from some rivers (Petrenko, 1983, cited from Andriyasheva, 1996; Nehlsen *et al.*, 1991; Riddell, 1993).

Negative consequences are associated with both commercial exploitation of biological resources and their artificial propagation, and the situation in the fishery is a typical example (Altukhov, 1995). As the corresponding adverse phenomena can only be revealed by monitoring the intraspecific genetic diversity, the problem of monitoring will be considered specifically in this chapter.

It should be noted here that, although the term 'monitoring' has been used widely in the international genetic literature, principles of corresponding studies have not yet been developed adequately. In particular, this concerns the judicious choice of the *reference point* necessary for evaluating the observed changes and understanding their significance. The purpose of this chapter is to specify the key problem of genetic monitoring, to review theoretical approaches to its solution, and to give examples of monitoring salmonid populations exposed to various anthropogenic influences. A forecast concerning the negative consequences of anthropogenic impact and the practical measures to prevent them will be described in the final section.

7.1 The problem of genetic monitoring and theoretical approaches to its solution

Genetic monitoring is *long-term surveillance over the state of population gene pools, intended to evaluate and predict their temporal and spatial dynamics and to determine the limits of permissible changes.* However, any prediction should be necessarily based on the concept of the *norm*, i.e. *normal state* or *normal process* (see section 6.4). Such an approach is a prerequisite to finding the necessary reference point, i.e. the baseline against which change can be estimated, and understanding the mechanisms of adverse anthropogenic influences on populations, species and entire ecosystems.

According to the theory of population genetics (Chapter 1), there are four basic processes that, separately or interacting with one another, determine evolutionary changes in populations:

(1) Random genetic drift
(2) Gene flow
(3) Gene mutations
(4) Natural selection.

Genetic drift resulting from a small N_e leads to a decrease in heterozygosity and, as a result, to inbreeding (F). The latter increases in a series of generations in the proportion $(1/2N_e)^t$, where t is the number of generations since the initial time t_0

when heterozygosity was H_0: $F_t = 1 - (1 - 1/2N_e)^t$. Correspondingly, $H_t = H_0\,(1 - 1/2N_e)^t$ (Chapter 1).

These simple equations show that the intensity of genetic drift, leading to the reduction of gene diversity and, most often, to degradation of populations, is inversely proportional to N_e: the lower the N_e value, the more intense the genetic drift, and vice versa. Such a relationship is critically important for animal and plant breeding, and managers concerned with manipulating the process of reproduction take special measures to prevent the negative consequences of closely related crossings (Falconer, 1960). In natural (native) populations, the effects of random drift are compensated by *gene flow* of a certain intensity (m). In other words, natural populations are usually non-homogeneous and represent historically formed *systems of subpopulations* simultaneously affected by both random drift and migration of genes, the processes that counterbalance one another (Chapters 4 to 6). Thus, under normal conditions, the negative effects of inbreeding are eliminated.

According to S. Wright, this balance for the island model of a subdivided population is estimated from the level of structural inbreeding of a subpopulation S relative to the entire subdivided population T, i.e. by the coefficient F_{ST} (section 1.6). The coefficient, which shows the local genetic differentiation expected in conditions of a stationary, selectively neutral process, can be designated F_e and compared with the actual standardized genetic variance F_0 estimated from allelic frequencies of different polymorphic loci (Chapters 4 to 6).

If a locus is selectively neutral, $F_0 \approx F_e$. This conclusion has considerable importance for population monitoring and evaluating the state of regional gene pools (Altukhov, 1995; Chapter 6). As will be shown below, any detectable anthropogenic influences disturbing the population structure of a species may lead to rapid changes in the ratio between different gene groups in the gene pool, and the empirical F_{ST} statistics become different from those expected for a selectively neutral process.

For example, anthropogenic influences leading to significant changes in the intensity or direction of gene migrations or in the N_e value can upset this evolutionarily established balance, which is characteristic for the native self-reproducing population systems. Thus, damage of spawning grounds results in the reduction and fragmentation of the reproductive range, usually followed by a drastic decrease in the effective population size and the rate of gene exchange. In such cases, an increase in interpopulation gene diversity (G_{ST}) and a decrease in intrapopulation gene diversity (H_S) should be expected. By contrast, excessive mixing and interaction of previously isolated gene pools may result in manifestations of outbreeding, i.e. in decreased viability of hybrid combinations. Such effects are particularly marked with respect to gene complexes that form the basis of the adaptive genetic structure of species, and are linked with adaptive morphophysiological traits and properties (Altukhov, 1984; Templeton, 1986).

Here, natural selection, which is usually regarded as the most systematic and important factor in the evolution of populations, again becomes prominent.

Viability of populations depends directly on the adaptive values (fitness) of genotypes. If the fitness of heterozygotes is increased (overdominance), each population has a cost to its adaptation to the particular environment by segregation of less viable homozygotes (the so-called segregational genetic load; Chapter 1). However, the cost of adaptation normally proves to be 'acceptable', because the ratio between homo- and heterozygotes is autoregulated via the non-random structure of gene migration, and is maintained at a stable level (section 6.4). Moreover, this level is *optimal* in the sense that *both decreased and excessively increased heterozygosity is unfavourable for normal functioning of a population* (Chapter 5; Altukhov, 1990, 1991).

The concept of optimal gene diversity as a prerequisite for successful existence of populations in a normally varying environment is particularly important in view of their systemic organization. Knowing *the ratio between intra- and interpopulation components of heritable variation under the conditions of normal reproduction or immediately before a certain anthropogenic influence, we have the unique possibility to plan an effective genetic monitoring of a population system, taking into account its gene pool structure*. To obtain reliable information, however, the following requirements should be met.

(1) The population should be well defined in terms of natural history and geography.
(2) It is necessary to describe the distribution of the subpopulation structure of the system in time and space, because these two factors are closely interrelated in the stationary genetic process (the ergodicity principle; see Altukhov, 1974).
(3) The material should be sampled in such a way as to characterize comprehensively the subpopulation structure by a complex of traits selected in advance.

Such traits should be selected taking into account the possibility of combining the methods and approaches of *population* and *quantitative* genetics, i.e. of analysing both mono- and polygenic traits. It became obvious during the last 10–15 years that the two types of hereditary variability – quantitative and qualitative – should not be opposed, but should be considered as 'two sides of one coin', within a common polygenic system of universal adaptive value with a particularly sensitive response to diverse external factors, including fishing pressure (Altukhov, 1983a; Zhivotovsky, 1984; Chapter 5). *In suboptimal environments or under sufficiently strong artificial selection, non-random correlations between individual heterozygosity for a number of loci and the values of adaptive quantitative traits are observed* (Altukhov, 1990, 1991; Chapter 5).

Thus, in addition to properly organized sampling (even distribution of samples in space and time), the analysis should also involve:

(1) Data on the demographic structure of each population, including sex, age, sex ratio, size, migrations, etc.

(2) Data on body weight, size, and metric proportions of individuals under study.
(3) Evaluation of individual genotypes for the maximum possible number of polymorphic gene loci, including various protein systems and, if necessary, systems of DNA polymorphism (both nuclear and non-nuclear).

This information makes it possible to analyse the distribution of poly- and monogenic traits and their joint variability, to estimate the ratio between components of gene diversity and understand the state of the genetic process in a certain population, and to estimate the contribution of random genetic drift, gene migration and selection, either separately or in combination (Chapters 4 to 6).

We omitted only one factor: mutation pressure. As the probability of mutation for individual genes is negligible (10^{-5}–10^{-6} per gene per generation), the effect of this factor may be disregarded under normal conditions. At the same time, pollution of the environment with radionuclides and chemical mutagens makes monitoring of the mutation load an important problem, especially with respect to human populations (Altukhov, 1981b, 1983a; Altukhov & Kurbatova, 1984; Altukhov *et al.*, 1985). However, this problem is beyond the scope of this book. Let us consider the results of applying the above principles of monitoring to salmon population systems. In recent years, several authors have used a similar approach in studies of different species of salmon and other fish (Kazakov & Titov, 1992; for reviews, see also Ryman & Utter, 1987; Hindar *et al.*, 1991; Carvalho, 1993; Carvalho & Hauser, 1995). One of the main findings in these monitoring programmes is *the loss of genetic diversity in artificially maintained populations, compared to native populations, revealed by means of biochemical markers* (Allendorf & Utter, 1979; Allendorf & Phelps, 1980; Ryman & Ståhl, 1980, 1981; Ståhl, 1983, 1987; Vuorinen, 1984; Krieg & Guyomard, 1985; Kazakov & Titov, 1992). Another important discovery concerns the negative effects of the *increase in intrapopulation gene diversity in the process of selective fishing and gene pool transplantation* from one part of the species range to another (Altukhov, 1974, 1981a, 1990; Altukhov & Salmenkova, 1981, 1987a, 1990, 1991).

7.2 Commercial fishing

7.2.1 General information

According to Food and Agriculture Organization (FAO) fisheries statistics, the annual catch of Pacific salmons averaged 0.72 million tons during 1981–93 (Kaeriyama, 1998). From data presented at the First Symposium of the North Pacific Anadromous Fish Commission (held in Sapporo, 1996),

> 'the present worldwide catch of salmon amounted to 1.5 million metric tons annually, hatchery and wild run fish combined. The North Pacific Rim catch amounted to almost 1.0 million metric tons of the total. Five salmon species make

up the bulk of the catch: pink – 34%, chum – 30%, sockeye – 27%, coho – 5% and chinook – 3%' (Low, 1998, p. 511).

It is interesting that this catch exceeded the previous historical high in the late 1930s and is thought to have coincided with favourable oceanic conditions for survival of salmon and industrial large-scale artificial propagation (Kaeriyama, 1998). The total trend of almost 80-year dynamics of Pacific salmon catches does correlate with global climatic changes (Beamish, 1995; Noakes *et al.*, 1998), and after an increase of catches in the 1990s, a decline is predicted (Chigirinsky, 1993).

The retrospective analysis of statistical data on commercial fishing in Russia, the United States, Canada and Japan – countries where the Pacific salmon species reproduce naturally in rivers – shows that catches of these fish vary considerably.

Before World War II and until the 1950s, the average annual catches in the Russian far east amounted to 167 200 tons. From the early 1960s to the mid-1970s, catches decreased sharply, averaging 34 800 tons per year. Since the mid-1970s, annual catches tended to increase, mostly on account of odd-year pink salmon generations, and during the 1990s varied in the range of 120 000–250 000 tons and even reached pre-war values in some years (Kazarnovsky, 1989a; Radchenko, 1998).

Throughout the history of commercial fishing, the bulk of salmon catches in eastern and north-eastern Russia consisted of chum and pink salmon; the former were abundant until the late 1960s but, since the early 1970s, pink salmon has become the main commercial species. Sockeye, chinook and coho salmon are of secondary commercial importance, although catches of the first have been increasing gradually since the early 1980s. Masu salmon catches are so small that official statistics on this species are absent in Russia. More than 80% of the total salmon catch in Russia is accounted for by the Kamchatka and Sakhalin–Kuril fishing regions (Kazarnovsky, 1989a; Radchenko, 1998).

In the United States, the annual catch of Pacific salmon is greater than in Russia, and at present, amounts to more than 400 000 metric tons (Low, 1998). Its structure is characterized by high proportions of highly valuable species: sockeye (about 30%), coho and chinook (slightly less than 20%) and chum salmon (about 15%); the rest is accounted for by pink salmon. The leading role in salmon fishing belongs to Alaska, followed by Washington, Oregon and California (Kazarnovsky, 1986b).

Salmon catches in Canada (in the post-war period) remain below 100 000 metric tons per year. The main commercial species is pink salmon, which accounts for 35% of the total catch, followed by sockeye (28%), chum (18%), coho (12%) and chinook (7%) salmon. Most Pacific salmon in Canada spawn in rivers of British Columbia, making it the main fishing region of the country (Kazarnovsky, 1989c).

In Japan, salmon fishing has long-established traditions. From the beginning of this century until World War II, the Japanese fishing fleet had been operating in all the coastal regions of the Far East. In some years, up to 90% of fishing grounds near Primorye, Sakhalin, the Sea of Okhotsk coast and Kamchatka were exploited exclusively by Japanese fishermen.

Annual salmon catches in this period exceeded 200 000 metric tons, reaching a peak of more than 529 000 tons in 1939. After World War II, Japan lost control over the main fishing grounds, and catches dropped sharply. However, after signing a peace treaty in San Francisco and entering into the Japanese–American–Canadian agreement on fishing, Japan regained the right of fishing in the open Pacific and, by 1955, developed its mobile salmon fishing industry. Since that time, the average annual catch of salmon in this country has never descended below that of 100 000 tons (Kazarnovsky, 1987). It ranged from 100 000 to 160 000 tons in 1965–82, increased to 170 000 tons after 1983, and exceeded 300 000 tons in 1995 (Hiroi, 1998). Since the early 1970s, catches consist mainly of chum salmon.

The average contributions of countries to the present North Pacific Rim catch of salmon, by weight, are: United States 48%, Japan 27%, Russia 20%, and Canada 5% (Low, 1998).

What are the effects that the pressure of commercial fishing produces on salmon stocks?

7.2.2 *Biological and genetic effects of fishing*

As to biological effects, Ricker (1981) investigated them over at least a 25-year period, from 1950 to 1975. Here we use two examples from his work. As the first illustration, we shall present data for pink salmon. Its populations are represented only by non-overlapping generations of even and odd years (Fig. 7.1). The second example will be data for chinook salmon, a species with a more complex age structure (Fig. 7.2). We can see clearly that the decline in the average body weight (length) in these species, fished with gill nets (pink salmon) and trolls (chinook salmon) along the coasts of British Columbia, appears in just six to twelve generations.

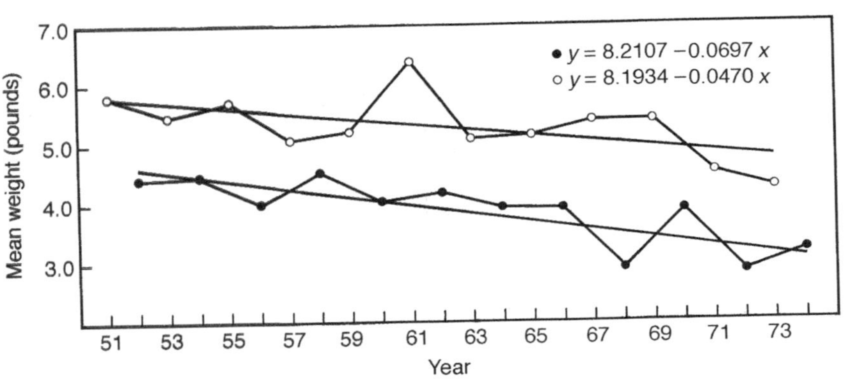

Fig. 7.1 Dynamics of mean weight of pink salmon caught by gill net in British Columbia waters. Hollow circles – odd-year generations; filled circles – even-year generations (from Ricker, 1981).

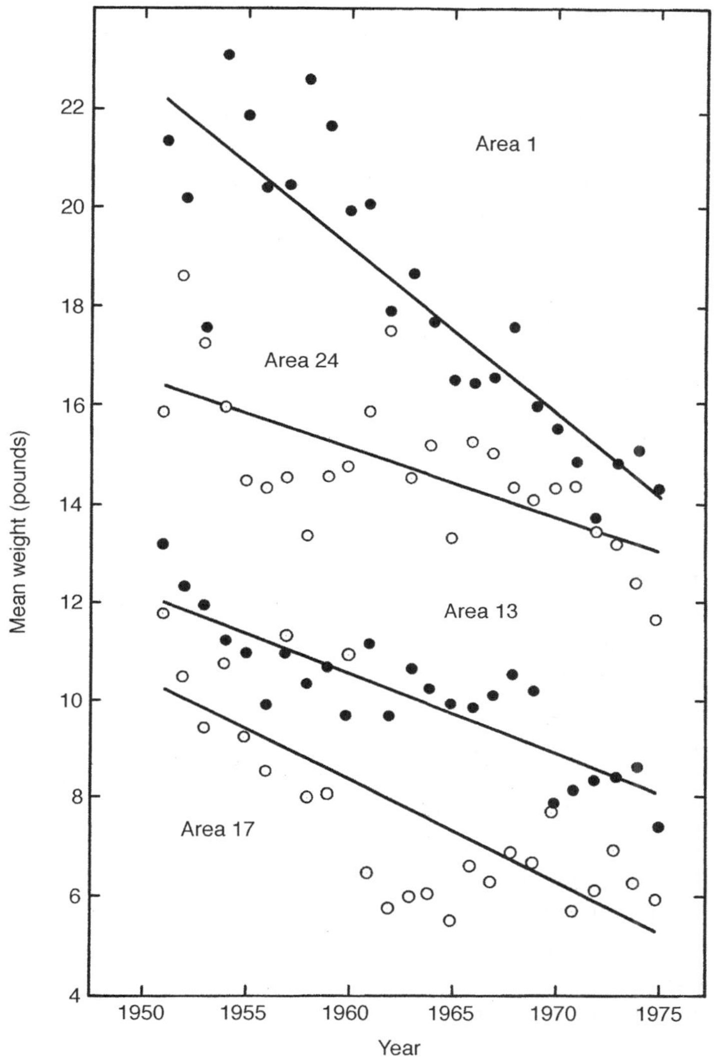

Fig. 7.2 Mean weight of troll-caught chinook salmon in four coastal areas of British Columbia (from Ricker, 1981)

Similar data were obtained by Ricker for two other species of Pacific salmon, chum and coho, although not in all regions was the response of individuals to fishing pressure a decrease of mean body weight and age. In some localities, body size increased or there was an absence of appreciable changes.

Ricker collected information on the physical and chemical environmental parameters (e.g. salinity, water temperature) for the entire period of observation, and showed that there was no reason to search for the cause of the observed dynamics in the influence of long-term environmental trends. The author concluded that it was fishing that was the main cause of the detected shifts and that this fishing had a definite selective character: in cases where, due to fishing management (fishing tools, and time and place of fishing), old and large fishes were removed first, an apparent tendency toward the diminution of the remaining individuals and to the acceleration of their growth rate was observed; in cases where smaller and younger fish groups were caught predominantly, the remaining fish became larger and older. In the absence of selectivity, the main biological parameters of a population were retained in time.

Using the methods of quantitative genetics and very good statistical records of catches, Ricker determined selection differentials for various fishing conditions and calculated coefficients of heritability (h^2) for such a polygenic characteristic as body weight according to the equation (Falconer, 1960):

$$R = sh^2$$

where R is a response to selection and s is a selection differential, i.e. the difference in the body size (weight) at the beginning and the end of a selection cycle.

The h^2 estimates obtained by Ricker for odd and even generations of pink salmon were around 0.20 and 0.30, respectively; they are in very good agreement with those known for other salmonid species from works where direct methods were used (Kirpichnikov, 1981; Gjedrem, 1983).

These data represent some of the first well-documented cases of genetic consequences of selective fishing. It should be stressed that here we deal with *intra-population selection*, examples of which are very few, as Nelson & Soulé (1987) note, although the first work discussing this subject in genetic terms was published by Miller (1957). Unfortunately, except for two or three examples of growth rate inheritance in rainbow trout and carp, this does not contain any reliable evidence, and the author's speculations about the impact of heavy exploitation on the species population structure are too general.

The studies of Favro *et al.* (1979, 1980, 1982) seem to be more advanced. We should admit that their assumption concerning the implementation of selective size-dependent fishing effects through the genetic system controlling individual growth rate differences is reasonable.

The object of investigation was a population of brown trout constantly exposed to fishing pressure since 1959. In 1977 its biomass was reduced by 36% due to fish diminution. Obtaining good agreement between the model and actual data on changes in the composition of catches toward smaller sizes and younger ages, Favro *et al.* pointed out a simultaneous decrease in the growth rate under fishing pressure, which is contradictory to numerous observations (including those mentioned above). Since empirical data on growth rate decrease in brown trout, as well as direct genetic proof, are absent in the work of Favro *et al.*, further interpretation of this study seems impossible (see also Nelson & Soulé, 1987).

To better understand the genetic processes occurring under the impact of commercial fishing, we will consider the results of our own studies, keeping in mind that natural salmon populations are complexly organized, heterogeneous population systems differentiated into smaller genetically differing subpopulations (Chapters 2 and 4).

The systemic organization of populations is of principal importance for the recognition of the selective effects of fishing pressure. It is apparent that exploitation of a population system will be rational only if this system is understood as a unity, with due consideration given to its internal structure. In commercial fishing, however, organization of groups at the subpopulation level is usually ignored, which results in destruction of population systems. Such consequences can be avoided if the spatial organization of the group as a whole, its subpopulation structure, is taken into account and fish are harvested uniformly from all its components.

Although this scheme appears to be abstract, it was confirmed by direct evidence obtained in studies on the consequence of commercial fishing for subdivided sockeye populations spawning in lakes Nachikinskoe, Blizhnee and Dal'nee in Kamchatka (Altukhov & Varnavskaya, 1983).

The choice of these populations was not arbitrary, and, along with the accessibility factor, was determined by the fact that these fish form a natural series reflecting the degree of their subjection to marine fishing in the order of its increase. In addition, these populations, especially the sockeye salmon of Lake Dal'nee, were the object of systematic observations for decades (Krogius, 1978, 1979), providing the researcher with unique biological information. Let us examine these data more closely.

Lakes Blizhnee and Dal'nee are small neighbouring lakes connected by channels with the Paratunka River, which flows into the Avachinsky Gulf of the Pacific Ocean (Fig. 7.3). In the 1930s, the population numbers were close to 100 000 spawners (Krogius, 1979). Sockeye salmon fry spent up to 3–4 years in Lake Dal'nee and 3 years in the sea; thus, the reproductive part of local populations was constituted by a large 'set' of up to 15 age groups. The available materials (Krogius, 1979) show that after 1947 the state of the migratory part of the sockeye shoal in Lake Dal'nee changed. Over the last 35 years, the sex ratio in the reproductive part of the population became uneven (with an excess of females), the production of small males ('jacks') increased, population numbers decreased catastrophically, and the assemblage became 'younger'. At the same time, relative numbers of the so-called 'residual' small form consisting of 90% of males increased and became dominant by 1976 (Table 7.1).

Considerable changes occurred in the sex–age structure of the sockeye salmon in Lake Blizhnee: while large males (2 and 3 years in the sea) predominated in the population in the 1930s, now their proportion is considerably reduced, whereas the proportion of jacks has significantly increased. Population numbers also dropped to, typically, 10 000–12 000 spawners during the years of the research.

Lake Nachikinskoe differs from lakes Blizhnee and Dal'nee in its larger size, and

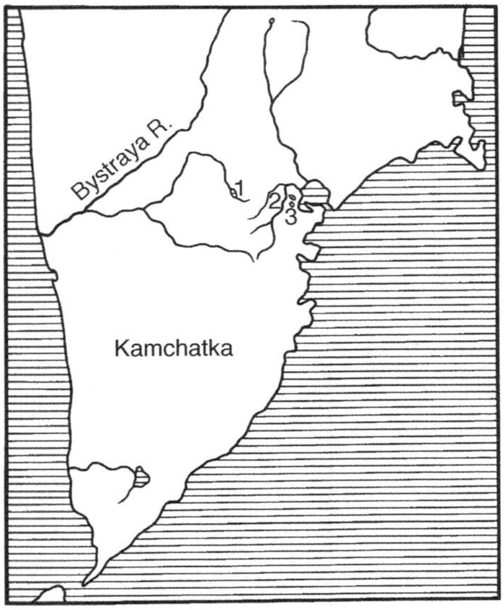

Fig. 7.3 The geographical location of sockeye salmon populations: 1 – Lake Nachikinskoe, 2 – Lake Blizhnee, 3 – Lake Dal'nee. The genetic structure at the *LDH-B2** and *PGM-2** loci was studied for 16, 5 and 3 subpopulations in Lake Nachikinskoe, Lake Blizhnee and Lake Dal'nee, respectively.

its morphology and hydrography; the juvenile Nachikinskoe sockeye salmon move along the Plotnikova and Bolshaya rivers to the Sea of Okhotsk. This population is a typical subdivided population of many spawning subpopulations, which form the structure of the spring and summer races. An analysis of the spawners' body length distributions with respect to their sex shows that noticeable changes in the last decades affected mainly the spring race, first of all due to increased frequency of small males. The numerical size declined, but did not undergo such a significant fall as in Lake Dal'nee's population.

Table 7.1 Average values of several biological characteristics of the sockeye salmon population system in Lake Dal'nee for periods from 1935 to 1976 (from Krogius, 1979).

Characteristics	Periods			
	1935–46	1947–56	1957–65	1966–76
Numbers of breeding fish entering the lake (thousands)	62.6	10.0	5.7	1.6
Proportion of females in the population migrating section (%)	52	54	59	68
Proportion of 'jacks' in the migrating males (%)	0.2	0.6	4.3	37.5
Numbers of 'residuals' in the lake (thousands)	14.9	7.6	5.8	5.3
Proportion of 'residuals' in all the males (%)	26.0	49.4	74.2	88.8

Genotypes of fishes were identified for the *LDH-B2** and *PGM-2** loci. The objective of the study was to compare the genetic structure of three sex–age (size) groups, spawning simultaneously: large and small males, and females. The expediency of this comparison follows from the data of section 5.4, where it was shown for Lake Azabachye sockeye salmon that the heterozygosity level is minimal in large males and maximal in small ones, while it is intermediate in females. A similar picture was found in the same three spawning groups of lakes Nachikinskoe, Blizhnee and Dal'nee (Table 7.2).

(1) Total distributions of the *LDH-B2** genotypes for the Lake Nachikinskoe sockeye salmon are characterized by a significant deficit of heterozygotes. We ascribe this situation to the Wahlund effect at a relatively neutral locus under conditions of genetic subdivision. The distributions of the *PGM-2** genotypes do not differ significantly from those expected according to Hardy–Weinberg equilibrium, or else there is an excess of heterozygotes. This picture results from selection in favour of heterozygotes, which was analysed in detail with the example of Lake Azabachye sockeye salmon.

(2) Division of the sample into groups distinguished by sex and age reveals fine-scale genetic differentiation: a statistically highly significant excess of heterozygotes is observed at the *PGM-2** locus in small males, with a clearly expressed deficit of these genotypes in large males, and a genotypic equilibrium in females. Such differentiation is not evident at the *LDH-B2** locus, which is more selectively neutral, but there is evidence of a deficit of heterozygotes in both females and large males, with Hardy–Weinberg equilibrium in small males. The frequency of heterozygous genotypes is maximal for the groups of small males, reaching 55% at the *PGM-2** locus.

(3) The same differences are typical for the distributions of the *PGM-2** genotypes in Lake Blizhnee sockeye salmon, providing evidence of heterozygotes' excess in small males, a deficit in large males, and an intermediate level of heterozygosity in females.

(4) A particularly significant excess of heterozygotes over the expected value at the *PGM-2** locus is found in Lake Dal'nee's small residual males. The observed frequency of heterozygous genotypes (62%) is highest for this group.

The observed differentiation is evidence of *optimal heterozygosity in females*, and simultaneously, of *its minimal and maximal levels in males*, according to their natural subdivision into two groups distinguished by age structure (Fig. 7.4 and see also Fig. 2.6). It should be stressed once again that male and female sockeye salmon represent two adaptive systems, with maximal variance of fitness in males.

At the same time, the fact that small males are the most heterozygous contradicts the previous conclusion concerning heterosis associated with the *PGM-2** locus: at first glance, it is illogical to discuss heterosis if higher heterozygosity is observed in small individuals. This inference, however, is premature, at least for as long as the

Table 7.2 Population genetic parameters for *LDH-B2** and *PGM-2** allozyme loci in different sex–size sockeye salmon groups in three lakes on the Kamchatka Peninsula.

Sockeye salmon lake populations and sex–size groups	Sample size	*LDH-B2**				*PGM-2**			
		q	*H*	*D*	χ^2	*q*	*H*	*D*	χ^2
Lake Nachikinskoe									
Small males (43–56 cm)	202	0.653	0.43	−0.06	0.72	0.648	0.55	+0.19	7.61*
Large males (57–73 cm)	296	0.667	0.39	−0.11	3.57	0.687	0.35	−0.19	10.68**
Females (50–69 cm)	579	0.683	0.38	−0.12	8.11*	0.657	0.46	+0.02	0.30
Total males and females	1077	0.673	0.39	−0.11	11.95**	0.664	0.45	0.00	0.00
Lake Blizhnee									
Small males (31–39 cm)	248	0.893	0.19	0.00	0.01	0.631	0.56	+0.20	10.30**
Large males (40–63 cm)	104	0.986	0.03	+0.04	0.03	0.711	0.36	−0.11	1.25
Females (43–59 cm)	191	0.937	0.11	−0.11	0.38	0.654	0.47	+0.04	0.33
Total males and females	543	0.926	0.13	−0.06	1.64	0.655	0.49	+0.08	3.42
Lake Dal'nee									
Anadromous form, spring race****	28; 30*****	0.946	0.11	+0.05	0.09	0.783	0.30	−0.12	0.41
Anadromous form, summer race****	38; 26*****	0.895	0.21	+0.12	0.52	0.769	0.31	−0.13	0.46
Small 'residuals'	158; 158*****	0.905	0.16	−0.04	0.28	0.646	0.62	+0.35	19.98**
Total	224; 214*****	0.908	0.17	−0.01	0.00	0.680	0.54	+0.23	11.77**

* $P<0.01$.
** $P<0.001$.**** From Kirpichnikov & Ivanova's (1977) data.
***** Number of fish analysed at loci *LDH-B2** and *PGM-2**, respectively.
q – frequency of the most frequent allele; *H* – observed heterozygosity; *D* – relative deviation of the observed heterozygosity from that expected; χ^2 – test for the correspondence of genotypic distributions to the Hardy–Weinberg equilibrium.

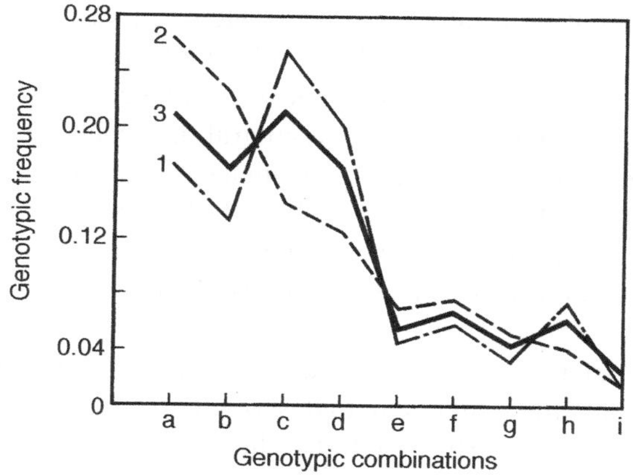

Fig. 7.4 The distribution of frequencies of paired genotype combinations at the *LDH-B2** and *PGM-2** loci in Lake Nachikinskoe spawning sockeye salmon. 1 – small males, 2 – large males, 3 – females; a – *100/100/100/100*; b – *100/120/100/100*; c – *100/120/100/135*; d – *100/100/135/135*; e – *100/120/100/100*; f – *100/120/135/135*; g – *120/120/100/135*; h – *120/120/100/100*; i – *120/120/135/135*. The probability of random differences between large and small males, and also between small males and females, is $P < 0.05$. The genotypic combinations are ranged in the order in which their theoretical frequencies diminish.

corresponding sex and age groups remain insufficiently studied with respect to specific features of their growth and maturation.

As shown earlier (Krogius, 1960, 1972, 1975), small males are distinguished from large fish by accelerated growth rate and earlier sexual maturation. Since the *PGM-2** locus is selectively more significant, it contributes most to the detected differences, and the excess and empirical level of heterozygosity are especially high in fast-growing and early-maturing small males. The three compared populations of Kamchatka's sockeye salmon can be organized in the following series according to the degree of increase in frequency of small and residual males: Nachikinskoe, Blizhnee and Dal'nee (18.7%, 45.7% and 73.8%, respectively). A similar situation is also observed for empirical heterozygosity at the *PGM-2** locus: 45%, 49% and 54%. As already noted above, these three populations are also arranged in the same sequence according to the impact of marine fishing.

This result needs special attention at least in two aspects. First, it reconfirms an earlier conclusion about heterotic selection at the *PGM-2** locus, and in part at the *LDH-B2** locus. Second, it enables us to comprehend many unclear cause–effect links that form the basis of the reaction of local salmon populations to fishing impact. In terms of population genetics, the detected differentiation in sex–age groups of sockeye salmon reflects nothing but the differences in the intensity (and direction) of selection. This selection favours small, more heterozygous males in the reproductive period, and, hence, is directed against large and more homozygous males. Occupying an intermediate position between these two extremes, females do not experience such strong differentiation pressure. This regularity can be explained easily by the fact that Pacific salmonids, reproducing in various river systems on both coasts of the North Pacific, have been subjected to intense fishing since the beginning of the twentieth century. The impact of fishing on the biological structure of populations is proven: in many cases, early-migrating stocks, especially the largest fish with a high frequency of males, are the primary subjects for heavy exploitation (Altukhov, 1974, 1981a,c; Krogius, 1979; Ricker, 1981). This led to a decrease in the numerical size of populations with delayed spawning migration periods, decline in body size, and rejuvenation of spawning groups (Vaughan, 1947; Krogius, 1979; Ricker, 1981).

Some authors associate a phenomenon of this kind with overfishing. However, with regard to the population genetic data described above, another conclusion can be drawn: the cause is not overfishing, but rather the distortion of the historically formed genetic structure of the groups and their systemic organization, or their subdivision into semi-isolated subpopulations, whose biological parameters (sex ratio, age structure, rates of growth and development, spawning migration periods, etc.) are derived from their historically formed gene pools (Altukhov, 1974, 1977, 1981a; Altukhov & Salmenkova, 1981). In particular, ordered behaviour, repeated from generation to generation, is observed in sockeye salmon during the spring migration period: large fish predominate in the beginning of the spawning migration of both races, and the level of heterozygosity changes in time, increasing toward the end of the spawning run (Altukhov & Varnavskaya, 1983; Varnavskaya, 1984).

Obviously, if fishing is selective, i.e. some subpopulations are harvested more than others, this will inevitably lead to changes in the genetic structure of a population as a whole.

Our findings, discussed here and in section 5.4, reveal that Pacific sockeye salmon has an adaptive genetic system, marked by allozyme genes, which is linked with intrapopulation differentiation for important polygenic traits such as sex, age, growth and maturation rates, life span, and spawning migration periods. The reorganization of this structure under fishing pressure also becomes clear.

The following picture emerges. Native sockeye salmon populations are characterized by historically formed genetic differentiation, which is correlated with differences in heterozygosity levels for an aggregate of Mendelian genes, in such a way that its maximal values are typical for small heterotic males ('jacks', residuals), its minimal values for large males, and its average, optimal levels are found in females. As already noted, large, more homozygous males have an adaptive advantage due to selective mating. The females are practically unaffected by directional and disruptive selection, and their genetic structure remains relatively unchanged over generations.

This is the 'adaptive norm' of a population. The small males are a *reserve of genetic diversity*; however, their portion in the native system of populations is small. Genetic structure of the male population as a whole corresponds to the structure of the female population. The genetic process in this system is stationary – selectively neutral on the average. The most important biological characteristics of a population system are highly stable in a normally fluctuating environment, and vary only within certain limits; they do not cross the bounds of a historically formed biological optimum, specific for each separate population. The sockeye populations of Lakes Nachikinskoe and partly of Lake Azabachye are an example of such a system.

Fishing influence causes changes in the vector (and intensity) of selection. The proportion of large, more homozygous fish in a population declines, while the proportion of small, more heterozygous fish increases. The optimal 1 : 1 sex ratio is disturbed, which leads to reduction of the effective population size. Maturation rate is accelerated because of heterosis, average life span is lower, and, consequently, generation time is reduced. This reaction to fishing pressure is possible because of the reserve of genetic variability carried by small males. However, the biomass of a population is reduced due to a decline in fertility. A gap grows between heterozygosity levels of a population as a whole and the population of females, whose genetic structure acts as a kind of 'memory' about the population's past state before it became an object of fishing. The sockeye salmon populations in Lake Blizhnee, and particularly Lake Dal'nee, are examples of such populations; Lake Dal'nee's heterozygosity excess at the *PGM-2** locus is especially high, reaching 23%. When reproductive size and capacity of a population are reduced even by fishing of constant intensity, perfectly compatible with a population's initial production potential, a numerical decline results. In this case, the excess of heterozygotes found from the relation of expected and empirical genotype numbers, and serving as an evidence of genotypic equilibrium distortion, acts as a marker of this process.

Indeed, if we use the estimate of mean population fitness for sockeye salmon of Lake Dal'nee at the *PGM-2** locus ($\bar{W} = 0.77$), we may show that the theoretically expected and observed numbers are almost the same in four consecutive generations (Table 7.3).

Data on the observed numbers in sockeye salmon of Lake Dal'nee are taken from the comprehensive work of Krogius *et al.* (1969). The expected number of the second generation is estimated by multiplying the observed number of the first generation by the $\bar{W}$ value, and in the same way for the next expected numbers.

It is clear that such an unfavourable process can lead finally to at least partial replacement of anadromous forms by residual (if not to complete degradation), and hence, to the loss of a commercially valuable resource (Table 7.1).

Table 7.3 Theoretically expected and observed numbers of sockeye salmon in Lake Dal'nee using the estimate of mean population fitness at the *PGM-2** locus.

Generations (years)	Average annual observed numbers (thousands)	Average annual expected numbers (thousands)
1 (1947–50)	9.93	—
2 (1951–4)	8.82	8.07
3 (1955–8)	7.30	6.56
4 (1959–62)	5.03	5.34
5 (1963–6)	4.84	4.34

However, as selective fishing is associated with increasing heterozygosity of stocks, the current situation can still be improved. To this end, it is necessary to revise the dates of fishing, making the fishing pressure more uniform in time, and to solve the problem of establishing the practice of fishing with nets of optimal mesh size, which should not remain the same throughout the period of the spawning run.

The genetic processes described above are characteristic not only of Pacific salmon populations, but also of other commercially important fish species (Altukhov, 1994a). In all the cases studied in detail, monitoring revealed the same pattern of changes in commercial fishes: a decrease in body size and individual age, an increase in the proportion of early-maturing small males, and the reduction of average life span. As the vector of selection remains unchanged (in favour of heterozygotes), the intrapopulation component of gene diversity increases, whereas the interpopulation component decreases, leading to a reduction in local genetic differentiation. Under such conditions, empirical F_{ST} (G_{ST}) values become significantly lower than expected. Thus, in the group of sockeye subpopulations from Lake Azabachye (Kamchatka), which were affected by selective fishing to the greatest extent, the value of local genetic differentiation (F_{ST}) was estimated at 0.008, compared with 0.059 expected in the case of the selectively neutral process (Altukhov & Salmenkova, 1994). Exposure to such strong artificial selection leads to the establishment of non-random relationships between polygenic (body length, growth rate and maturation rate) and monogenic traits (allozymes) (Altukhov & Varnavskaya, 1983; Altukhov, 1991; Chapter 5).

Genetic data on selective effects of commercial fishing on Atlantic salmon populations are absent, but there are numerous observations concerning a decrease in body size and individual age of fish in commercial stocks (Schaffer & Elson, 1975; for reviews, see Dirin, 1987; Thorpe, 1993). A general trend is an increase in the frequency of grilse as illustrated by the dynamics of their proportion in Atlantic salmon catches in Scotland (Fig. 7.5), which we estimated from the data published by Williamson (1988).

It should also be emphasized that an optimal harvest rate should be distributed among populations in proportion to their size and reproductive ability. Such practice is especially important while harvesting mixed stocks on prespawning migration routes. Hence, it is necessary to have information about the stock composition of mixed fisheries, which presumes reliable identification of stocks.

7.2.3 Genetic stock identification

The problem of managing a mixed-stock fishery is urgent because in many regions artificial propagation on an industrial scale has been progressing in recent years (Heard, 1996; Mahnken *et al*, 1998; see section 7.3), and commercial fishing in these regions should be largely, oriented toward exploitation of hatchery populations. Therefore, it is necessary to know the population composition of mixed stocks at sea, its dynamics during the fishing season and in different years, the time of

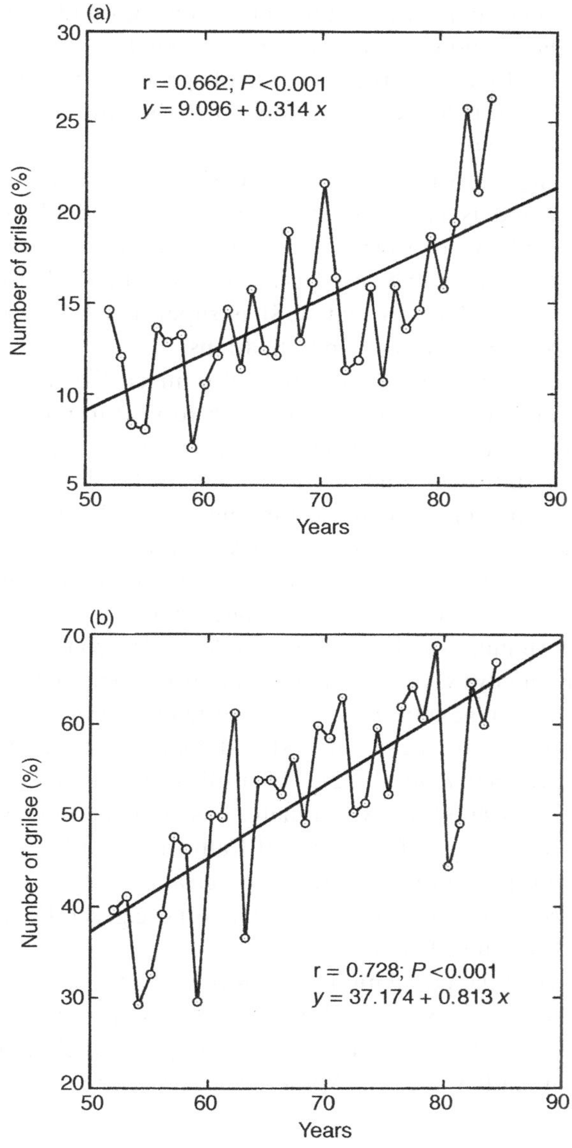

Fig. 7.5 Dynamics of the proportion of Atlantic salmon grilse in Scottish catches in 1952–84. (a) Rod-and-line fishing, (b) net-and-coble fishing (calculated by the authors based on the data from Williamson, 1988).

migration of particular populations through fishing grounds, etc. To this end, specialists use methods allowing identification of populations by natural criteria, such as the structure of scales or the presence of specific parasites, or by means of marking young and adult fish in different ways, including the use of coded wire tags (for review see Utter, 1991).

The method of genetic stock identification (GSI), proposed in the early 1980s and subsequently progressed, is based on the existence of interpopulation differences in allele frequencies of allozyme or DNA loci: as a rule, they are stable and specific for each population and, hence, can be used as markers (Grant *et al.*, 1980; Milner *et al.*, 1985; Utter *et al.*, 1987). Due to that, GSI is superior to the previously known methods and suitable for large-scale investigations.

To determine the population composition of a mixed fishery, it is necessary to have information on genotypic distributions with respect to numerous loci both in potential constituent stocks (populations) and in a representative sample from the stock mixture. The statistical analysis of these data by the method of maximal likelihood provides an approach for estimating the individual contributions of populations or their regional groups to stock mixtures (Pella & Milner, 1987). Different variants of such statistical analyses and corresponding computer programs have been developed (Wood *et al.*, 1987; Millar, 1990).

Accuracy in determining stock composition depends strongly on the number of loci included in the analysis, the probable number of components in the mixture, and the degree of their genetic divergence. Studies showed that estimations of mixed stock composition by means of GSI and coded wire tagging (CWT) are in good agreement (Brodziak *et al.*, 1992). This is illustrated by the data of Gall *et al.* (1989) cited by Utter (1991) for GSI estimates based on a blind test of coded wire-tagged fish taken in fisheries off the coast of California and Oregon (Table 7.4).

It has also been proposed that different biological characteristics of populations should be used in combination with genetic markers (e.g. see Fournier *et al.*, 1984; Wood *et al.*, 1989; Koljonen & Pella, 1997).

The first estimates of the population composition of a harvested stock mixture by a set of allozyme loci were obtained for sockeye salmon returning to Cook Inlet (Alaska) (Grant *et al.*, 1980) and for artificially propagated chinook salmon populations in the Columbia River basin (Milner *et al.*, 1981, cited from Pella & Milner,

Table 7.4 Mixed stock composition by GSI estimates and CWT measurements.

Stock group	Actual % based on CWT	Estimated % by GSI
Sacramento River	21.9	24.2
California coast	3.7	3.8
Klamath River	48.9	48.5
South Oregon/Smith River	24.7	24.7
Mid Oregon	0.9	0.0

1987). Since then, the GSI method has been applied to analysing mixed fisheries of different salmonid species in many regions (see Shaklee & Phelps, 1990).

Currently, a large-scale programme for genetic stock identification is being implemented, in which some governmental agencies of the United States and Canada are involved and several research laboratories participate. Researchers have already created a vast genetic database on chinook, chum, sockeye and pink salmon populations inhabiting the region extending from California to Alaska (Shaklee & Phelps, 1990; Shaklee *et al.*, 1990b).

Within the framework of this programme, many new loci have been studied in salmonid fishes (Beacham *et al.*, 1985a,b; Shaklee *et al.*, 1990b, 1991; Shaklee & Varnavskaya, 1994; Varnavskaya *et al.*, 1994a; Wilmot *et al.*, 1994, 1998), and recent investigations describe the use of polymorphic DNA markers (mtDNA, micro- and minisatellite loci) for genetic stock identification (Beacham *et al.*, 1996, 1998; Miller *et al.*, 1996; Carney *et al.*, 1997; Seeb *et al.*, 1998).

These studies are obviously important in a practical sense as they reveal the pattern of distribution of fishing pressure on salmonid populations, which is necessary for fisheries management aimed at the protection of natural population complexes and for interception accounting (i.e. US vs. Canadian stocks). On the other hand, their scientific significance is no less obvious: specialists have already obtained and are currently obtaining abundant information on gene pools of salmonid populations in all areas of their range, which is necessary for understanding evolutionary processes occurring in the populations and the genetic dynamics of populations in different environmental conditions and under anthropogenic impact.

We used the GSI method for determining the population origin of immature chum salmon by-catch in the course of commercial pink salmon fishing off south-western Sakhalin. The results showed that these chum salmon are more similar to fish from the Amur River than to other potential 'candidates', and it is the Amur population that accounts for approximately 90% of the by-catch (Salmenkova *et al.*, 1990).

Chum salmon commercially harvested in the sea off north-western Sakhalin were also identified genetically. These fish spawn in both continental and insular (Sakhalin) rivers flowing into the Amur lagoon. The proportions of Amur and Sakhalin chum salmon in sea catches in 1990–1992 were determined by analysing allele frequencies at allozyme loci and comparing them with those in samples from the Langry and Chingai rivers – two Sakhalin rivers draining into the Amur lagoon – and in samples of summer and autumn chum salmon from the Amur River. The results showed that, according to genetic characteristics and the time of migration, the summer run harvested in this region consisted of summer chum salmon from the Amur. In the autumn chum salmon catches made in the sea, fish from the Amur strongly prevailed in 1990 and 1991, whereas fish from Sakhalin prevailed in 1992. The variable composition of autumn chum salmon stocks migrating along the Sakhalin coast of the Amur lagoon, apparently associated with changes of hydrologic conditions in the latter, should be taken into account in establishing commercial quotas on chum salmon harvested in this region (Omelchenko *et al.*, 1994).

7.3 Artificial propagation

7.3.1 General information

The first attempts to cultivate salmon (trout) were made as early as in the fourteenth century in France and, later, in Germany and Scotland (Wilkins, 1989; Laird, 1996). Biotechnology of Pacific salmon cultivation was first mastered in Canada and then in the United States, from where it was conveyed to Japan in the 1870s. The main objects of cultivation in the North American continent included sockeye, chinook and, to a lesser extent, chum salmon. Japanese fish breeders concentrated their efforts on autumn chum salmon, organizing hatcheries on islands of the Japanese archipelago and the southern Kuril Islands. The first hatchery in southern Sakhalin was built by the Japanese in 1912; by 1943, 21 hatcheries operated there, incubating 1 to 30 million eggs. The first three Russian hatcheries were built in the years 1909–15, one in the Amur River basin and two in the Kamchatka Peninsula, on the Bol'shaya and Kamchatka rivers (Rukhlov, 1982).

As biotechnology of cultivation improved, the number of hatcheries increased in all countries. Today, the greatest contribution to propagation of salmonid species in hatcheries is made by Japan and the United States. Since the early 1980s, Japan has been releasing each year about 2 billion fry, mainly chum (>90%) and, in a small part, pink and masu salmon, which are cultivated on 250 rivers of Hokkaido and Honshu (Isaksson, 1994; Kaeriyama, 1998). In the United States, more than 1 billion fry of pink and chum salmon are currently released each year in Alaska alone, where artificial fish cultivation has been developed intensively since the late 1970s. In total, hatcheries of the United States and Canada release annually over 2 billion juvenile chum, pink, chinook, sockeye and coho salmon and steelhead trout annually.

In Russia, where Pacific salmon propagation is concentrated mainly in Sakhalin, an average of 500–550 million individuals of pink and chum salmon fry are released each year (Radchenko, 1998).

In 1995, a total of 5.2 billion juvenile salmonids were released from hatcheries into the North Pacific, i.e. about 20% of the total number involved in seaward migration. Of these fish, 41% were released in Japan, 39% in the United States, 10% in Canada and 10% in Russia (Low, 1998). Data on hatchery production in detail are shown in Table 7.5.

Atlantic salmon are propagated artificially in hatcheries located in the territory of the former Soviet Union, which produce a total of approximately 2.5 million smolts. About 5 million are produced and released each year in Sweden and Finland, 1 million in Norway, 6 million in Iceland, and small hatcheries also operate in Ireland, the United States and Canada (Kazakov & Titov, 1992; Heggberget *et al.*, 1993; Isaksson, 1994; Heard, 1996) (Table 7.5).

Table 7.5 Salmon hatchery releases in different countries and regions.

Country or region	Species	Approximate annual releases of juveniles (in millions)	Reference
Norway	Atlantic salmon	1	Heard, 1996
Sweden } Finland }	Atlantic salmon	5	"
Iceland	Atlantic salmon	6	"
Russia	Pink salmon	300–350	Radchenko, 1998
	Chum salmon	200–220	"
	Atlantic salmon	2.5	Kazakov & Titov, 1992
Japan	Chum salmon	2000	Hiroi, 1998
	Pink salmon	120–150	"
	Masu salmon	3.1–8.0	"
	Sockeye salmon	0.03–2.0	"
British Columbia*	Chinook salmon	64.5	Heard, 1996
	Coho salmon	20.3	"
	Sockeye salmon	303.5	"
	Chum salmon	230.4	"
	Pink salmon	107.2	"
	Steelhead trout	2.27	"
	Cutthroat trout	0.29	"
Oregon & Washington**	Chinook salmon	~ 240	Allee, 1988, cited by Laird, 1996
	Coho salmon	~ 115	
	Chum salmon	~ 95	"
	Sockeye salmon	0.04	"
	Steelhead trout	16	"
Alaska***	Pink salmon	~ 800	Heard, 1996
	Chum salmon	300–400	"
	Sockeye salmon	70–75	"
	Coho salmon	14–19	"
	Chinook salmon	7–10	"

* Data for 1989.
** Data for 1987.
*** Data for 1988–92.

Rapid progress in cage culture (mariculture) of salmonid fishes has been a growing trend in recent years. This applies mainly to Atlantic salmon, although there is experience in commercial culture of coho and chinook salmon and Arctic charr. Achievements in cage culture of Atlantic salmon are impressive: in 1996, its production reached 400 000 metric tons, compared to about 10 000 tons harvested annually from natural populations of this species. The main producer of Atlantic salmon reared in cage culture is Norway, followed by Chile, Scotland, and some other European countries. The next important species in this respect is coho salmon, which is cultivated mainly in Chile and the United States (Lem & Shehadeh, 1997).

Rainbow trout is also cultivated on a large scale. The world harvest of this species, largely accounted for by commercial culture, varied in the 1980s from 40 000 to 85 000 metric tons per year (Alexeev, 1988).

Recently the great variety of fish farming aspects, including various genetic questions (Jorstad & Naevdal, 1996), have been reviewed in the comprehensive book *Principles of Salmonid Culture* (Pennell & Barton, 1996). This saves us from having to consider the problems of mariculture technology, and we refer the reader to this book for corresponding information.

Our next objective is to concentrate attention on the artificial propagation of salmonid stocks in hatcheries.

7.3.2 Monitoring of genetic and biological parameters of salmon populations reproduced in hatcheries

As previously emphasized, salmon populations are systems with a complex structure consisting of multiple discrete subpopulations. To reproduce such systems artificially in hatcheries, we must collect eggs and sperm throughout the spawning run rather than limiting ourselves to using only a part of the differentiated gene pool (Altukhov, 1974). The more notable the subpopulation structure of a system, the less the chance of reproducing the whole from a part. Unfortunately, this fact is often ignored in hatcheries and, consequently, the genetic diversity of artificially reproduced populations gradually decreases. Allendorf & Phelps (1980) showed a decrease in genetic diversity of a hatchery population of cutthroat trout (Montana, USA) compared with the native donor population: the portion of polymorphic loci was reduced by 57%, the average number of alleles per locus reduced by 29%, and the level of individual heterozygosity decreased by 21%. Ståhl (1987) noted that increased homozygosity for a set of allozyme loci is characteristic for several Baltic, East and West Atlantic populations of *Salmo salar* and he suggested that it reflects inbreeding resulting from a small number of spawners in the hatcheries.

For the last several years, we have monitored three adjacent pink salmon populations of south Sakhalin: two native, from the Firsovka and Bakhura rivers, and one artificially produced in the Naiba River (Fig. 6.8; Altukhov *et al.*, 1989). The latter is maintained by Sokolovsky and Bereznyakovsky hatcheries that had produced the Naiba stock of chum salmon but, after a sharp decrease in its size in 1972, began to work with pink salmon. Judging from yields, artificial fish breeding resulted in a several-fold increase in the size of the local population. In recent years, however, the biological structure of the hatchery stock has changed: the fish became considerably larger, the proportion of males increased, and the size of the groups began to decrease.

To reveal the mechanisms responsible for these changes, work was organized as follows:

(1) We compared the genetic characteristics of spawners selected for breeding with those of rejected ones, using a set of electrophoretically identified allozyme loci.
(2) We analysed alterations of sex ratio and body length in a series of generations of the artificially produced Naiba population.
(3) We compared genetic and biological parameters of this population with those of two native groups reproducing in the neighbouring rivers.

We were also interested in determining the relationship between individual heterozygosity and biologically important traits. To this end, besides body size we analysed males for the frequency of abnormal gill rakers (bent, split or fused), using this frequency as a parameter of developmental stability. We distinguished weak (only one gill arch affected) and severe (two or more gill arches affected) abnormalities, but only the latter were taken into account during comparative analysis.

The dynamics of body length and sex ratio in the Naiba population was estimated for the whole period of systematic artificial reproduction of pink salmon from records kept at the hatchery since 1972–3. Corresponding data on the self-reproducing Firsovka and Bakhura populations have been collected in the course of our studies. Genotypes were identified for six allozyme loci.

At the hatchery, individual batches of eggs, each obtained from 50 females, are fertilized by sperm obtained from 20–30 males. Moreover, gametes are collected usually at the beginning of the spawning run, when males prevail, and in the middle of it, when the sex ratio approaches equilibrium. Comparison of males selected for breeding with rejected (control) males showed that fish breeders prefer large fish (Table 7.6). No such selectivity was revealed for females. A particularly important

Table 7.6 Morphological and genetic differences between two groups of pink salmon males.

Parameter	Males used for breeding	Males not used for breeding	Significance of differences (t_d-test)
Body length (cm)	52.6 ± 0.2	47.8 ± 0.1	18.02***
σ	3.87	2.64	6.33***
N	300	293	
Proportion of fish homozygous for all loci	0.443 ± 0.029	0.362 ± 0.028	2.01**
N	300	293	
Proportion of fish with severe abnormalities of gill rakers	0.17 ± 0.03	0.09 ± 0.02	2.08**
N	203	210	

Loci *MDH-B1,2**, *G3PDH**, *PGDH**, *PGM-2**, *MEP-2**, *PGI-A** were analysed; N – number of fish studied.
** $P < 0.05$.
*** $P < 0.01$.

fact is that males used at the hatchery also differed from the control males in the level of allozyme homozygosity and in the frequency of abnormal gill rakers (Table 7.6): *the large males selected for breeding were more homozygous and had a higher frequency of gill raker anomalies than did the control group*. Assuming that this selection was performed more or less regularly throughout the period of artificial reproduction of the Naiba population, and that the 'rearguard' proportion with an excess of females was not involved in breeding, we could expect definite changes in the biological structure of this population. This applied, in particular, to a decrease in heterozygosity for allozyme loci, an increase in body length, and deviation from the optimal sex ratio toward an increase in the proportion of males, because eggs are collected from early-migrating groups with a high frequency of males.

Two groups of facts confirmed this suggestion: the first was data obtained during biological monitoring of the Naiba population for several generations, and the second was the results of comparing the biological and genetic characteristics of this population with those of two native populations spawning in the Firsovka and Bakhura rivers. We found that, in consecutive generations of the Naiba population, the proportion of females decreased (Fig. 7.6) and body length increased in the more 'productive' odd years, in which artificial selection among males was more stringent (Fig. 7.7).

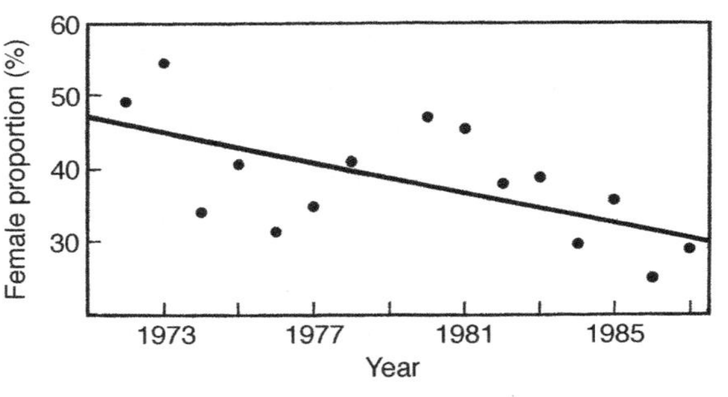

Fig. 7.6 Variability of female proportion for the generations of pink salmon population in the Naiba River after the beginning of artificial reproduction. $y = 46.2 - 0.92x$, $r = -0.556$, d.f. $= 13$, $P < 0.05$.

Knowing the difference between the values of average body length at the beginning and end of the selection cycle (selection differential) and the heritability coefficient of the trait (for pink salmon, approximately 0.27 (Kirpichnikov, 1981; Ricker, 1981; section 7.2.2)), we can use Falconer's equation to estimate the expected shift in the value of the trait and to compare it with the shift actually observed in a series of generations. The results (5.8 and 5.3 cm, respectively) are so similar that any further comments seem unnecessary: fish breeding is obviously a process of the selective type. As under fishing pressure (see section 7.2.3.), a non-random correlation appears between variability of a polygenic trait (body length)

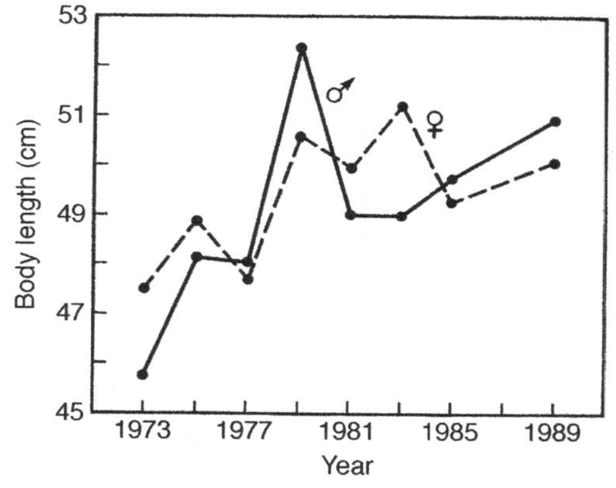

Fig. 7.7 Variability of mean body length in pink salmon population in the Naiba River for odd-year generations after beginning of hatchery reproduction. The points on the plot are mean values of fish body length from the data of the hatchery biological analyses. Continuous line – males, broken line – females. Standardized deviation from respective total average values (for nine generations) for males and females increases significantly for this period ($r = 0.63$, d.f. $= 16$, $P < 0.05$).

and the integral genotype structure for a complex of allozyme loci (see also Altukhov, 1990).

Important data are also shown in Table 7.7. These are estimates of sex ratio and average body length for five 'odd' generations of pink salmon from the naturally reproducing Firsovka population. The stationary state of this population is quite obvious in comparison with the Naiba population, which undergoes marked time-dependent changes. It is also seen that the body length of both males and females in the Naiba population is significantly greater and that variance of the trait is lower than in the adjacent Firsovka and Bakhura populations (Table 7.8). Correspondingly, fish from the River Naiba have a decreased level of multi-locus allozyme heterozygosity (Tables 7.8, 7.9).

Table 7.7 Sex ratio and average body length of pink salmon spawners from the native Firsovka River population in generations of odd years.

Year	Number of fish studied	Proportion of females (%)	Mean body length and variance			
			Females		Males	
			$\bar{x}$	σ	$\bar{x}$	σ
1979	416	51.2	46.6	2.88	47.3	3.88
1981	349	48.3	48.0	2.20	47.9	2.47
1985	250	50.0	45.7	2.51	47.3	4.10
1987	100	61.0	47.1	2.18	47.4	4.54
1989	216	51.8	46.3	4.27	45.9	5.91

Table 7.8 Body length ($\bar{x}$) and average heterozygosity (H) for allozyme loci in three spawning populations of pink salmon.

River (population)	Mean body length (cm) and variance (σ)										Heterozygosity			
	Females					Males								
	N	$\bar{x}$	σ	Significance of differences		N	$\bar{x}$	σ	Significance of differences		N	$H \pm m$	Significance of differences	
				Populations compared	t_d-test				Populations compared	t_d-test			Populations compared	t_d-test
Bakhura (1)	175	45.3	2.8	1–2	1.32	225	46.4	4.5	1–2	1.86	400	0.821 ± 0.019	1–2	2.57*
Firsovka (2)	135	45.7	2.5	1–3	16.88**	115	47.3	4.1	1–3	11.02**	250	0.892 ± 0.020	1–3	3.67*
Naiba (3)	480	49.3	2.4	2–3	14.64**	593	50.2	4.1	2–3	6.97**	785	0.730 ± 0.016	2–3	6.33*

* $P < 0.05$.
** $P < 0.01$.
Heterozygosity was estimated for the same loci as at Table 7.6. N – number of fish studied.

Table 7.9 Genotypic diversity index (μ) in samples of spawning pink salmon populations with natural (Firsovka and Bakhura rivers) and artificial (Naiba River) reproduction.

Year	River	Numbers analysed		μ (average per locus)	Significance of differences (F-test)
		Individuals	Polymorphic loci		
1977	Naiba	234	4	1.920 ± 0.039	3.58**
	Firsovka	250		2.129 ± 0.045	
1979	Naiba	214	4	1.992 ± 0.040	1.39
	Firsovka	243		2.073 ± 0.043	
1981	Naiba	474	4	2.148 ± 0.047	2.78**
	Firsovka	449		2.300 ± 0.050	
1985	Naiba	318	5	2.157 ± 0.040	3.87***
	Firsovka	250		2.394 ± 0.046	
1985	Naiba	318	5	2.157 ± 0.040	2.54*
	Bakhura	400		2.295 ± 0.037	
1985	Firsovka	250	4	2.394 ± 0.046	1.67
	Bakhura	400		2.295 ± 0.037	

* $P < 0.05$.
** $P < 0.01$.
*** $P > 0.001$.
μ – Zhivotovsky's index (1984).

Effects of this kind are obviously equivalent to the consequences of inbreeding, which seemingly can be ignored in artificial propagation of Pacific salmon, as dozens or hundreds of thousands of male and female spawners are involved in crossings. In fact, genetic variation decreases under the effect of selective breeding at such a rate as if the effective population size N_e was only dozens of individuals. This conclusion cannot be verified experimentally because we have no data on the initial heterozygosity of pink salmon from the Naiba River prior to the onset of artificial propagation. However, assuming that it was intermediate between the levels of heterozygosity in the neighbouring native fish of the Bakhura and Firsovka rivers, we obtain $N_e \approx 25$ provided the process is selectively neutral. But we know that the process of reduction of gene diversity in the Naiba population is not neutral: on the contrary, it is maintained by directional selection in favour of homozygotes, and the intensity of selection reaches 1.5–2% per generation. This means that in the absence of compensatory mechanisms, the Naiba population of pink salmon could have become fully homozygous at the studied loci during no more than 40 generations.

Such a rapid loss of genetic variation has an extremely deleterious effect on population fitness (Cross & King, 1983). On the other hand, the increase of heterozygosity under the impact of selective fishing, which leads to elimination of more large (homozygous) individuals, as in the case of sockeye salmon (Altukhov & Varnavskaya, 1983), also results in degradation of stocks.

There is only one way out of this impasse: *in the normally fluctuating environment, in the course of commercial fishing and artificial propagation, it is necessary to preserve the specific levels of genetic diversity established in each population in the course of its natural history, which are indispensable for successful adaptation.* This opinion, first expressed more than 25 years ago (Altukhov, 1973), is additionally confirmed by the data discussed above, and is well supported (e.g. see Kirpichnikov, 1981; Thorpe, 1993; Allendorf & Waples, 1996; Andriyasheva, 1996).

We should emphasize that the existence of the *optimal level of genetic diversity corresponding to the adaptive maximum of a population* is reliably marked by genotypes of allozyme loci whose polymorphism is often interpreted as selectively neutral (e.g. Kimura, 1983). The data discussed here and in Chapters 4 and 5 show that this interpretation is erroneous at least in cases when the environment is not neutral, i.e. as it is in nature. In most cases, it is possible to reveal stable relationships between individual genotypes at allozyme loci and the expression of adaptively significant morphofunctional traits. Moreover, there are convincing reasons for accepting the model of fitness with overdominance as one of the main mechanisms providing for the maintenance of allozyme variation in several salmon species (Altukhov, 1983a, 1990, 1991; Altukhov *et al.*, 1987; chapter 5).

Comparing natural and artificially maintained populations of Atlantic salmon and trout reveals two opposite processes associated with redistribution of intra- and interpopulation components of gene diversity (Table 7.10). Thus, *Salmo salar* bred in hatcheries demonstrate a higher level of interpopulation gene diversity but lower intrapopulation variation than under natural conditions. A directly opposite but even more prominent trend is characteristic of Spanish and French stocks of brown trout.

Table 7.10 Genetic diversity of natural and hatchery salmon populations.

Species, region	Natural			Hatchery			Reference
	H_T	H_S	G_{ST}	H_T	H_S	G_{ST}	
Salmo trutta							
Spain	0.069	0.027	0.615	0.092	0.089	0.028	Garcia-Marin *et al.*, 1991
	(4 populations)			(4 populations)			
France	0.111	0.050	0.550	0.077	0.072	0.063	Krieg & Guyomard, 1985
	(12 populations)			(8 populations)			
Salmo salar, within the Baltic sea, eastern and western Atlantic basins, average	0.041	0.038	0.64	0.037	0.030	0.196	Ståhl, 1987
	(29 populations)			(24 populations)			

In the case of Atlantic salmon, artificial propagation leads increasingly to inbreeding, resulting from a small number of spawners involved in this process (Ståhl, 1987). In the case of brown trout, the increase of intrapopulation heterozygosity and homogenization of genetic differentiation between populations (Table 7.10) is either the consequence of the mixing gene pools of brood stocks differing in their origin, or as a result of selection in favour of heterozygotes (Krieg & Guyomard, 1985; Garcia-Marin *et al.*, 1991). A similar effect may be assumed for *S. salar* reared on fish farms. Comparison of farmed strains with five native strains for three allozyme loci (*AAT-3**, *MEP-2**, *SDH-1**) clearly shows a higher level of heterozygosity in a strain selected for farming purposes (Mork, 1991). Thus, salmon stocks suffer from both inbreeding and outbreeding (Altukhov & Salmenkova, 1994).

Similar effects associated with artificial propagation were observed in our analysis of data on Pacific salmon species. Thus, the assessment of G_{ST} in chum salmon at the intraregional level showed that this parameter in natural populations is similar and exceeds that in artificially propagated fish prevailing in Hokkaido; in the latter, G_{ST} is significantly lower because fertilized eggs are transported between hatcheries (Fig. 7.8). The reviews by Waples (1991) and Hindar *et al.*, (1991) provide examples of the decrease of interpopulation genetic diversity in other salmonid species.

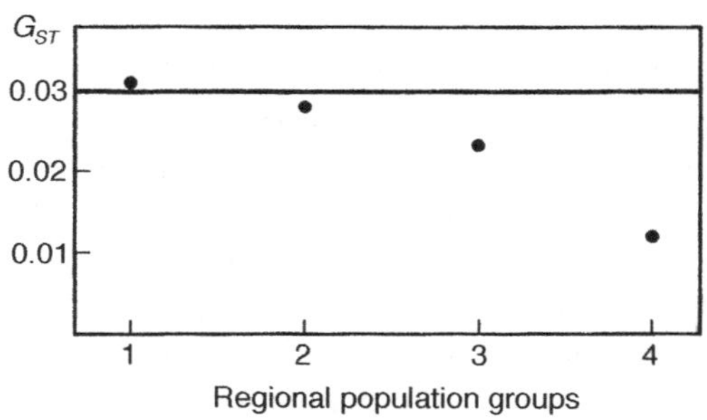

Fig. 7.8 Values of interpopulation genetic differentiation (G_{ST}) averaged for regional groups of chum salmon spawning populations in (1) north-east of Russia, (2) western Alaska, (3) British Columbia, (4) Hokkaido (using data from Table 6.1.). The line parallel with the x-axis indicates the average regional G_{ST} level for natural population systems (1 and 2).

Conversely, chinook salmon populations from the regions of intensive artificial propagation (Oregon–California) differ from natural population systems by higher G_{ST} values. This is a consequence of genetic drift explained by the fact that a small number of spawners is used in breeding (Fig. 7.9; see also Waples & Teel, 1990; Bartley *et al.*, 1992a).

On the whole, the analysis of adverse genetic processes associated with artificial propagation provides direct evidence that they involve disturbances of equilibrium between inbreeding and outbreeding, which counterbalance one another under conditions of normal reproduction of differentiated gene pools (i.e. in subdivided populations). Negative consequences of inbreeding are well known, whereas the

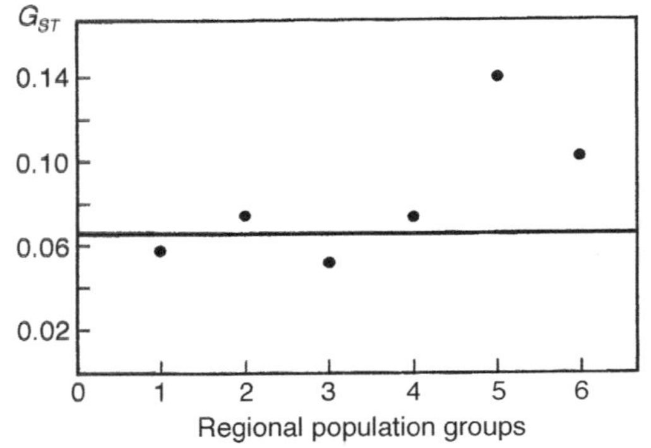

Fig. 7.9 Values of interpopulation genetic differentiation (G_{ST}) for regional groups of chinook salmon spawning populations in (1) Alaska, (2, 3) Yukon basin, (4) British Columbia, (5) California, (6) the region of Oregon–California (using data from Table 6.7). The line parallel with the x-axis indicates the average regional G_{ST} level for natural population systems (1–4).

corresponding data concerning outbreeding are as yet scarce, and further investigations in this field are required. Nevertheless, several known examples are fairly illustrative (Table 7.11).

Negative consequences involving reduction of fitness and productivity of populations can result from measures taken at hatcheries in order to increase hatchery efficiency by accelerating the growth of young fish in species with a long freshwater period. Such observations were made on hatchery populations of coho and chum salmon in the United States (Washington & Koziol, 1993). An increase in freshwater growth rate of young fish leads to accelerated maturation and a decrease in average age and body size of adult fish in the population. Hence, the 'biomass potential of adult fish' becomes lower, i.e. the number of individuals reaching maximum potential adult size within the population decreases. In addition, female fertility, i.e. egg quantities, reduce (Fig. 7.10).

One more condition principally important for the efficient functioning of salmon hatcheries consists in determining the optimal ratio between the release of juveniles and the return of spawners.

7.3.3 Monitoring of an optimal ratio between release and return

The ratio between the release of juveniles and the return of spawners (return rate) is specific for each local population and reflects the relationship between its size and the carrying capacity of the environment (see Chapter 1, Fig. 1.3), which depends primarily on food supply to sea-run juveniles in the coastal waters. If a certain optimum is exceeded, the carrying capacity of the environment decreases, which inevitably results in the reduction of population size.

To avoid such consequences, it is necessary to determine the dependence between the return rate η and the number of juveniles released x, based on the theory of

Table 7.11 Examples of negative consequences of population deviation from genetic optimum (outbreeding and inbreeding).

Species	Effect	Reference
Outbreeding		
Oncorhynchus gorbuscha	Homing of hybrids (native × transplanted) is less than in native fish	Bams, 1976
	Hybrids of genetically isolated odd- and even-year pink salmon have increased F_1 variation in size, and low F_2 return and increased bilateral asymmetry	Gharrett & Smoker, 1991
O. mykiss	Survival of hybrids (hatchery × wild) is reduced	Reisenbichler & McIntyre, 1977
O. nerka	Increase of heterozygosity following selective fishing results in decrease of fish body size, age of maturity and abundance	Altukhov & Varnavskaya, 1983
Salmo trutta	Hybridization of hatchery and natural stocks disturbs reproductive isolation of populations.	Ryman, 1983
Inbreeding		
O. mykiss	Increased mortality of eggs and fry, increased numbers of abnormal fry, reduced performance of fry and young growth and development	Kincaid, 1976, 1983; Gjerde *et al.*, 1983
O. mykiss	Decrease of heterozygosity on inbreeding is accompanied by a decrease in development stability and by a higher level of fluctuating asymmetry	Leary *et al.*, 1985
O. gorbuscha	Hatchery selection of large males (they proved to be more homozygous than small ones) results in increased male proportion, reduction of population gene diversity and production	Altukhov *et al.*, 1989
Salmo salar	Decrease of sea survival of hatchery juveniles from inbred families	Ryman, 1970

population number dynamics of fish (Ricker, 1954). To a first approximation, this dependence is described by the formula $\eta = Ae^{-\lambda x}$, where λ is the proportionality factor inversely relating the natural logarithm of η to the number of released juveniles, and A is the maximal return corresponding to the minimal release. Under such conditions, the dependence between the number of returning spawners y and the number of juveniles released x is determined from the equation $y = Axe^{-\lambda x}$, and the greatest value of return corresponds to the optimal release.

We performed such work for the salmon hatcheries of the Russian far east while implementing genetic recommendations that gametes should be collected consistently throughout the period of the spawning run in the river, taking into

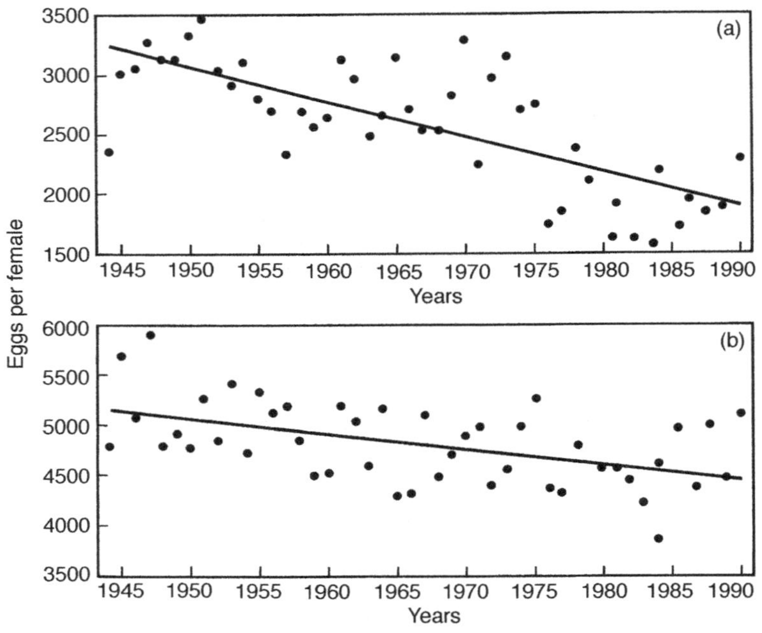

Fig. 7.10 Eggs per female by year (1944–90) for (a) coho and (b) chinook escaping to Soos Creek Hatchery, Washington. (a) $y=3235.8-28.6x$; $r=-0.732$; $P<0.001$. (b) $y=5105.2-13.5x$; $r=-0.480$; $P<0.001$ (our calculations). (Reprinted from Washington & Koziol (1993), with permission from Elsevier Science Publishers.)

account its subpopulation structure (Altukhov *et al.*, 1972). This scheme, introduced for the first time in 1973, soon proved to be inconsistent with the existing Soviet system of planned economy: personnel of each hatchery were interested in progressively increasing the numbers of released fish because this parameter was the only formal criterion for assessing the efficiency of the propagation process, irrespective of return rate. In cases of a deficiency in local spawners, it was common practice to transfer artificially fertilized eggs from other hatcheries (Altukhov & Salmenkova, 1987a,b, 1990; see section 7.4). As a consequence, although the total release of chum and pink salmon fry reached approximately 900 million fish (vs. 600 million in 1969), the return of spawners to many rivers decreased considerably. Two examples showing a negative correlation between the number of released fry and return rate for chum and pink salmon (Fig. 7.11) are illustrative of this situation. The above data were obtained in the course of long-term genetic and biological monitoring of spawning populations actively supported by specialists of the Sakhalinrybvod.

For each hatchery, information about the numbers of released fry and returning spawners (excluding those caught in the sea) was processed to calculate the optimal scale of release. The corresponding recommendations were accepted by the Ministry of Fish Industry of the former USSR and, since 1985, have been implemented

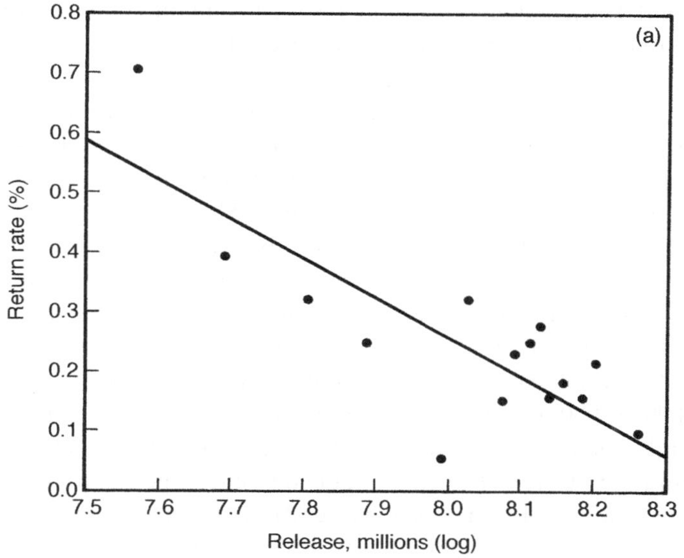

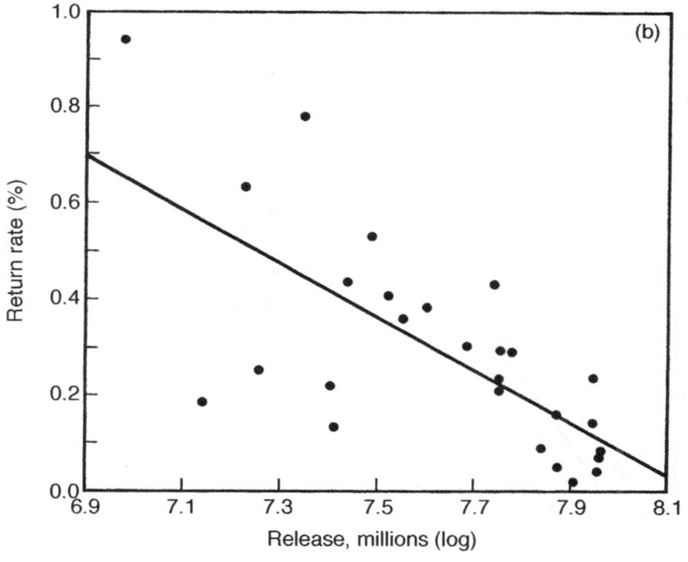

Fig. 7.11 Relationship between release (million fry, log) and return rate (%) at Sakhalin salmon hatcheries. (a) Naiba River pink salmon, 1968–86; $y = 5.5267 - 0.6587 \log x$; $r = -0.8035$; $P < 0.001$. (b) Kalininka River chum salmon, 1952–84; $y = 4.5485 - 0.5577 \log x$, $r = -0.7009$; $P < 0.001$.

at Sakhalin and Kuril hatcheries. As a result, the number of released fish was reduced to approximately 500 million individuals, and this soon produced a favourable effect on artificially propagated fish.

Let us consider the results of this work using an example of a chum salmon population from the small (about 4 km long) Kalininka River on the south-western Sakhalin coast (Fig. 6.2). This population has been kept under observation since the hatchery on this river was put into operation in 1951.

Figure 7.12 shows the dynamics of the ratio between the release of fry and the return of spawners. At the initial stage, the latter parameter increased proportionally to the former. However, when the number of released fry exceeded approximately 50 million, that of returning spawners began to decrease fairly rapidly: at 80–90 million fry released, the number of spawners in 1981, 1983 and 1984 was only 20 000–40 000, i.e. similar to that at the initial stage of artificial propagation. Meanwhile, using Ricker's function for approximating the empirical data on the period from 1952 to 1984 (seven to eight consecutive generations, as generation length in chum salmon is 4 years; see Chapter 2), we determined the optimal rate of release at $x = 46$ million fry per year, which corresponds to the expected annual return of about 160 000 spawners (to the 'natal' river alone). In fact, at the annual average release of 45 million fry, the annual return of spawners in the years 1985–8 averaged 280 500.

It should be emphasized that analysing the relationship between release and return in the Kalininka population of chum salmon, we excluded the data on the extremely high numbers of spawners returning to the river in 1978, 1979 and 1982, i.e. immediately after the 200 nautical mile zone was commissioned in 1977. This

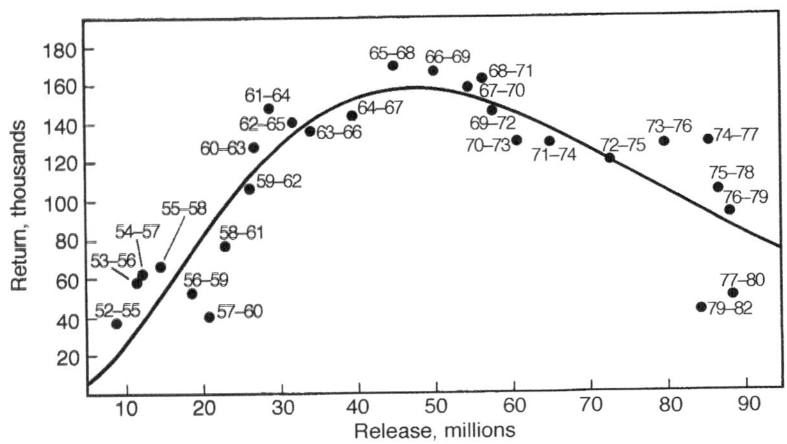

Fig. 7.12 Relationship between release (million fry) and return (thousand adults) in the Kalininka River in 1952–82. Approximation by modified Ricker's function (1954): $y = 0.177x^{2.4} \times e^{(0.052)x}$; $x_{max} = 46.3$; $y_{opt} = 158.1$. Points are mean values for every four years indicated.

immediate increase in population size clearly demonstrates that chum salmon of south-western Sakhalin had been intensively harvested for a long time by Japanese fishermen, who removed approximately two-thirds of the reproductive fish from the hatchery population created and maintained by Russian specialists. Thus, throughout the preceding period of relatively steady production rate at the Kalininka hatchery (1964–71), the annual return of spawners averaged 158 400 fish, whereas in the years 1978–82 this value increased to 303 000 fish. The number of spawners decreased in subsequent years, but this was accounted for by overproduction of fry at the hatchery up to 1986.

Let us now consider the results of observations on the pink salmon population from the Naiba River (Fig. 6.8).

This population has also been kept under long-term biological monitoring since the very beginning of its artificial propagation, when 77.5 million fry were released in 1968; this was followed by 64.5 million released in 1970, and 36.9 million in 1973. These figures reflect sharp differences in the abundance of spawners of even- and odd-year generations: in 1972, 10 800 adult fish were captured in the river, whereas in 1973 this number was about 200 000, i.e. almost 20 times as great. In subsequent years, the annual rate of release increased, reaching a peak of 182.5 million fry in 1981 and averaging 141.2 million fry over the period from 1975 to 1984. Since the beginning of artificial propagation, no more than 20% of spawners were allowed to migrate to their natural spawning grounds.

This practice initially provided for a rapid increase in population size and the levelling of corresponding differences between even- and odd-year generations, so that approximately 300 000 spawners entered the river in 1976 and 1977. Their number reached a peak of 369 000 in 1978 and was only slightly lower (345 000) in 1980. During the period from 1975 to 1980, no less than 260 000 spawners were recorded each year in the river only (excluding fish caught in coastal waters using trap nets). From 1981, however, this number began to decrease rapidly, so that no more than 51 700 spawners returned in 1985, and only 16 000 in 1986. This was accompanied by changes reflected in Figs 7.6 and 7.7 and Tables 7.8 and 7.9: the proportion of females in the spawning population became reduced, the average body size of fish increased, and the level of genetic diversity reduced.

After 1987, the situation with the Naiba population of pink salmon began to improve. The proportion of females increased to 42% in 1991 (vs. 21% in 1987) and to 34% in 1992 (vs. 15% in 1986); in the odd-year generations, the number of fish increased to 242 000 in 1989 and reached a peak of about 300 000 in 1991. Although the size of the odd-year population subsequently decreased slightly, this was counterbalanced by a considerable increase in the number of returning even-year spawners: from 16 000 in 1986 to 230 000 in 1996; in other words, sizes of the even- and odd-year populations became similar again. These data demonstrate how great can be the role of artificial propagation in salmon population management.

An important condition is that this management should be based on fundamental principles of genetics and population biology, which imply the maintenance of (1) intrinsic genetic diversity of populations and (2) balance between the number of released fish and their potential food supply. Fish biologists know that food deficiency in the sea areas where young fish feed can be compensated by supplementary feeding. This approach is widely used at Japanese hatcheries breeding chum salmon on Hokkaido and Honshu: they release fry into the sea after the latter reach a body weight of 500–600 mg. In the case of pink salmon, supplementary feeding is not so effective because the presmolt stage in this species is very short, and, hence, food supply in the estuary of the natal river is critically important for the survival of young fish. This applies fully to the Naiba population. As in the case of chum salmon from the Kalininka River, the graph of the quantitative relationship between the release of fry and the return of spawners in the Naiba population is distinctly curvilinear, showing that the correlation between these parameters is initially direct but subsequently becomes inverse. By approximating the empirical data by Ricker's function, the optimal rate of release was estimated at $x = 91.7$ million fry, which is 1.5–2 times as low as the number of young fish annually released according to production plans from the Sokolovsky and Bereznyakovsky hatcheries, which maintained the Naiba population during the period from 1975 to 1984.

In recent years, the mass release of juvenile salmon from hatcheries in Japan, Canada, the United States and Russia has apparently created an unusual situation, with disturbances of ecological balance between the size of salmonid populations and sufficiency of food supply for them. Instead of being local and limited to coastal waters, these disturbances are now observed in the open ocean ecosystem, over the vast water area of the North Pacific. Such a tendency was demonstrated for chum salmon propagated at Japanese hatcheries (Ishida *et al.*, 1993; Kaeriyama, 1998) and, almost simultaneously, for salmonid species of the north-east Pacific (Bigler *et al.*, 1996).

In essence, this tendency differs principally from the aforementioned effects of selective fishing and artificial propagation.

If selective fishing leads to the removal of large fish, which are more homozygous, the level of heterozygosity in a population increases and, as a consequence, fish maturation becomes accelerated, the age of fish in spawning stocks decreases and, hence, their body size decreases as well. Conversely, if selective fishing or fish propagation promotes the survival of large (= more homozygous) fish, as in the case of pink salmon from the Naiba River, an increase in intrapopulation homozygosity is observed. In salmonid species characterized by the presence of different age groups in populations, such selective removal leads to a reduction in intrapopulation polymorphism, retardation of sexual maturation, 'ageing' of spawning stocks, and an increase in average body size. Apparently such consequences of selective harvest of small, early-maturing fish were revealed by Ricker (1981) in the course of his long-term (1940–75) observations on chum, sockeye and coho salmon populations from several regions of British Columbia.

The situation described in recent publications (Ishida *et al*, 1993; Helle & Hoffman, 1995; Bigler *et al.*, 1996; Kaeriyama, 1998) is entirely different: the observed decrease of fish body size occurs against the background of retarded sexual maturation and increasing age of spawning stocks, which has nothing to do with the presently well-known polygenic system of relationships between heterozygosity of the genotype and the rates of growth and sexual maturation. It is apparent that the factors responsible for these new phenomena are the reduced mortality of fry at early stages due to intensive feeding at hatcheries and, correspondingly, insufficient food supply to salmon in the ocean. However, the density-dependent consequences of the excessive increase in numerical size of salmon populations in the ocean may prove to be as negative for their fitness as those of imbalance between population size and food supply in estuaries of rivers used for fish breeding.

Now we will look at the genetic consequences of stock transplantation.

7.4 Stock transplantation

The high food value of salmonid fishes, their rapid growth during sea feeding migrations and a strong homing instinct make them attractive objects for acclimatization. Success in transplantation and acclimatization to new water bodies was achieved in experiments with species such as rainbow trout and brown trout, and mainly with their freshwater (fluvial and lacustrine) forms (MacCrimmon & Marshall, 1968; MacCrimmon *et al.*, 1970; MacCrimmon, 1971; Karpevich *et al.*, 1991; Krueger & May, 1991). On the other hand, almost all attempts to acclimatize Pacific salmon species, especially beyond their historically established range, have failed (Karpevich, 1975; Withler, 1982; McDowall, 1990; Lever, 1996; Allendorf & Waples, 1996). Rare exceptions to this rule include successful transplantation of North American chinook salmon to New Zealand, where this species formed relatively small but steadily reproducing populations in rivers of the South Island (McDowall, 1990). There are also the data on successful introductions of sockeye salmon (kokanee) to lakes of the United States and Japan (Hokkaido; Seeley & McCammon, 1966) and of pink, coho and chinook salmon to the Great Lakes (see Withler, 1982).

For almost 30 years, attempts have been made to introduce pink salmon to the waters of northern Europe. Their outcome, however, is as yet unclear, especially in view of the fact that numerous previous experiments on transplanting this species to the Atlantic waters of North America proved to be unsuccessful (see Karpevich *et al*, 1991). During the period from 1956 to 1986, fertilized pink salmon eggs were transplanted from Sakhalin to some hatcheries on rivers of the Kola Peninsula almost every year with some breaks. The returns were usually small, except for several abundant odd-year generations of the 1970s, when pink salmon demonstrated extensive straying both westward – along the Scandinavian coast to the British Isles – and eastward from the Kola Peninsula. It was concluded that deviation of new environmental conditions, primarily temperature, from those optimal for pink salmon was the main factor responsible for a low return rate

(Dyagilev & Markevich, 1979; Yves, 1992). However, beginning from the late 1980s (when egg transplantations had already been discontinued), in the White Sea basin, the progressive increase of pink salmon returns occurred in five consecutive odd-year generations. They were possibly the progeny of naturally spawning fish that were transferred from more northern regions of the Far East in the mid-1980s (Yakovenko, 1998). A steady increase in the returns in a series of generations provides evidence that pink salmon acclimatization in northern Europe may be successful. However, to be certain of this, more long-term observations, will be necessary.

Compared with attempts at transplantation of fish beyond the species range, interpopulation transfer is much more common: fertilized eggs or fry are often transferred from one hatchery to another, usually in cases when the number of available spawners is insufficient for artificial propagation. We will discuss the problem of this stock transfer from the population genetic aspect in view of the data obtained in our own research.

The studies were carried out in 1969–81 and involved several populations of autumn-spawning chum salmon artificially reproduced in the hatcheries of the Russian far east. The numbers of artificially fertilized eggs transplanted from one river basin to another are shown in Table 7.12 (see Fig. 6.2).

Data on return rates, i.e. ratios of the numbers of returning adult spawners to numbers of fry released from hatcheries, for populations from the Kalininka, Naiba and Kurilka rivers have been collected for many years. Prior to mass transplantations of artificially fertilized eggs, the averaged values of these coefficients for hatchery fish reared in natal waters amounted to 0.4–0.5% for Naiba, 0.4% for Kalininka and 0.7% for Kurilka (for escapement in rivers only; Altukhov, 1974; Efanov *et al.*, 1979). Our investigations included:

(1) Systematic analysis of the genetic structure of these populations for a set of allozyme (protein) loci, the number of which was increased during the experiments up to ten polymorphic loci.
(2) Recording numbers of released fry and returning adult spawners, to determine the return rate to each river.
(3) Genetic evaluation of spawning adults to determine the proportions of native fish and 'aliens' derived from transplanted eggs, as a measure of the effectiveness of transplantations.

To identify fish genotypes for the allozyme loci, starch and polyacrylamide gel electrophoresis was used. The proportion of 'alien' fish was evaluated quantitatively by calculating the impact of alleles with respect to separate loci (Neel & Schull, 1958; Altukhov, 1974), and then by averaging the obtained values. Genetic distances between populations were estimated according to Nei (1975).

To identify different subpopulations in a mixed population after transplanting eggs reliably, it is important to ensure that the genetic characters of each of the

Table 7.12 Acclimatization transfers of chum salmon eggs in the Sakhalin–Kuril–Amur region.

Year	Number of transferred eggs, in thousands
	Kurilka–Naiba
1972	30 000
1973	3 500
	Naiba–Amur
1964	22 000
1967	9 000
1968	98 000
1969	62 000
1970	17 600
	Naiba–Tym
1967	9 487
	Naiba–Buyuklinka
1968	5 390
1976	3 000
	Kalininka–Naiba
1959	4 830
1964	70 610
1965	70 000
1966	54 123
1967	25 665
1968	26 310
1969	35 563
1970	18 820
1971	56 000
	Iturup Island rivers–Kalininka
1976	~ 30 000
	Tym–Naiba
1966	10 208

The name of the 'donor' river is followed by the name of the 'recipient' river.

native stocks remain constant (Altukhov *et al.*, 1980; Altukhov, 1981a; Salmenkova *et al.*, 1986). Each chum salmon population has its own 'genetic face' (Fig. 7.13). Table 7.13 gives values for genetic distances between populations. The data given emphasize the genetic uniqueness of chum salmon spawning in the south-west of Sakhalin Island (Fig. 6.8); this population differs greatly from all the other stocks in its allele frequencies at the *MDH-B1,2** and *ESTD** loci. Other populations have greater genetic similarity. However, practically every one has its own specific combination of allele frequencies at one or more loci. For example, the chum salmon from the Kurilka River differ in their allele frequencies at the *mIDHP-1** and *sIDHP-2** loci, while those from the Naiba River differ in the *MDH-B1,2** and

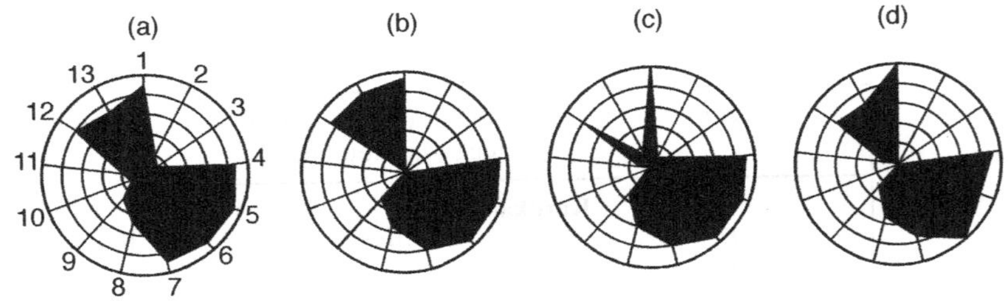

Fig. 7.13 Polygons of the allele frequencies of allozyme loci in chum salmon populations. (a) Naibi River; (b) Amur River; (c) Kalininka River; (d) Kurilka River. In the centre of a circle, allele frequency = 0; on the perimeter, allele frequency = 1. Radii: 1–3 – *MDH-B1,2 *100*, **74*, **122* alleles; 4 – *AAT-1,2 *100* allele; 5 – *LDH-A1 *100* allele; 6 – *PGDH *100* allele; 7 – *mIDHP-1 *100* allele; 8–11 – *sIDHP-2 *100*, **43*, **25*, **83* alleles, respectively; 12 – *mMEP-2 *100* allele; 13 – *ESTD *100* allele.

*mIDHP-1** loci, and those from the Amur River (autumn-spawning race) differ in the *ESTD** and *sIDHP-2** loci.

Three examples given below show the fate of fish that developed from transplanted eggs, and were thus exposed to novel environmental conditions.

Transfer of Naiba (Sokolovsky hatchery) chum salmon eggs to the Amur River (Teplovsky hatchery)

The *MDH-B1,2** and *LDH-A1** loci were first examined in Amur and Naiba chum salmon in 1974. At that time, after several years of transplanting approximately 200 million eggs (Table 7.12), one could expect to observe the return of fish of various age groups belonging to the first and second generations after transfer. In 1974, 60% of chum salmon from the Teplovsky hatchery were in age group 3+. Nine samples (each containing 46–50 individuals) were taken for electrophoresis at intervals of 1–3 days throughout the spawning migration from 24 September to 10 October. No samples contained a significant proportion of Naiba chum salmon (Altukhov *et al.*, 1980). Fifteen years later, a sample of Amur chum salmon from the Teplovsky hatchery was again analysed for seven polymorphic enzymes encoded by ten loci.

Table 7.13 Genetic distances ($\times 10^4$) between chum salmon populations related by transplantations of artificially fertilized eggs. Calculations were made from allele frequencies of the polymorphic loci *MDH-B1,2**, *sAAT-1,2**, *LDH-A1**, *PGDH**, *mMEP-2**, *mIDHP-1**, *sIDHP-2** and *ESTD**, averaged for a number of years.

Rivers	1	2	3	4
1. Naiba	0	514	52	104
2. Kalininka	—	0	492	913
3. Kurilka	—	—	0	135
4. Amur**	—	—	—	0

** Autumn race.

The Teplovsky fish had not altered and still differed from the Naiba fish, further evidence of the inefficiency of this transplantation.

Transfer of Kurilka chum salmon eggs to the Naiba river (Sokolovsky hatchery)

Eggs were transplanted in 1972–3 (Table 7.12). Analysis for several allozyme loci were carried out in chum salmon, returning for spawning to the Naiba River in 1976, 1980 and 1987. In 1976 one should expect the admixture of fish in age groups 3+ and 2+ from the transfers of 1972 and 1973 respectively. In 1976 three samples were collected from a spawning run for electrophoretic analysis. Allele frequencies for the *LDH-A1** locus, which differ significantly in the natural Naiba and Kurilka populations, had moved towards values typical of Kurilka fish (for details see Altukhov *et al.*, 1980). Therefore adults derived from the Kurilka population, via transported eggs, were present in the spawning run in the Naiba River. The proportion of these alien fish in the Naiba, calculated from allele frequencies for the *LDH-A1** locus, was 60% although this may have been an over estimate because the 1976 samples were taken only from the middle part of the chum salmon spawning run. The return rate for Kurilka chum salmon to the Naiba river was 0.17%, significantly smaller than the 0.7% return rate for the native Kurilka fish. A return rate was also estimated by examining scale structure (Efanov *et al.*, 1979), which differs in the Kurilka and Naiba chum salmon stocks. The value was 0.11%, approximately the same as the coefficient determined from enzyme analysis.

Observations on the fish of the Naiba River were continued. In 1980, allele frequencies averaged for a series of samples again showed a slight shift towards the value typical of Kurilka fish for a number of loci (Salmenkova *et al.*, 1986). The proportion of Kurilka fish in the Naiba River was estimated at 35% and the return rate was 0.05% for this second generation of adults derived from the Kurilka stock. Therefore, compared with the first adult return, subsequent returns had significantly decreased. In the late 1970s and early 1980s the Sokolovsky hatchery received chum salmon eggs from various hatcheries on Sakhalin, and so data for the period following these transplantations cannot be interpreted unambiguously. However, in 1987, the last year of the study, samples of Naiba fish were even less like the Kurilka fish; in 1980 the genetic distance between them was 50×10^{-4}, and in 1987 it was 70.4×10^{-4}.

The low returns of 'alien' fish, which decreased in successive generations, show that the transplantation of eggs was ineffective, and that transplanted fish appear to be intolerant of novel environmental conditions.

Transfer of Kurilka chum salmon eggs to the Kalininka River hatchery

In 1976, 34 million artificially fertilized eggs were transferred from the Kurilka hatchery to the Kalininka hatchery (Table 7.12). These two populations differed in their allele frequencies for a number of loci (Table 7.14) (Salmenkova *et al.*, 1983;

Table 7.14 Frequencies of the most common alleles of the polymorphic loci in the Kalininka and Kurilka chum salmon populations.

River, year (number of samples)	Number of fish	Allele frequencies						
		*MDH-B1,2**	*sAAT-1,2**	*LDH-A1**	*mMEP-2**	*mIDHP-1**	*sIDHP-2**	*PGDH**
Kalininka river								
1969 (3)	191	0.919	—	—	—	—	—	—
1970 (4)	349	0.920	—	—	—	—	—	—
1971 (4)	187	0.924	—	0.985	—	—	—	—
1974 (4)	392	0.902	—	0.996	—	—	—	—
1975 (2)	99	0.909	—	1.0	—	—	—	—
1976 (3)	125	0.908	0.918	1.0	—	—	—	—
1977 (3)	150	0.923	0.908	0.983	—	—	0.650	0.997
1978 (3)	148	0.910	0.935	0.993	—	—	0.685	0.997
1980 (6)**	585	0.910	0.937	0.980	0.905	0.908	0.677	0.996
1981 (5)**	492	0.910	0.927	0.989	0.883	0.898	0.662	0.990
Total	2718	0.912	0.928	0.989	0.895	0.904	0.669	0.994
χ^2-test of heterogeneity		3.08	3.82	17.59***	2.74	0.62	1.38	—
d.f.		9	4	7	1	1	3	3
Kurilka river								
1970 (1)	95	0.979	—	0.845	—	—	—	—
1971 (2)	81	0.985	—	0.810	—	—	—	—
1972 (2)	94	0.987	—	0.830	—	—	—	—
1977 (1)	31	0.968	—	0.871	—	—	0.484	1.0
1978 (1)	97	0.995	0.954	0.845	—	0.768	0.569	1.0
1979 (1)	95	0.992	0.958	0.898	—	0.801	0.506	1.0
1980 (1)	97	0.977	0.952	0.874	0.745	0.737	0.505	1.0
Total	590	0.984	0.955	0.852	0.745	0.768	0.522	1.0
χ^2-test of heterogeneity		4.48	0.08	7.26	—	2.23	2.50	—
d.f.		6	2	6	—	2	3	—

* Allele frequencies for isoloci *MDH-B1,2** and *sAAT-1,2** are calculated on the basis of the assumption of equal frequency for each locus of the pair.
** Average allele frequencies are given for samples, in which admixture of Kurilka river fish is absent.
*** $P < 0.01$. In this case the significant value χ^2 can be explained by the statistical properties of the test (by the low expected frequencies of the rare allele).

Altukhov & Salmenkova, 1987a). Adult chum salmon returning to spawn at the Kalininka hatchery were sampled throughout the whole of the spawning runs in 1980 and 1981, and nine allozyme loci were analysed.

In 1980, allele frequencies in the first six samples, taken from 3 September to 3 October, were typical of the Kalininka fish, but those of the last two samples, on 13 and 20 October, were practically identical to those of Kurilka fish (Table 7.15). The conclusion that these late-run spawning fish of 1980, in age group 3+, were from transplanted eggs of the Kurilka population is also supported by their later mean time of returning to spawn, characteristic of the Kurilka native stock. This is in good agreement with results obtained by Okazaki (1982a), who reported allele frequencies and time of return typical for the one native stock after transplanting fish into the Tokachi River (Hokkaido).

In 1981, the return of age 4+ adults of Kurilka origin to the Kalininka River was very insignificant, registered only in a sample on 12 October 1981. The return corresponds to a very small proportion (2–19%) of 5-year-old fish in the native spawning run to the Kurilka River.

Analysis of the allele frequencies confirmed that the last part of the 1980 Kalininka run of spawning adults, on 12 and 20 October, consisted of about 85% Kurilka fish, whereas on 12 October 1981 only 39% were Kurilka fish. In other samples taken in 1980 and 1981, adults of the Kurilka population were not found.

Hatchery statistics on the size of runs, coupled with the results of this study, indicate that only 0.03% of Kurilka fish released into the Kalininka returned to spawn in the river. This value is more than one order of magnitude smaller than the corresponding return rate to their native rivers for the Kalininka (0.4%) and Kurilka (0.7%) populations.

Thus the results obtained in these experiments are in accord with those in the first two examples given above, confirming that transplanting eggs from one population to the river basin of another population is not an effective way of establishing a new self-reproducing population of fish in territory already occupied by another.

From the viewpoint of genetics, successful acclimatization after transplanting eggs would result in adaptation to a new environment. Its efficiency can only be determined when a self-reproducing population is established with a relatively stable gene pool and capable of infinitely long self-reproduction. Unfortunately, the evidence for such phenomena is scarce and difficult to obtain with fish. Nevertheless, indirect evidence from numerous examples testify that it occurs rarely as a result of experimental transplants of fish eggs. Ricker (1972) summarized a great volume of data on transplantations of North American salmonid populations and showed that the return of the first generation to the 'alien' river is rather common, but is less pronounced than the return to the native river; in the following generations the return to the 'alien' river decreases sharply or does not occur at all.

Out of numerous transplantations of chum salmon in Japan, only a few within Hokkaido were successful (at least some of them were observed during three generations after transplantation (Okazaki, 1978, 1982a)). In the review of literature

Table 7.15 Frequencies of the most common alleles of polymorphic loci in Kalininka river chum salmon population.

Date	Number of fish	Allele frequencies						
		*MDH-B1,2**	*sAAT-1,2**	*LDH-A1**	*mMEP-2**	*mIDHP-1**	*sIDHP-2**	*PGDH**
1980								
3 September	98	0.888	0.940	0.970	0.914	0.890	0.697	1.0
13 September	97	0.919	0.955	0.977	0.914	0.910	0.655	1.0
19 September	98	0.920	0.940	0.995	0.885	0.905	0.711	0.980
25 September	96	0.913	0.895	0.975	0.880	0.919	0.662	1.0
30 September	96	0.903	0.951	0.978	0.944	0.909	0.655	0.995
3 October	100	0.917	0.943	0.985	0.895	0.900	0.680	1.0
13 October	98	0.947	0.942	0.861	0.804	0.830	0.545	1.0
20 October	100	0.967	0.958	0.876	0.795	0.795	0.475	1.0
1981								
3 September	96	0.920	0.967	0.979	0.872	0.870	0.655	0.985
10 September	98	0.913	0.929	0.995	0.885	0.900	0.640	1.0
20 September	100	0.905	0.890	0.990	0.891	0.920	0.660	0.985
29 September	98	0.897	0.933	0.980	0.914	0.890	0.657	0.985
6 October	100	0.910	0.905	0.990	0.865	0.910	0.697	0.995
12 October	99	0.925	0.927	0.950	0.853	0.830	0.568	0.995
20 October	100	0.903	0.936	0.936	0.865	0.899	0.604	1.0

See notes for Table 7.14.

on transplanting of Pacific salmon, Withler (1982) concludes that transplantations carried out within a species' natural area fail to produce any new anadromous stocks.

All these facts are evidence for the unique and conservative character of local adaptations formed by natural selection during thousands of generations under the environmental conditions intrinsic to the natural history of a population. When we transfer some part of a population into a new environment, the reserve of genetic resistance that is sufficient under some environmental conditions is usually insufficient under others. *The fact that transplanted fish return to spawn at the same time as they would in the native river, and the fact that the genetic structure of transplanted fish remains unchanged, despite the low values of the return rate, testifies to the catastrophic consequences of selection under the new conditions.*

An important question now arises: in which periods of the life cycle is natural selection acting to render transplants ineffective? Unfortunately there are no direct observations on the mortality rates of transplanted eggs during their incubation in hatcheries, although it is believed their mortality rate is no different from that of eggs from native fish at the same hatchery. Moreover, the mortality rates of native and alien fry in the estuaries of rivers are not known. It is quite probable that selection is most intense during this period of the life cycle, and that the native fish are more successful at competing for food.

Can we increase the efficiency of transplantation if it is necessary to restore a destroyed population or to create a new one? We should take into account the fact that a single transplantation represents only a minute part of the population gene pool. Moreover, the species of fish considered here are not single, homogeneous panmictic populations; they are composed of distinct subpopulations, characterized by complicated genetic structures (Altukhov, 1974). Therefore, the restoration of a population that has been dangerously depleted, or the creation of a new population where one has been extirpated, can be carried out only with fish that are similar in origin or are genetically similar (Altukhov & Salmenkova, 1987a). Only then are they likely to be relatively interchangeable because the similarity of the gene pools, selected over long periods of time, would be expressed in terms of similar biological and ecological characteristics, such as sex ratio and age structure, time of spawning runs, physical and chemical requirements for embryonic development and comparable levels of resistance to diseases.

As a rule, genetic similarity between populations correlates directly with geographic distance between them (e.g. see Kijima & Fujio, 1982; Chapter 6), and the smaller this distance, the more probable is the similarity in their adaptations. Indeed, consider the example of two transplantations of chum salmon from the Kurils (the Kurilka River) to Sakhalin (the Naiba and Kalininka rivers), which were studied in the greatest detail: as estimated by ten allozyme loci, the genetic distance between fish from the Kurilka and Naiba rivers proved to be ten times smaller than that between fish from the Kurilka and Kalininka rivers (Table 7.13). In agreement with these data, the return rate to the Naiba River in the first generation of Kuril chum salmon was 0.17%, and that to the Kalininka River was 0.03% (Salmenkova *et al*, 1986; Altukhov & Salmenkova, 1990). In experiments on coho salmon, Reisenbichler (1988) also observed that the rate of return to an 'alien' river among transplanted fish was inversely proportional to the distance of transfer from their 'natal' river.

Another fact that should be taken into account is that only a small part of the genetic diversity existing in a population system is usually employed in the work on stock transfer. As already noted, salmon should not be regarded as panmictic populations: they are complex systems of subpopulations, and each of the latter has a specific gene pool. Hence, material transplanted to a new range should be collected from the greatest possible number of populations (gene pool donors) and from a series of generations. Analysing rare examples of more or less successful transplantations of salmonid fishes, e.g. sockeye salmon (kokanee) to Lake Shikotsu in Honshu (Seeley & McCammon, 1966) or chinook in New Zealand (McDowall, 1990), it becomes clear that success was determined by the fact that fish at early stages of individual development were transplanted for a long time from different geographically separated populations.

In the rare cases where transplantations appear to have been successful, we should not underestimate the following important circumstances. First, we still lack a reliable series of long-term observations for the fate of transplants (e.g. pink salmon to the Great Lakes in North America), bearing in mind that in nature time is measured in generations and not in years. Second, the so-called successful acclimatizations, for example with *S. trutta* or *O. mykiss*, have unfavourable genetic and ecological consequences for native populations that are depressed and frequently displaced.

Negative genetic consequences are associated with changes in gene pools of local species, which result from hybridization with introduced species followed by introgression of their genes. Hybridization between different populations upon transfer or contact between hatchery and natural stocks leads to outbreeding depression and the loss of locally adapted gene complexes. These processes reduce fitness of local populations, making them more vulnerable to environmental stress. Adverse ecological effects of stock transfer are attributed to competition for food and spawning grounds, predation of local fish by introduced species, unfavourable transformation of the environment, and other interactions between introduced forms and the local fauna (for reviews, see Hindar *et al.*, 1991; Krueger & May, 1991; Rhymer & Simberloff, 1996).

Among possible consequences of stock transfer, especially dangerous is the spread of infections that can be lethal to local species because they are not immune to them. The history of salmon aquaculture in Norway provides an impressive example of such a situation (Heggberget *et al.*, 1993). Beginning from the mid-1970s, *S. salar* populations of more than 30 Norwegian rivers have been perishing from a disease caused by the ectoparasite monogenean *Gyrodactylus salaris*, supposedly brought to Norway from Swedish hatcheries together with transplanted salmon eggs or juveniles. Subsequent studies showed that Baltic *S. salar* populations are far less sensitive to this parasite than Norwegian populations. Populations of this species in Norway were also seriously damaged by furunculosis acquired from rainbow trout transferred from Denmark.

We must admit that despite important genetic data confirming the inefficiency of numerous transplantations, and notwithstanding the clear attitude of the International Union for Conservation of Nature and Natural Resources (IUCN, 1978), which regards the acclimatization of organisms in new localities as a major form of 'biological pollution' of the environment, so far no consensus has been reached among scientists and industrial managers. Meanwhile the destruction of unique gene pools in natural populations has continued. No wonder the number of species referred to as 'rare' has been increasing drastically. Taking into account the vital necessity of conserving ancient biological diversity and pursuing a strategy for survival, we must make every effort to stop the process of destruction through 'biological pollution' and develop a more sophisticated means of solving the vitally important problem of restoration of vanished populations. This approach must employ the principles of population genetics.

Conclusion

Monitoring natural populations of commercially important fish species reveals a rather uniform pattern of changes caused by anthropogenic pressure. In virtually all cases, these influences result in unfavourable genetic processes, i.e. *to types of reproduction of species gene pools that lead to deviation from the optimal ratio between intra- and interpopulation components of gene diversity and to the loss of 'memory' about the previous state of the population.* This situation is based on the fact that people involved in the industrial employment of natural resources pay no attention to the historically developed subpopulation structure of affected populations. Even fish breeding, apparently serving the cause of maintenance and artificial reproduction of biological resources, can lead to negative consequencies. They are associated with a redistribution of gene diversity in such a way that its intrapopulation component H_s decreases, whereas the interpopulation component G_{ST} increases. The situation is typical for salmon hatcheries, at which breeders either use an insufficient number of spawners, thus provoking inbreeding, or perform unintentional selection in favour of homozygotes (the effect is the same in practice). This process is *maladaptive* and can lead to irreversible degradation of populations even after the corresponding influence is terminated.

Redistribution of the components of gene diversity via an increase in the level of intrapopulation heterozygosity was revealed during monitoring of harvested self-reproducing populations, and also in Atlantic salmon populations reproduced in hatcheries and in cage culture (Altukhov & Salmenkova, 1994). Although this process is *adaptive*, it also leads to degradation of populations, because the cost of adaptation proves to be excessive. For example, highly productive anadromous populations are substituted by less productive resident forms, or mortality at early developmental stages increases (Ryman & Utter, 1987; Altukhov, 1991; Altukhov *et al.*, 1991). At the same time, the opportune termination of an adverse external influence provides the possibility of the damaged subpopulation structure being restored and the genetic processes in the system returning to the norm.

It is important to emphasize a very significant fact, touched on only partly in Chapter 6, when we discussed similarity and differences of genetic processes in population systems of biological species on the one hand, and human populations on the other. It becomes evident that humans, as both the creators and subjects of history, affecting the environment by their activities, induce in plant and animal species the same processes as those currently observed in human populations: the population system evolves to its extreme states, i.e. to panmixia or to maximal subdivision, when interpopulation differences disappear or become excessive. The former is characteristic for the European human population, in which smoothing of interpopulation differences is accompanied by an increase in intrapopulation polymorphism ($\bar{F}_0 = 0.0279$). The latter process is observed in the indigenous Mongoloid populations of northern Asia and America ($\bar{F}_0 = 0.081$ and $\bar{F}_0 = 0.082$, respectively) that have already approached the limit of spatial differentiation (Rychkov *et al.*, 1982; Rychkov & Yashchuk, 1983). Obviously, both processes are the evolutionary cost paid for 'civilization' and the 'primitive way of life', respectively. Which of the two states is the more favourable for human society should be discussed separately. But, as for wild animal species, the answer is unambiguous: *both processes lead to decreased fitness or even degradation of population systems, bringing them to the limits of maintaining their integrity.*

In order to overcome these negative effects, at least in salmon fisheries, it is necessary to follow some basic principles of management.

(1) Local salmon populations are historically formed systems of subpopulations that interact through gene exchange and differ in their gene pools. Adaptive genetic differences at the subpopulation level are reflected by the fact that each subpopulation has its own sex–age structure, physiological rhythms, numerical size, productivity, resistance to diseases, etc. This means that the inner structure of a population system, as mentioned above, is ordered in time and space in a certain way, which is a guarantee of the genetic stability of a subdivided population and the regulation mechanism of intra- and inter-population gene diversity components.

(2) Successful fishing and artificial reproduction of a population system are impossible if its inner subpopulation structure is ignored. In order to preserve this structure, economic efforts must be distributed evenly and proportionally among all elementary populations (that is, subpopulations or stocks) comprising the system. At the same time, fishing and fish breeding have a selective character, since they differentially remove (or reproduce) distinct genotypes. This leads to irreversible changes in the historically formed ratio between

intra- and interpopulation components of gene diversity, which is normally maintained at a certain optimal level.

(3) Predominant removal of large, old, late-maturing homozygous fish by selective fishing increases intrapopulation gene diversity. Fish become smaller in size because of early sexual maturation, the rate of succession of generations increases, and the optimal sex ratio is disturbed. As we have seen, such a picture is especially typical for Pacific salmon and, apparently, for many other fish species. The same effects are highly probable for breeding programmes dealing with fish farming of Atlantic salmon (especially in Norway). This process is adaptive, but if it is not stopped, the adaptive cost can be too high (for example, replacement of highly productive anadromous forms by residual forms of less value). Still, timely termination of unfavourable external pressure leaves a chance for the restoration of a subpopulation structure and the normalization of genetic processes in a population system.

(4) Genetic diversity increases with various transplantations when breeding of native populations with introduced stocks is carried out; in this case, adaptive gene combinations are destroyed as a result of outbreeding. On the contrary, hatchery selection of large homozygous fishes will lead to consequences similar to the negative effects of random genetic drift and concurrent inbreeding. The intrapopulation diversity will decrease, while the interpopulation diversity will increase. This process is maladaptive and most likely leads to an irreversible degradation of populations.

(5) For rational management:
   (a) It is necessary to harvest populations throughout the whole period of the spawning migration with compulsory regulations for the distribution of fishing gear and their mesh size in time (space), especially for gill nets.
   (b) To maintain broodstocks in aquaculture, it is necessary to avoid both inbreeding and outbreeding, reproducing each group separately as an independent integral unit; it is necessary to select parents characterized by average levels of heterozygosity for a set of allozyme loci in production of smolts.
   (c) Eggs should be collected during the whole spawning run in hatchery reproduction of the populations, and disproportionate usage of reproductive products and preference for large fish over small ones (and vice versa) should be avoided.

These minimal guidelines provide the possibility for not only the maximal economic gain, but also the long-term conservation of natural populations. It is obvious that only such a type of fishery can be called sustainable, i.e. non-exhaustive. We need to enhance contacts between research groups working in this area and fishery managers employed in exploitation and reproduction of salmon genetic resources in fisheries. International cooperation is also of great importance.

Summary

Studying populations of salmonid fishes, we concentrated our efforts on revealing their biological features and trends in population genetic processes that are characteristic particularly of natural communities not yet destroyed irreversibly because of anthropogenic impact. Interacting with the environment, such populations are maintained in a steady state.

The most important condition for this balance, and therefore for the stability of populations, is their systemic organization which is linked by subdivision into semi-isolated subpopulation units. Since the biological features of populations are essentially derived from their historically evolved gene pools, a genetic stability of the population in time reflects the stability of its important parameters such as total and genetically effective numbers, sexual and age structure, and reproduction coefficient. Clearly, a system of this kind should comprise also a balanced genetic load value – harmful mutations and unfavourable combinations of genotypes will inevitably be segregated and eliminated by natural selection in the lengthy process of historical development. It is evident as well that a population system's genetic parameters, and the biological ones which they determine, lend themselves to long-term forecasting and in principle are controllable.

What is most important for the existence of population systems? As follows from the sum of data discussed in this book, it is a balance between the processes of differentiation and integration of their gene pools. In nature, under normally fluctuating environments, this balance is achieved owing to the existence of feedback in relationships between the parameters of a population system, such as the effective size of its structural components and the intensity of gene migration to them from the common gene pool. This mechanism helps to maintain genetic diversity of the system at the optimal level providing for adaptation to the environment. The optimal level of genetic diversity in a population system implies the existence of not only a definite ratio between homo- and heterozygous genotypes in its gene pool, but also a steady (i.e. autoregulated) ratio between the intra- and interpopulation components of this diversity. In some salmonid species, mechanisms maintaining and stabilizing the interpopulation component of gene diversity manifest themselves at different hierarchical levels of population structure. This fact can apparently be regarded as evidence that gene pools of some population systems still remain in the normal state acquired in the course of their natural history.

The ratio of intra- and intergroup components of hereditary variation that is characteristic of a population system under conditions of the *normal* genetic process serves as a reference point for revealing unfavourable genetic changes in populations, e.g. those caused by selective fishing, selective artificial propagation or stock transfer. Unfavourable genetic processes occurring in salmonid populations under the effect of different human activities can lead to a degradation in the evolutionarily developed structure and a decrease in the volume of genetic 'memory' in the corresponding population systems.

Indeed, the results obtained during genetic monitoring of commercially important species are discouraging. Gene diversity at various evolutionarily established levels is disturbed due to fishing and, moreover, during actions undertaken with good intentions, such as artificial reproduction of salmon in hatcheries. In all cases, disproportionate removal of certain genotypes and insufficient utilization or non-uniform reproduction of other genotypes give rise to unfavourable processes that can lead to a decrease in the fitness of populations. The mechanisms underlying these phenomena involve both a decrease and an increase in gene diversity, relative to the optimum established in the course of natural history.

We emphasize this conclusion, since researchers concerned with biosphere problems concentrate their attention on the reduction of biological and genetic diversity. An artificial increase of gene diversity relative to the historical optimum may also be unfavourable for a population. However, the latter process is adaptive and, if the corresponding external pressure is stopped in time, it is possible to recover the genetic structure and to normalize the process of reproduction of the population gene pool. The opposite process relating to the loss of population gene diversity is maladaptive and irreversible.

There are good reasons to believe that, in many cases (at least in those when intrapopulation polymorphism decreases, whereas non-adaptive spatial differentiation increases), the permissible level of genetic changes has already been exceeded. This conclusion is fundamental for our long-term forecast concerning the consequences of anthropogenic pressure on natural and artificially propagated populations. This forecast is obviously unfavourable. *Hence, contrary to some current concepts that evolution of the biosphere is still in progress, we would rather conclude that the biosphere is deteriorating.* To prevent this process, it is necessary to revise the strategy of human interactions with nature, as mentioned at the beginning of Chapter 6.

The facts discussed in this book suggest that human interactions with salmon resources should not disturb the systemic organization of populations and should maintain intra- and interpopulation gene diversity at the optimal level. Such an approach envisages:

(1) The maintenance of gene diversity in surviving population systems during their exploitation and artificial reproduction (non-exhaustive wildlife management).
(2) Rehabilitation of systems with a disturbed structure.

(3) Creation of new population systems in regions with appropriate natural historical and economical conditions.

These principles of conservation of the optimal diversity developed during evolution should pertain to all levels of biological integration, including the level of ecological systems. Anthropogenic pressure affects the biosphere as a whole, but only populations, i.e. self-reproducing intraspecific groups of organisms formed in the course of natural history, serve as a target for direct external influences. We believe that the principal condition ensuring the stability of any ecosystem is the same as in the case of the population system of individual species: self-regulation via interaction of relatively independent structural components that exchange information about their own state and the state of the environment. It is important to maintain, restore and, if required, imitate the historically conditioned direction and intensity of these information flows. Otherwise, neither the long-term existence of a conserved or newly formed population nor its ability to respond adequately to external influences, not exceeding the historical optimum, would be possible.

This approach is based on understanding principal differences between the historical and evolutionary processes. The history of humans, at least at the phase of genetic differentiation of the species, is a quasi-stationary process with an uninterrupted course. Populations involved in it can maintain equilibrium only dynamically. As their genetic relatedness weakens, they diverge in time and lose identical genes by descent.

The situation with natural populations is quite different. Rapidly undergoing the dynamic stage of development, they subsequently remain in a steady state, being included in a biogeocenosis. Their gene diversity and the ratio of its intra- and intergroup components remain unchanged, thus maintaining a memory of ancestral gene pools. Hence, it is quite natural that no strictly directed microevolutionary processes occur in native population systems under normally fluctuating environments, and only various stabilizing mechanisms preventing accidental gene fixation operate in them. Any directed reorganization of the genetic population structure is primarily the consequence of powerful anthropogenic impact.

In other words, it is totally wrong to place the sign of equality between natural variability, which is intrinsic to the biota, and changes that occur in it because of anthropogenic impact. Without taking this into account, it is impossible to comprehend the current situation with gene pools of natural populations. The absence of a clear conception of the norm (optimum) – normal state or normal process – deprives us of the memory of the previous state as a reference point necessary for understanding the significance of genetic changes and their immediate and long-term consequences. Conversely, the finding of such a baseline makes it possible to discover and formulate the principles of conservation and rational management of natural population gene pools, as discussed here.

Taking into account the successes won in the genetics of natural populations, we can introduce into our practical activities methods for industrial utilization, artificial reproduction and acclimatization that will make it possible to change to an intensive type of fisheries. But this is only practicable if one makes allowances for the special features of the internal structure of populations, preserving their historically formed genetic heterogeneity, and if the autoregulatory mechanisms that ensure their effective adaptation under conditions of a normally fluctuating environment are maintained. Only in this way can an optimal policy of management that maximizes not just economic advantage but also the natural preservation of the populations be realized in practice.

These principles laid the foundation for *conservation genetics*, a new scientific field developing before our eyes. If this book may be useful in promoting such a strategy for salmon management, the authors will believe their task accomplished.

References

Adams, N.S., Spearman, W.J., Burger, C.V., Currens, K.P., Schreck, C.B. & Li, H.W. (1994) Variation in mitochondrial DNA and allozymes discriminates early and late forms of chinook salmon (*Oncorhynchus tshawytscha*) in the Kenai and Kasilof rivers, Alaska. *Canadian Journal of Fisheries and Aquatic Sciences* **51** (Suppl. 1), 172–81.

Alberts, B., Bray, D., Raff, M., Roberts, K. & Watson, J.D. (1983) *Molecular Biology of the Cell*. Garland, New York.

Alexandrov, S.M. (1973) *Sakhalin Island. History of the Relief Development of Siberia and the Far East*. Nauka, Moscow.

Alexeev, A.P. (1988) World fishery and aquaculture of salmonid fishes. In *Modern State of Research of Salmonid Fishes* (Konovalov, S.M., ed.), pp. 7–9. Institute of Ecology of the Volga River Basin, Tol'yatti.

Allendorf, F.W. (1975) *Genetic variability in a species possessing extensive gene duplication: genetic interpretation of duplicate loci and examination of genetic variation in populations of rainbow trout*. DPhil thesis, University of Washington.

Allendorf, F.W. & Leary, R.F. (1986) Heterozygosity and fitness in natural populations of animals. In *Conservation Biology: The Science of Scarcity and Diversity* (Soule, M.E., ed.), pp. 57–76. Sinauer, Sunderlend, MA.

Allendorf, F.W. & Leary, R.F. (1988) Conservation and distribution of genetic variation in a polytypic species, the cutthroat trout. *Conservation Biology* **2**, 170–84.

Allendorf, F.W. & Phelps, S.R. (1980) Loss of genetic variation in hatchery stock of cutthroat trout. *Transactions of the American Fisheries Society* **109**, 537–43.

Allendorf, F.W. & Thorgaard, G.H. (1984) Tetraploidy and evolution of salmonid fishes. In *Evolutionary Genetics of Fish* (Turner, B.J., ed.), pp. 1–53. Plenum Press, New York.

Allendorf, F.W. & Utter, F.M. (1973) Gene duplication within the family Salmonidae: disomic inheritance of two loci reported to be tetrasomic in rainbow trout. *Genetics* **74**, 647–54.

Allendorf, F.W. & Utter, F.M. (1976) Gene duplication in the family Salmonidae III. Linkage between two duplicated loci coding for aspartate aminotransferase in the cutthroat trout (*Salmo clarki*). *Hereditas* **82**, 19–24.

Allendorf, F.W. & Utter, F.M. (1979) Population genetics. In *Fish Physiology* (Hoar, W.S., Randall, D.J. & Brett, J.R., eds), vol. 8, pp. 407–54. Academic Press, New York.

Allendorf, F.W. & Waples, R.S. (1996) Conservation and genetics of salmonid fishes. In *Conservation Genetics: Case Histories from Nature* (Avise, J.C. & Hamrick, J.L., eds), pp. 238–80. Chapman & Hall, New York.

Allendorf, F.W., Utter, F.M. & May, B.P. (1975) Gene duplication within the family Salmonidae: detection and determination of the genetic control of duplicate loci through inheritance studies and the examination of populations. In *Isozymes IV. Genetics and Evolution* (Market, C.L., ed.), pp. 415–31. Academic Press, New York.

Allendorf, F.W., Mitchel, N., Ryman, N. & Ståhl, G. (1977) Isozyme loci in brown trout (*Salmo trutta* L): detection and interpretation data. *Hereditas* **86**, 179–90.

Allendorf, F.W., Knudsen, L.L. & Phelps, S.R. (1982) Identification of a gene regulating the tissue expression of a phosphoglucomutase locus in rainbow trout. *Genetics* **102**, 259–68.

Allendorf, F.W., Knudsen, L.L. & Leary, R.F. (1983) Adaptive significance of differences in the tissue-specific expression of a phosphoglucomutase gene in rainbow trout. *Proceedings of the National Academy of Sciences of the USA* **80**, 1397–1400.

Allendorf, F.W., Gellman, W.A. & Thorgaard, G.H. (1994) Sex-linkage of two enzyme loci in *Oncorhynchus mykiss* (rainbow trout). *Heredity* **72**, 498–507.

Alm, G. (1959) Connection between maturity, size and age in fishes. *Reports of the Institute for Freshwater Research, Drottningholm* **40**, 5–145.

Altukhov, Yu.P. (1969a) Immunogenetic approach to the problem of intraspecific differentiation in fish. *Advances in Modern Genetics (Russia)* **2**, 161–95.

Altukhov, Yu.P. (1969b) On the relation between monomorphism and polymorphism of haemoglobins in fish microevolution. *Proceedings of the USSR Academy of Sciences* **189**, 1115–17.

Altukhov, Yu.P. (1971) Recent physiological, biochemical and immunological studies on the problem of intraspecific differentiation in marine fish. *Rapports et proces verbaux, Conceil international pour l'exploration de la mer* **161**, 103–8.

Altukhov, Yu.P. (1973) Local fish communities as genetically stable population systems. In *Fish Biochemical Genetics* (Troshin, A.S., ed.), pp. 43–53. Nauka, Leningrad.

Altukhov, Yu.P. (1974) *Fish Population Genetics*. Pishchevaya Promyschlennost, Moscow.

Altukhov, Yu.P. (1977) Problems of population genetic organization of species in fish. *Journal of General Biology (Russia)* **38**, 893–906.

Altukhov, Yu.P. (1981a) Hereditary heterogeneity of Pacific salmon stocks in the theory and practice of fishery management. In *Genetics and Reproduction of Marine Organisms* (Kasyanov, V.L. & Pudovkin, A.I., eds), pp. 20–8. DVNTS, Vladivostok.

Altukhov, Yu.P. (1981b) Genetic monitoring as related to environmental conditions. In *Genetics and Human Welfare* (Vartanyan, M.E., ed.), pp. 205–20. Nauka, Moscow.

Altukhov, Yu.P. (1981c) The stock concept from the viewpoint of population genetics. *Canadian Journal of Fisheries and Aquatic Sciences* **38**, 1523–38.

Altukhov, Yu.P. (1982) Biochemical population genetics and speciation. *Evolution* **36**, 1168–81.

Altukhov, Yu.P. (1983a) *Genetic Processes in Populations*, 1st edn. Nauka, Moscow.

Altukhov, Yu.P. (1983b) Systemic organization of populations and its significance for conservation and rational exploitation of the genetic pool of the biosphere. In *Applied Aspects of the program 'Man and Biosphere'* (Gvishiany, D.M. & Sokolov, V.E., eds), pp. 97–121. VINITI, Moscow.

Altukhov, Yu.P. (1984) Adaptive norm concept and the problem of outbreeding. *Bulletin of the USSR Academy of Medical Sciences* **7**, 16–21.

Altukhov, Yu.P. (1985a) Two kinds of genetic variability and the problem of speciation. *Evolution* **39**, 223–6.

Altukhov, Yu.P. (1985b) Molecular evolution of populations. In *Molecular Mechanisms of Genetic Processes* (Sozinov, A.A., ed.), pp. 100–31. Nauka, Moscow.

Altukhov, Yu.P. (1989a) *Genetic Processes in Populations*, 2nd edn. Nauka, Moscow.

Altukhov, Yu.P. (1989b) Balancing selection as the factor of maintenance of allozyme polymorphism. *Advances in Modern Biology (Russia)* **107**, 323–40.

Altukhov, Yu.P. (1990) *Population Genetics: Diversity and Stability*. Harwood Academic Publishers, London.

Altukhov, Yu.P. (1991) The role of balancing selection and overdominance in maintaining allozyme polymorphism. *Genetica* (Netherlands) **85**, 79–90.

Altukhov, Yu.P. (1994a) Genetic consequences of selective fishing. *Genetika* **30**, 5–21.

Altukhov, Yu.P. (1994b) The principles of population monitoring for conservation genetics. In *Conservation Genetics* (Loeschcke, V., Tomiuk, J. & Jain, S.K., eds), pp. 337–50. Birkhauser Verlag, Basel.

Altukhov, Yu.P. (1995) Intraspecific genetic diversity: monitoring and conservation. *Genetika* **31**, 1331–57.

Altukhov, Yu.P. & Blank, M.L. (1991) Computer modelling of genetic processes in subdivided populations. *Proceedings of the USSR Academy of Sciences* **219**, 1467–72.

Altukhov, Yu.P. & Blank, M.L. (1992) Genetic dynamics of population systems with variable parameters of structure and selection. *Proceedings of the USSR Academy of Sciences* **326**, 1068–72.

Altukhov, Yu.P. & Dubrova, Yu.E. (1981) Biochemical polymorphism of populations and its biological significance. *Advances in Modern Biology (Russia)* **91**, 467–80.

Altukhov, Yu.P. & Kalabushkin, B.A. (1974) Stable polymorphism in modern and ancient populations of molluscs, *Littorina squalida*. *Proceedings of the USSR Academy of Sciences* **215**, 477–80.

Altukhov, Yu.P. & Kurbatova, O.L. (1984) Human heredity and environment. In *Human Heredity and Environment* (Altukhov, Yu.P., ed.), pp. 7–34. Nauka, Moscow.

Altukhov, Yu.P. & Kurbatova, O.L. (1990) The problem of the adaptive norm in human populations. *Genetika* **26**, 583–98.

Altukhov, Yu.P. & Pobedonostseva, E.Yu. (1978) Experimental modelling of genetical processes in subdivided populations. *Proceedings of the USSR Academy of Sciences* **238**, 466–9.

Altukhov, Yu.P. & Pobedonostseva, E.Yu. (1979a) Biological dynamics in an experimental population system of *Drosophila melanogaster*. *Journal of General Biology (Russia)* **40**, 368–76.

Altukhov, Yu.P. & Pobedonostseva, E.Yu. (1979b) Genetical dynamics in an experimental population system of *Drosophila melanogaster*. *Journal of General Biology (Russia)* **40**, 916–23.

Altukhov, Yu.P. & Rychkov, Yu.G. (1970) Population systems and their structural components: genetic stability and variability. *Journal of General Biology (Russia)* **31**, 507–26.

Altukhov, Yu.P. & Rychkov, Yu.G. (1972) Genetic monomorphism of species and its biological significance. *Journal of General Biology (Russia)* **33**, 281–300.

Altukhov, Yu.P. & Salmenkova, E.A. (1981) Applications of the stock concept to fish populations in the USSR. *Canadian Journal of Fisheries and Aquatic Sciences* **38**, 1591–600.

Altukhov, Yu.P. & Salmenkova, E.A. (1987a) Stock transfer relative to natural organization, management, and conservation of fish populations. In *Population Genetics and Fishery Management* (Ryman, N. & Utter, F., eds), pp. 333–43. University of Washington Press, Seattle.

Altukhov, Yu.P. & Salmenkova, E.A. (1987b) Population genetics of coldwater fish. In

Proceedings of the World Symposium on Selection, Hybridization, and Genetic Engineering in Aquaculture (Tiews, K., ed.), vol. 1, pp. 3–29. Heinemann, West Berlin.

Altukhov, Yu.P. & Salmenkova, E.A. (1990) Introduction of distinct stocks of chum salmon, *Oncorhynchus keta* (Walbaum), into natural populations of the species. *Journal of Fish Biology* **37** (Suppl. A), 25–33.

Altukhov, Yu.P. & Salmenkova, E.A. (1991) The genetic structure of salmon populations. *Aquaculture* **98**, 11–40.

Altukhov, Yu.P. & Salmenkova, E.A. (1994) Straying intensity and genetic differentiation in salmon populations. *Aquaculture and Fishery Management* **25** (Suppl. 2), 99–120.

Altukhov, Yu.P. & Varnavskaya, N.V. (1983) Adaptive genetic structure and its connection with intrapopulational differentiation for sex, age and growth rate in sockeye salmon, *Oncorhynchus nerka* (Walb.). *Genetika* **19**, 796–807.

Altukhov, Yu.P., Apekin, V.S. & Limansky, V.V. (1964) The basic principles of serological investigations of intra- and interspecific differentiation in fishes. In *Problems of Fish Physiology of the Black Sea and the Sea of Azov* (Shulman, G.E., ed.), pp. 53–71. Pishchevaya Promyshlennost, Moscow.

Altukhov, Yu.P., Limansky, V.V., Payusova, A.N. & Truveller, K.A. (1969a) Immunogenetical analysis of intraspecific differentiation of the European anchovy from the Black and Azov seas. 1. The blood groups and possible mechanism of genetic control. *Genetika* **5**, (4), 50–64.

Altukhov, Yu.P., Limansky, V.V., Payusova, A.N. & Truveller, K.A. (1969b) Immunogenetical analysis of intraspecific differentiation of the European anchovy from the Black and Azov seas. 2. Elementary populations of anchovy and its place in population genetical structure of species. *Genetika* **5**, (5), 81–94.

Altukhov, Yu.P., Salmenkova, E.A. & Sachko, G.D. (1970) Duplication and polymorphism of lactate dehydrogenase genes of the Pacific salmon. *Proceedings of the USSR Academy of Sciences* **195**, 711–14.

Altukhov, Yu.P., Salmenkova, E.A., Omelchenko, V.T., Sachko, G.D. & Slynko, V.I. (1972) Number of mono- and polymorphous loci in the population of tetraploid salmon species *Oncorhynchus keta* (Walb.). *Genetika* **8** (2), 67–75.

Altukhov, Yu.P., Salmenkova, E.A., Konovalov, S.M. & Pudovkin, A.I. (1975a) Stationary distributions of the frequencies of lactate dehydrogenase and phosphoglucomutase genes in the system of subpopulations of sockeye salmon *Oncorhynchus nerka* (Walb.). 1. Stability of the system over generations with simultaneous variability of the component subpopulation. *Genetika* **11** (4), 44–53.

Altukhov, Yu.P., Pudovkin, A.I., Salmenkova, E.A. & Konovalov, S.M. (1975b) Stationary distributions of the frequencies of lactate dehydrogenase and phosphoglucomutase genes in the system of subpopulations of sockeye salmon *Oncorhynchus nerka* (Walb.). 2. Random genetic drift, migration and selection as factors of stability. *Genetika* **11** (4), 54–62.

Altukhov, Yu.P., Bernashevskaya, A.G., Milishnikov, A.N. & Novikova, T.A. (1979) Experimental modelling of genetic processes in the populational system of *Drosophila melanogaster*, corresponding to the circular stepping-stone model. 1. The approach to the problem and local differentiation of *6-Gpd* and *Est-6* allele frequencies. *Genetika* **15**, 646–55.

Altukhov, Yu.P., Salmenkova, E.A., Ryabova, G.D. & Kulikova, N.I. (1980) Genetic differentiation of chum salmon *Oncorhynchus keta* (Walb.) populations and effectiveness of some acclimatization measures. *Marine Biology (Russia)* **3**, 23–38.

Altukhov, Yu.P., Salmenkova, E.A., Omelchenko, V.T. & Efanov, V.N. (1983) Genetic differentiation and population structure in the pink salmon of the Sakhalin–Kuril region. *Marine Biology (Russia)* **3**, 46–51.

Altukhov, Yu.P., Naidich, V.A. & Zhivotovsky, L.A. (1984) Computer simulation of the interaction between random genetic drift and gene migration in a system of partially isolated populations. *Genetika* **20**, 605–9.

Altukhov, Yu.P., Suskov, I.I., Afanasyev, K.I., Malinina, T.V., Shurhkal, A.V. & Rakitskaya, T.A. (1985) Frequency of rare protein variants in normal and congenitally malformed infants. *Genetika* **21**, 2031–43.

Altukhov, Yu.P., Salmenkova, E.A., Omelchenko, V.T., Rubtzova, G.A. & Dubrova, Yu.E. (1987) Balancing selection as a possible factor maintaining uniformity of allele frequencies of enzyme loci in populations of pink salmon, *Oncorhynchus gorbuscha* (Walb.). *Genetika* **23**, 1884–96.

Altukhov, Yu.P., Mezhzherin, S.V., Salmenkova, E.A. & Omelchenko, V.T. (1989) Effect of selective breeding on adaptive genetic and biological structure of *Oncorhynchus gorbuscha* (Walb.) population. *Genetika* **25**, 1843–53.

Altukhov, Yu.P., Salmenkova, E.A. & Kartavtzev, Yu.F. (1991) Relationship of allozyme heterozygosity with viability and growth rate of pink salmon. *Tsitologiya i Genetika* **25**, 47–51.

Altukhov, Yu.P., Korochkin, L.I. & Rychkov, Yu.G. (1996) Hereditary biochemical diversity in evolution and development. *Genetika* **32**, 1450–73.

Anas, R.E. (1959) Three-year-old pink salmon. *Journal of the Fisheries Research Board of Canada* **16** (1), 92–3.

Anderson, L., Ryman, N. & Ståhl, G. (1983) Protein loci in the arctic charr, *Salvelinus alpinus*: electrophoretic expression and genetic variability patterns. *Journal of Fish Biology* **23**, 75–93.

Anderson, S., Bankier, A.T., Barrel, B.G. *et al.* (1981) Sequence and organization of the human mitochondrial genome. *Nature* **190**, 457–65.

Andriyasheva, M.A. (1996) *Conception of Gene Pool Conservation in Wild Fish Populations.* Nauchniye Tetradi, St Petersburg.

Antonova, V.P. (1976) Biological estimation of the spawning Atlantic salmon in the Pechora River. In *Ecology and Systematics of Salmonid Fishes* (Nizovtsev, G.P., ed.), pp. 5–8. Nauka, Leningrad.

Antonova, V.P. & Chuksina, N.A. (1985) Influence of fishing on catches, return and structure of spawning stocks of Atlantic salmon in the Pechora River. In *The Population Biology and Ecology Investigations of the Salmonid Fishes in the Northern Waters* (Skarlato, O.A., ed.), pp. 5–10. ZIN AN SSSR, Leningrad.

Apostolidis, A., Karakousis, Y. & Triantaphyllidis, C. (1996a) Genetic divergence and phylogenetic relationships among *Salmo trutta* L. (brown trout) populations from Greece and other European countries. *Heredity* **76**, 551–60.

Apostolidis, A., Karakousis, Y. & Triantaphyllidis, C. (1996b) Genetic differentiation and phylogenetic relationships among Greek *Salmo trutta* L. (brown trout) populations as revealed by RFLP analysis of PCR-amplified mitochondrial DNA segments. *Heredity* **77**, 608–18.

Arnason, E. & Rand, D.M. (1992) Heteroplasmy of short tandem repeats in mitochondrial DNA of Atlantic cod, *Gadus morhua*. *Genetics* **132**, 211–20.

Arnheim, N. (1983) Concerted evolution of multigene families. In *Evolution of Genes and Proteins* (Nei, M. & Koehn, R.K., eds), pp. 38–61. Sinauer, Sunderland, MA.

Aro, K.V. & Shepard, M.P. (1967) Salmon of the North Pacific Ocean. 5. Pacific Salmon in Canada. *Bulletin of the International North Pacific Fisheries Commission* **23**, 225–322.

Aspinwall, N. (1973) Inheritance of α-glycerophosphate dehydrogenase in pink salmon, *Oncorhynchus gorbuscha* (Walb.). *Genetics* **73**, 639–43.

Aspinwall, N. (1974a) Genetic analysis of the North American population of pink salmon (*Oncorhynchus gorbuscha*), possible evidence for the neutral mutation random drift hypothesis. *Evolution* **28**, 295–305.

Aspinwall, N. (1974b) Genetic analysis of duplicate malate dehydrogenase loci in the pink salmon, *Oncorhynchus gorbuscha*. *Genetics* **76**, 65–72.

Avise, J.C. (1994) *Molecular Markers, Natural History and Evolution*. Chapman and Hall, New York.

Avise, J.C. & Vrijenhoek, R.C. (1987) Mode of inheritance and variation of mitochondrial DNA in hybridogenetic fishes of the genus *Poeciliopsis*. *Molecular Biology and Evolution* **4**, 514–25.

Avise, J.C., Arnold, J., Ball, R.M. *et al.* (1987) Intraspecific phylogeography: the mitochondrial DNA bridge between population genetics and systematics. *Annual Review of Ecological Systems* **18**, 489–522.

Ayala, F.J. (ed.) (1976) *Molecular Evolution*. Sinauer, Sunderland, MA.

Ayala, F.J. & Kiger, J.A. (1984) *Modern Genetics*. Benjamin/Commings Publishing Company, Menlo Park, MA.

Ayala F.J., Tracey, M.L., Barr, L.G., McDonald, J.E. & Peres-Sales, S. (1974a) Genetic variation in natural populations of five *Drosophila* species and the hypothesis of the selective neutrality of protein polymorphism. *Genetics* **77**, 343–84.

Ayala, F.J., Tracey, M.L., Hedgecock, D. & Richmond, R.C. (1974b) Genetic differentiation during the speciation process in *Drosophila*. *Evolution* **28**, 576–92.

Azbelev, V.V. (1960) Materials on biology of Atlantic salmon from the Kola Peninsula and its survival. *Scientific Papers of the Research Institute of Fisheries and Oceanography* **12**, 5–70.

Bachevskaya, L.T. (1990) *Interpopulation differences and intrapopulation differentiation in chum salmon of the North Okhotsk sea*. PhD thesis, Magadan.

Bailey, G.S. & Wilson, A.C. (1968) Homologies between isoenzymes of fishes and those of higher vertebrates. Evidence for multiple H4 lactate dehydrogenases in trout. *Journal of Biological Chemistry* **243**, 5843–53.

Bailey, G.S., Cock, G.T. & Wilson, A.C. (1969) Gene duplication in fishes: malate dehydrogenase of salmon and trout. *Biochemical and Biophysical Research Communications* **34**, 605–12.

Bailey, G.S., Wilson, A.C., Halver, J. & Johnson, C. (1970) Multiple forms of supernatant malate dehydrogenase in salmonid fishes: biochemical, immunological and genetic studies. *Journal of Biological Chemistry* **245**, 5927–40.

Bailey, G.S., Tsuyuki, H. & Wilson, A.C. (1976) The number of genes for lactate dehydrogenase in salmonid fishes. *Journal of the Fisheries Research Board of Canada* **33**, 760–67.

Baker, R. (1978) Demystifying vertebrate migration. *New Scientist* **80**, 526–8.

Bakkala, R.G. (1970) Synopsis of biological data on the chum salmon, *Oncorhyncus keta* (Walbaum) 1792. *FAO Fish Synopsis 41; US Fisheries and Wildlife Service Circular* **315**, 1–89.

Bakshtansky, E.L. (1962) Pink salmon in a lake. *Scientific–Technical Bulletin of the Polar Research Institute of Fisheries and Oceanography* **4**, 46–7.

Bakshtansky, E.L. (1980). The introduction of pink salmon into Kola peninsula. In: *Salmon Ranching* (Thorpe, J., ed.), pp. 245–59. Academic Press, London.

Bakshtansky, E.L., Klovach, N.V. & Lepskaya, V.A. (1991) Migration of Atlantic salmon. In *Biological Bases of the Distribution of Commercial and Food Marine Animals* (Neiman, A.A., Karpinsky, M.G. & Tarverdieva, M.N., eds.), pp. 94–116. All-Union Research Institute of Fisheries and Oceanography, Moscow.

Balanovskaya, E.V. & Rychkov, Yu.G. (1990) Ethnogenetics: adaptive structure of the mankind gene pool according to data on human polymorphic genetic markers. *Genetika* **26**, 739–48.

Bams, R.A. (1976) Survival and propensity for homing as affected by presence or absence of locally adapted paternal genes in two transplanted populations of pink salmon (*Oncorhynchus gorbuscha*). *Journal of the Fisheries Research Board of Canada* **33**, 2716–25.

Baranova, L.S. & Biske, S.F. (1964) *North-east of the USSR. History of Relief Development of Siberia and the Far East.* Nauka, Moscow.

Bartley, D.M. & Gall, G.A.E. (1990) Genetic structure and gene flow in chinook salmon populations of California. *Transactions of the American Fisheries Society* **119**, 55–71.

Bartley, D.M., Bagley, M., Gall, G.A.E. & Bentley, B. (1992a) Use of linkage disequilibrium data to estimate effective size of hatchery and natural fish populations. *Conservation Biology* **6**, 365–71.

Bartley, D.M., Bentley, B., Brogziak, J., Gomulkiewicz, R., Mangel, M. & Gall, G.A.E. (1992b) Geographic variation in population genetic structure of chinook salmon for California and Oregon. *Fishery Bulletin US* **90**, 77–100.

Bartley, D.M., Bentley, B., Olin, P.G. & Gall, G.A.E. (1992c) Population genetic structure of coho salmon (*Oncorhynchus kisutch*) in California. *California Fish and Game* **78**, 88–104.

Beacham, T.D. (1984) Age and morphology of chum salmon in southern British Columbia. *Transactions of the American Fisheries Society* **113**, 727–36.

Beacham, T.D. (1985) Meristic and morphometric variation in pink salmon (*Oncorhynchus gorbuscha*) in southern British Columbia and the Puget Sound. *Canadian Journal of Zoology* **63**, 366–72.

Beacham, T.D. (1988) A genetic analysis of early development in pink (*Oncorhynchus gorbuscha*) and chum (*Oncorhynchus keta*) salmon at three temperatures. *Genome* **30**, 89–96.

Beacham, T.D., Withler, R.E. & Gould, A.P. (1985a) Biochemical genetic stock identification of chum salmon (*Oncorhynchus keta*) in southern British Columbia. *Canadian Journal of Fisheries and Aquatic Sciences* **42**, 437–48.

Beacham, T.D., Withler, R.E. & Gould, A.P. (1985b) Biochemical genetic stock identification of pink salmon (*Oncorhynchus gorbuscha*) in southern British Columbia and the Puget Sound. *Canadian Journal of Fisheries and Aquatic Sciences* **42**, 1474–83.

Beacham, T.D., Gould, A.P., Withler, R.E., Murray, C.B. & Barner, L.W. (1987) Biochemical genetic survey and stock identification of chum salmon (*Oncorhynchus keta*) in British Columbia. *Canadian Journal of Fisheries and Aquatic Science* **44**, 1702–13.

Beacham, T.D., Withler, R.E., Murray, C.B. & Barner, L.W. (1988) Variation in body size, morphology, egg size, and biochemical genetics of pink salmon in British Columbia. *Transactions of the American Fisheries Society* **117**, 109–26.

Beacham, T.D., Murray, C.B. & Withler, R.E. (1989) Age, morphology, and biochemical genetic variation of Yukon River chinook salmon. *Transactions of the American Fisheries Society* **118**, 46–63.

Beacham, T.D., Withler, R.E. & Stevens, T.A. (1996) Stock identification of chinook salmon (*Oncorhynchus tshawytscha*) using minisatellite DNA variation. *Canadian Journal of Fisheries and Aquatic Sciences* **53**, 380–94.

Beacham, T.D., Margolish, L. & Nelson, R.J. (1998) A comparison of methods of stock identification for sockeye salmon (*Oncorhynchus nerka*) in Barkley Sound, British Columbia. In *Assessment and Status of Pacific Rim Salmonid Stocks, NPAFC Bulletin No. 1* Welch, D.W., Eggers, D.M., Wakabayashi, K. & Karpenko, V., eds.), pp. 227–39. North Pacific Anadromous Fish Commission, Vancouver.

Beamish, R.J. (ed.) (1995) Climate change and northern fish population. *Canadian Special Publications on Fisheries and Aquatic Science* **121**, 739.

Beckenbach, A.T. (1991) Rapid mtDNA sequence analysis of fish populations using the polymerase chain reaction (PCR). *Canadian Journal of Fisheries and Aquatic Sciences* **48** (Suppl. 1), 95–8.

Behnke, R.J. (1972) The systematics of salmonid fishes of recently glaciated lakes. *Journal of the Fisheries Research Board of Canada* **29**, 639–71.

Behnke, R.J. (1980) A systematic review of genus *Salvelinus*. In *Salmonid Fishes of the Genus Salvelinus* (Balon, E.K., ed.), pp. 441–80. Junk Publishers, The Hague.

Bentzen, P., Harris, A.S. & Wright, J.M. (1991) Cloning of hypervariable minisatellite and simple sequence microsatellite repeats for DNA fingerprinting of important aquacultural species of salmonids and tilapia. In *DNA Fingerprinting Approaches and Applications* (Burke, T., Dolf, G., Jeffreys, A.J. & Wolf, R., eds), pp. 243–62. Birkhauser Verlag, Basel.

Berg, L.S. (1935) Materials on Atlantic salmon biology. *Bulletin of the All-Union Research Institute on Lake and River Fisheries* **20**, 3–114.

Berg, L.S. (1940) System of modern and fossil Pisces and fish-like vertebrates. *Proceedings of the Zoological Institute of the USSR Academy of Sciences* **5** (2), 87–517.

Berg, L.S. (1948a) The origin of trouts and other freshwater salmonid fishes. In *In Memory of the Academician S.A. Zernov* (Pavlovsky, Ye.N., ed.), pp. 159–72. USSR Academy of Sciences, Moscow

Berg, L.S. (1948b) *Freshwater Fishes of the USSR and Adjacent Countries* Part 1. USSR Academy of Sciences Moscow.

Berg, W.J. & Ferris, S.D. (1984) Restriction endonuclease analysis of salmonid mitochondrial DNA. *Canadian Journal of Fisheries and Aquatic Sciences* **41**, 1041–7.

Berg, W.J. & Gall, G.A.E. (1988) Gene flow and genetic differentiation among California coastal rainbow trout populations. *Canadian Journal of Fisheries and Aquatic Sciences* **45**, 122–31.

Berger, E.M. & Weber, L. (1974) The ribosomes of *Drosophila*. 2. Studies on intraspecific variation. *Genetics* (US) **78**, 1173–83.

Bermingham, E., Forbes, S.H., Friedland, K. & Pla, C. (1991) Discrimination between Atlantic salmon (*Salmo salar*) of North American and European origin using restriction analyses of mitochondrial DNA. *Canadian Journal of Fisheries and Aquatic Sciences* **48**, 884–93.

Bernatchez, L. & Dodson, J.J. (1990) Allopatric origin of sympatric populations of lake whitefish (*Coregonus clupeaformis*) in the Allegash basin, northern Maine. *Evolution* **44**, 1263–71.

Bernatchez, L., Guyomard, R. & Bonhomme, F. (1992) DNA sequence variation of the mitochondrial control region among geographically and morphologically remote European brown trout, *Salmo trutta*, populations. *Molecular Ecology* **1**, 161–73.

Bespalyi, V.G. & Glushkova, O.Yu. (1987) *North-east. Quarternary Glaciations of the USSR Territory*. Nauka, Moscow.

Bickham, J.W., Wood, C.C. & Patton, J.C. (1995) Biogeographic implications of cytochrome *b* sequences and allozymes in sockeye (*Oncorhynchus nerka*). *Journal of Heredity* **86**, 140–44.

Bielak, A. (1992) Canadian Atlantic salmon recovery efforts. *Proceedings of Scientific Papers of the State Research Institute on Lake and River Fisheries* **304**, 85–99.

Bigler, B.S., Welch, D.W. & Helle, J.H. (1996) A review of size trend among North Pacific salmon (*Oncorhynchus* spp.). *Canadian Journal of Fisheries and Aquatic Sciences* **53**, 455–65.

Billington, N. & Hebert, P.D.H. (1991) Mitochondrial DNA diversity of fishes and its implications for introductions. *Canadian Journal of Fisheries and Aquatic Sciences* **48** (Suppl. 1), 80–94.

Birman, I.B. (1981) Sea migrations and origin of season races of the anadromous salmonid. *Journal of Ichthyology (Russia)* **21**, 37–48.

Birman, I.B. (1985) *Sea Life Period and Problems of Dynamics of Pacific Salmon Stocks*. Agropromizdat, Moscow.

Birt, T.P., Green, J.M. & Davidson, W.S. (1991) Mitochondrial DNA variation reveals genetically distinct sympatric populations of anadromous and non-anadromous Atlantic salmon, *Salmo salar*. *Canadian Journal of Fisheries and Aquatic Sciences* **48**, 577–82.

Blanco, G., Sanchez, J.A., Vazquez, E., Rubio, J. & Utter, F.M. (1992) Genetic differentiation among natural European populations of Atlantic salmon, *Salmo salar* L., from drainages of the Atlantic ocean. *Animal Genetics* **23**, 11–18.

Blank, M.L. & Altukhov, Yu.P. (1992) Genetic dymamics in populations with complex migration structure. *Proceedings of the USSR Academy of Sciences* **324**, 122–6.

Blank, M.L. & Altukhov, Yu.P. (1995) Disequilibrium genetic dynamics of population systems. *Proceedings of the USSR Academy of Sciences* **344**, 136–9.

Bodmer, W.F. & Cavalli-Sforza, L.L. (1968) A migration matrix model for the study of random genetic drift. *Genetics* **59**, 565–92.

Bowen, B.W., Meylan, A.B., Ross, J.P., Limpus, C.J., Balazs G.H. & Avise, J.C. (1992) Global population structure and natural history of the green turtle (*Chelonia mydas*) in terms of matriarchal phylogeny. *Evolution* **46**, 865–81.

Boyd, F.C. (1964) Return of pink salmon to Robertson Creek shows promise of success. *Canadian Fish Culturist* **32**, 59–62.

Braitseva, O.A. (1968) *The Stratigraphy of Quaternary Deposits and Glaciation in Kamchatka*. Nauka, Moscow.

Brodziak, J., Bentley, B., Bartley, D., Gall, G.A., Gomulkiewicz, R. & Mangel, M. (1992) Test of genetic stock identification using coded wire tagged fish. *Canadian Journal of Fisheries and Aquatic Sciences* **49**, 1507–17.

Brooker, A.L., Cook, D., Bentzen, P., Wright, J.M. & Doyle R.W. (1994) Organization of microsatellites differs between mammals and coldwater teleost fishes. *Canadian Journal of Fisheries and Aquatic Sciences* **51**, 1959–66.

Brown, A.J.L. & Langley, C.H. (1979) Re-evaluation of the level of genetic heterozygosity in natural populations of *Drosophila melanogaster* by two-dimensional electrophoresis. *Proceedings of the National Academy of Sciences of the USA* **76**, 2381–4.

Brown, W.M., George, M. & Wilson, A.C. (1979) Rapid evolution of animal mitochondrial DNA. *Proceedings of the National Academy of Sciences of the USA* **76**, 1967–71.

Brunner, P.C., Douglas, M.R. & Bernatchez, L. (1998) Microsatellite and mitochondrial DNA assessment of population structure and stocking effects in Arctic charr *Salvelinus alpinus* (Teleostei, Salmonidae) from Central Alpine lakes. *Molecular Ecology* **7**, 209–23.

Brykov, V.A., Polyakova, N., Skurihina, L.A. & Kukhlevsky, A.D. (1996) Geographical and temporal mitochondrial DNA variability in populations of pink salmon. *Journal of Fish Biology* **48**, 899–909.

Brykov, V.A., Polyakova, N., Skurihina, L.A. *et al.* (1999a) Population genetic structure of pink salmon, *Oncorhynchus gorbuscha* (Walbaum), by the results of restrictase analysis of mitochondrial DNA: dynamics of diversity in generations. *Genetika* **35**, 657–65.

Brykov, V.A., Polyakova, N., Skurihina, L.A. *et al.* (1999b) Population genetic structure of pink salmon, *Oncorhynchus gorbuscha* (Walbaum), by the results of restrictase analysis of mitochondrial DNA: temporal heterogeneity during the spawning run. *Genetika* **35**, 666–72.

Bumpus, H.C. (1899) The elimination of the unfit as illustrated by the introduced sparrow, *Passer domesticus*. *Biological Lectures of the Marine Biological Laboratory, Woods Hole* **11**, 209–26.

Burger, C.V., Spearman, W.J. & Cronin, M.A. (1997) Genetic differentiation of sockeye salmon subpopulations from geologically young Alaskan lake system. *Transactions of the American Fisheries Society* **126**, 926–38.

Burgner, R.L. (1991) Life history of sockeye salmon (*Oncorhynchus nerka*). In *Pacific Salmon Life Histories* (Groot, C. & Margolis L., eds), pp. 3–117. University of British Columbia Press, Vancouver.

Buri, P. (1956) Gene frequency in small populations of mutant *Drosophila*. *Evolution* **10**, 367–402.

Bushuyev, V.P. (1973) The two-component haemoglobins in salmon are associated with its allotetraploid origin. In *Fish Biochemical Genetics* (Troshin, A.S., ed.), pp. 62–6. Nauka, Leningrad.

Bushuyev, V.P., Omelchenko, V.T. & Salmenkova, E.A. (1975) Species-specificity and intraspecific constancy of electrophoregrams and thermal resistance of haemoglobins in some fishes (Clupeiformes). *Journal of General Biology (Russia)* **36**, 569–78.

Campton, D.E. & Johnston, J.M. (1985) Electrophoretic evidence for a genetic admixture of native and non-native rainbow trout in the Yakima River, Washington. *Transactions of the American Fisheries Society* **114**, 782–93.

Campton, D.E. & Utter, F.M. (1987) Genetic structure of anadromous cutthroat trout (*Salmo clarki clarki*) populations in the Puget Sound area: evidence for restricted gene flow. *Canadian Journal of Fisheries and Aquatic Sciences* **44**, 573–82.

Cann, R.L., Brown, W.M. & Wilson, A.C. (1984) Polymorphic sites and the mechanism of evolution in human mitochondrial DNA. *Genetics* **106**, 479–99.

Carl, L.M. & Healey, M.C. (1984) Differences in isozyme frequencies and body morphology among three juvenile life history types of chinook salmon (*Oncorhynchus tshawytscha*) in the Nanaimo River, British Columbia. *Canadian Journal of Fisheries and Aquatic Sciences* **41**, 1070–77.

Carlin, B. (1969) The migration of salmon. *Swedish Salmon Research Institute Report* **2–4**, 14–22.

Carney, B.L., Gray, A.K. & Gharrett, A.J. (1997) Mitochondrial DNA restriction site variation within and among five populations of Alaskan coho salmon (*Oncorhynchus kisutch*). *Canadian Journal of Fisheries and Aquatic Sciences* **54**, 940–49.

Carr, A. (1986) Rips, fads and little loggerheads. *Bioscience* **86**, 78–85.

Carr, A., Carr, M.H. & Meylan, A.B. (1978) The ecology and migration of sea turtles. 7. The West Caribbean green turtle colony. *Bulletin of the American Museum of Natural History* **162**, 1–46.

Carvalho, G.R. (1993) Evolutionary aspects of fish distribution: genetic variability and adaptation. *Journal of Fish Biology* **43** (Suppl. A), 53–73.

Carvalho, G.R. & Hauser, L. (1995) Molecular genetics and stock concept in fisheries. In *Molecular Genetics in Fisheries* (Carvalho, G.R. & Pitcher, T.J., eds.), pp. 55–79. Chapman & Hall, London.

Carvalho, G.R. & Pitcher, T.J. (eds) (1995) *Molecular Genetics in Fisheries*. Chapman & Hall, London.

Cavalli-Sforza, L.L. & Bodmer, W.F. (1971) *The Genetics of Human Populations*. Freeman, San Francisco.

Cavalli-Sforza, L.L., Barrai, J. & Edwards, A.W.F. (1964) Analysis of human evolution under random genetic drift. *Cold Spring Harbor Symposium on Quantitative Biology* **29**, 9–20.

Chebanov, N.A. (1979) Behaviour and mate pairing of chum salmon spawners of different sizes with dominance of males at a spawning ground. *Ekologiya* **2**, 72–9.

Chebanov, N.A. (1982) On spawning behaviour of pink salmon, *Oncorhynchus gorbuscha* (Walb.), with different sex ratio at a spawning ground. *Ekologiya* **1**, 57–65.

Chereshnev, I.A. (1996) *Biological Diversity of Freshwater Fish Fauna in the Russian North-east*. Dalnauka, Vladivostok.

Chereshnev, I.A. & Skopetz, M.B. (1990) *Salvethymus svetovidovi gen. et sp. nova* – New endemic fish of the salmonid (*Salmonidae*) subfamily from Elgigitgin Lake (Central Chukotka). *Journal of Ichthyology (Russia)* **30**, 201–13.

Chesser, R.K. (1983) Genetic variability within and among populations of the black-tailed prairie dog. *Evolution* **37**, 320–31.

Chetverikov, S.S. (1926) Some aspects of the evolutionary process from the viewpoint of modern genetics. *Journal of Experimental Biology (Russia)* **1**, 3–54.

Chigirinsky, A.I. (1993) Global natural factors, commercial fishing and numbers of Pacific salmon. *Fisheries (Russia)* **2**, 19–22.

Clifton, D.R. & Rodriguez, R.J. (1997) Characterization and application of a quantitative DNA marker that discriminates sex in chinook salmon (*Oncorhynchus tshawytscha*). *Canadian Journal of Fisheries and Aquatic Sciences* **54**, 2647–52.

Cope, O.B. (1957) Races of cutthroat trout in Yellowstone lake. *US Fish and Wildlife Service Special Report – Fisheries* **208**, 74–84.

Coulthart, M.B., Rhomberg, L.R. & Singh, R.S. (1984) The nature of genetic variation for species formation. *Evolution* **38**, 689–92.

Coyne, J.A. (1992). Genetics and speciation. *Nature* **355**, 511–15.

Crane, A., Seeb, L.W. & Seeb, J.E. (1994) Genetic relationships among *Salvelinus* species inferred from allozyme data. *Canadian Journal of Fisheries and Aquatic Sciences* **51** (Suppl. 1), 182–97.

Cronin, M.A., Spearman, W.J., Wilmot, R.L., Patton, J.C. & Bickham, J.W. (1993). Mitochondrial DNA variation in chinook (*Oncorhynchus tshawytscha*) and chum salmon (*O. keta*) detected by restriction enzyme analysis of polymerase chain reaction (PCR) products. *Canadian Journal of Fisheries and Aquatic Sciences* **50**, 708–15.

Cross, T.F. & King, J. (1983) Genetic effects of hatchery rearing in Atlantic salmon. *Aquaculture* **33**, 33–40.

Cross, T.F. & Payne, R.H. (1971) NADP-isocitrate dehydrogenase polymorphism in the Atlantic salmon, *Salmo salar. Fish Biology* **11**, 493–6.

Cross, T.F. & Ward, R.D. (1980) Protein variation and duplicate loci in the Atlantic salmon, *Salmo salar* L. *Genetic Research* **36**, 147–65.

Cross, T.F., Ward, R.D. & Abreu-Grobois, A. (1979) Duplicate loci and allelic variation for mitochondrial malic enzyme in the Atlantic salmon *Salmo salar* L. *Comparative Biochemistry and Physiology* **62B**, 403–6.

Crow, J.F. (1954) Breeding structure of populations. 2. Effective population number. In *Statistics and Mathematics in Biology* (Bancroft, T.A., Gowen, J.W. & Lush, J.L., eds), pp. 543–56. Iowa State College Press, Ames.

Crow, J.F. (1958) Some possibilities for measuring selection intensities in man. *Human Biology* **30**, 1–13.

Crow, J.W. & Morton, N.E. (1955) Measurement of gene frequency drift in small populations. *Evolution* **9**, 202–14.

Crozier, W.W. & Moffett, I.J.J. (1990) Inheritance of allozymes in Atlantic salmon (*Salmo salar* L.). *Aquaculture* **88**, 253–62.

Danzmann, R.G., Ferguson, M.M. & Allendorf, F.W. (1985) Allelic differences in initial expression of parental alleles at an isocitrate dehydrogenase locus in rainbow trout (*Salmo gairdneri*). *Developmental Genetics* **5**, 117–27.

Danzmann, R.G., Ferguson, M.M. & Allendorf, F.W. (1988) Heterozygosity and components of fitness in strains of rainbow trout. *Biological Journal of the Linnean Society* **33**, 285–304.

Danzmann, R.G., Ihssen, P.E. & Hebert, P.D.N. (1991) Genetic discrimination of wild and hatchery populations of brook charr, *Salvelinus fontinalis* (Mitchill), in Ontario using mitochondrial DNA analysis. *Journal of Fish Biology* **39** (Suppl. A), 69–77.

Datta, U., Datta, P. & Mandal, R.K. (1988) Cloning and characterization of highly repetitive fish nucleotide sequence. *Gene* **62**, 331–6.

Davidson, W.S., Birt, T.P. & Green, J.M. (1989a) Organization of the mitochondrial genome from Atlantic salmon (*Salmo salar*). *Genome* **32**, 340–2.

Davidson, W.S., Birt, T.P. & Green, J.M. (1989b) A review of genetic variation in Atlantic salmon, *Salmo salar* L., and its importance for stock identification, enhancement programmes and aquaculture. *Journal of Fish Biology* **34**, 547–60.

Davisson, M.T., Wright, J.E. & Atherton, L.M. (1973) Cytogenetic analysis of pseudolinkage of LDH loci in the teleost genus *Salvelinus*. *Genetics* **73**, 645–58.

Dempson, J.B., Verspoor, E. & Hammar, J. (1988) Intrapopulation variation of the esterase-2 polymorphism in the serum of anadromous Arctic charr, *Salvelinus alpinus*, from a northern Labrador river. *Canadian Journal of Fisheries and Aquatic Sciences* **43**, 463–8.

Devlin, R.H., McNeil, B.K., Groves, D.D. & Donaldson, E.M. (1991) Isolation of a Y-chromosomal DNA probe capable of determining genetic sex in chinook salmon (*Oncorhynchus tshawytscha*). *Canadian Journal of Fisheries and Aquatic Sciences* **48**, 1606–12.

Dirin, D.K. (1980) Basic parameters of size–age diversity of salmonid fishes. In *Salmonid Fishes* (Skarlato, O.A., ed.), pp. 133–58. ZIN AN SSSR, Leningrad.

Dirin, D.K. (1987) Dynamics of structure and fishing of salmonid fishes. In *Ecological and Physiological Investigations of Commercial Fishes of the North Basin* (Matishov, G.G., ed.), pp. 39–61. Nauka, Leningrad.

Dobzhansky, T. (1970) *Genetics of the Evolutionary Process*. Columbia University Press, New York.

Dobzhansky, T., Ayala, F.J., Stebbins, G.L. & Valentine, J.W. (1977) *Evolution*. W.H. Freeman, San Fransisco.

Domanico, M.J. & Phillips, R.B. (1995) Phylogenetic analysis of the Pacific salmon (genus *Oncorhynchus*) based on mitochondrial DNA sequence data. *Molecular Phylogenetics and Evolution* **4**, 366–71.

Donaldson, L.R. & Allen, G.H. (1957) Return of silver salmon *O. kisutch* to point of release. *Journal of the Fisheries Research Board of Canada* **87**, 13–22.

Donnelly, R.F., Johnson, K.B., Hersberger, W.K. & Bevan, D.E. (1979) *Identification of Kodiak Island pink salmon population based on biochemical genetic variation*. Final report, University of Washington, Seattle.

Dorofeeva, E.A. (1982) Classification and phylogeny of Pacific salmon (subfamily *Salmoninae*). In *Systematics and Ecology of Bony Fishes* (Korovina, V.M., ed.), pp. 15–22. ZIN AN SSSR, Leningrad.

Dorofeeva, E.A. (1985) Principles of classification of salmonid fishes. In *Morphology and Systematics of Salmonid Fishes* (Korovina, V.M. & Dorofeeva, E.A., eds), pp. 4–12. ZIN AN SSSR, Leningrad.

Dorofeeva, E.A. (1994) Modern views on classification of Pacific trouts (genus *Parasalmo*) and salmon (genus *Oncorhynchus*). In *Systematics, Biology and Biotechnology of Breeding of Salmonid Fishes* (Reshetnikov, Yu.S., ed.), pp. 55–7. GOSNIORKH, St Petersburg.

Dowling, T.E., Moritz, C. & Palmer, J.D. (1990) Nucleic acids II: restriction site analysis. In *Molecular Systematics* (Hillis, D.M. & Moritz C., eds), pp. 250–317. Sinauer, Sunderland, MA.

Du, S.J., Devlin, R.H. & Hew, C.L. (1993) Genomic structure of growth hormone genes in chinook salmon (*Oncorhynchus tshawytscha*): presence of two functional genes, GH-I and GH-II, and a male-specific pseudogene, GH-Y. *DNA and Cell Biology* **12**, 739–51.

Dubinin, N.P. (1931) Genetic automatic processes and their influence on the mechanisms of organic evolution. *Journal of Experimental Biology (Russia)* **7**, 463–79.

Dubinin, N.P. (1948) Experimental analysis of the integration of hereditary systems in the evolution of populations. *Journal of General Biology (Russia)* **9**, 203–44.

Dubinin, N.P. & Romashov, D.D. (1932) The genetic structure of a species and its evolution. *Biological Journal (Russia)* **1**, 52–95.

Dubinin, N.P., Romashov, D.D. & Geptner, M.A. (1937) Aberrant polymorphism in *Drosophila fasciata* Meig. *Biological Journal (Russia)* **6**, 311–84.

Dubrova, Yu.E. (1980) Stabilizing selection and distribution of allele frequencies of the loci coding of protein structure. *Proceedings of the USSR Academy of Sciences* **252**, 472–5.

Dubrova, Yu.E. & Gavrilets, S.Yu. (1989) Epistatic interaction of genes in progeny from distant marriages within Russian population. *Proceedings of the USSR Academy of Sciences* **309**, 211–15.

Dubrova, Yu.E., Karaphet, T.M., Sukernik, R.I. & Goltzova, T.V. (1990) Heterozygosity and fertility relationship in the forest nentzy and nganasans. *Genetika* **26**, 122–9.

Dubrova, Yu.E., Dambueva, L.K., Kholod, O.N., Prokhorovskaya, V.D. & Pushkina, E.I. (1991) Influence of the mother's heterozygosity on the variation of anthropometric traits in the newborn. *Genetika* **27**, 2128–76.

Dubrova, Yu.E., Salmenkova, E.A., Altukhov, Yu.P., Kartavtsev, Yu.F., Kal'kova, E.V. & Omelchenko, V.T. (1995) Family heterozygosity and progeny body length in pink salmon *Oncorhynchus gorbuscha* (Walbaum). *Heredity* **75**, 281–9.

Dvinin, P.A. (1952) Salmonid fish of south Sakhalin. *Bulletin of the Pacific Research Institute of Fisheries and Oceanography* **37**, 69–108.

Dyagilev, S.E. & Markevich, N.B. (1979) Time differences in the maturation of *Oncorhynchus gorbuscha* (Walb.) generations of odd and even years as the main factor influencing their acclimatization in the north of the European part of the USSR. *Journal of Ichthyology (Russia)* **19**, 230–45.

Efanov, V.N. (1989) Population structure of pink salmon reproducing in the rivers of Sakhalin region. In *Reserves of Salmon Fisheries of the Far East* (Glubokovsky, M.K., ed.), pp. 52–65. DVO AN SSSR, Vladivostok.

Efanov, V.N. & Kochneva, Z.P. (1980) On the second catch of sex-mature males of pink salmon in the age smaller than one year. *Marine Biology (Russia)* **2**, 88–9.

Efanov, V.N., Kaev, A.M. & Kovtun, A.A. (1979) Results of an introduction of fall chum from the Kurilka River into the Naiba River. *Bulletin of the Pacific Research Institute of Fisheries and Oceanography* **103**, 86–93.

Ehrlich, P.R. & Holm, R.W. (1963) *The Process of Evolution*. McGraw-Hill, New York.

Ehrlich, P.R. & Raven, P.H. (1969) Differentiation of populations. *Science* **165**, 1228–32.

Elo, K. (1993) Gene flow and conservation of genetic variation in anadromous Atlantic salmon (*Salmo salar*). *Hereditas* **119**, 149–59.

Elo, K., Vuorinen, J.A. & Niemela, E. (1994) Genetic resources of Atlantic salmon (*Salmo salar* L.) in Teno and Naatamo rivers, north-eastern Europe. *Hereditas* **120**, 19–28.

Elo, K., Ivanoff, S., Vuorinen, J.A. & Piironen, J. (1997) Inheritance of RAPD markers and detection of interspecific hybridization with brown trout and Atlantic salmon. *Aquaculture* **152**, 55–65.

Engel, W., Schmidtke, J. & Wolf, U. (1971) Genetic variation of α-glycerophosphate dehydrogenase isoenzymes in clupeoid and salmonid fish. *Experientia* **27**, 1489–91.

Engel, W., Schmidtke, J. & Wolf, U. (1975) Diploid-tetraploid relationships in teleostean fishes. *Isozymes* **4**, 449–62.

Enyutina, R.I. (1972) Amur's pink salmon [Fishery – biological essay]. *Bulletin of the Pacific Research Institute of Fisheries and Oceanography* **77**, 3–126.

Ermolenko, L.N. & Rudminaitis, E.A. (1984) Polymorphism of esterase D in salmonid fishes. *Genetika* **20**, 1204–10.

Ermolenko, L.N., Rudminaitis, E.A. & Makoedov, A.N. (1988) Genetics of NADP-dependent malate dehydrogenase isozymes of the salmonid fishes. *Genetika* **24**, 528–38.

Ershov, P.N. (1985) On biology of Atlantic salmon from the Pila and Tshchoguy rivers of the Kola peninsula. In *Studies on Population Biology and Ecology of the Salmonid Fishes in the North Water Bodies* (Skarlato, O.A., ed.), pp. 94–119. ZIN AN SSSR, Leningrad.

Estoup, A., Presa, P., Krieg, F., Vaiman, D. & Guyomard, R. (1993) $(CT)_n$ and $(GT)_n$

microsatellites: a new class of genetic markers for *Salmo trutta* L. (brown trout). *Heredity* **71**, 488–96.

Estoup, A., Rousset, F., Michalakis, Y., Cornuet, J.M., Adriamanga, M. & Guyomard, R. (1998) Comparative analysis of microsatellite and allozyme markers – a case study investigating microgeographic differentiation in brown trout (*Salmo trutta*). *Molecular Ecology* **7**, 339–53.

Evsyukov, A.N., Zhukova, O.V. & Rychkov, Yu.G. (1999) The geography of genetic processes in the human population: gene migration in Northern Eurasia (the Central Asian Region). *Genetika* **35**, 83–94.

Excoffier, L., Smouse, P.E. & Quattro, J.M. (1992) Analyses of molecular variance inferred from metric distances among DNA haplotypes: application to human mitochondrial DNA restriction data. *Genetics* **131**, 479–91.

Falconer, D.S. (1960) *Introduction to Quantitative Genetics*. Oliver & Boyd, Edinburgh.

Favro, L.D., Kuo, P.K. & McDonald, J.F. (1979) Population genetic study of the effects of selective fishing on the growth rate of trout. *Journal of the Fisheries Research Board of Canada* **36**, 552–61.

Favro, L.D., Kuo, P.K. & McDonald, J.F. (1980) Effects of unconventional size limits on the growth rate of trout. *Canadian Journal of Fisheries and Aquatic Sciences* **37**, 873–6.

Favro, L.D., Kuo, P.K., McDonald, J.F., Favro, D.D. & Kuo, A.D. (1982) A multi-locus genetic model applied to the effects of selective fishing on the growth rate of trout. *Canadian Journal of Fisheries and Aquatic Sciences* **39**, 1540–3.

Felsenstein, J. (1992) *PHYLIP (Phylogeny inference package) Version 3.4*. Department of Genetics, SK-50, Washington University, Seattle.

Ferguson, A. (1989) Genetic differences among brown trout, *Salmo trutta*, stocks and their importance for conservation and management of the species. *Freshwater Biology* **21**, 35–46.

Ferguson, A. & Fleming, C.C. (1983) Evolutionary and taxonomic significance of protein variation in the brown trout (*Salmo trutta* L.) and other salmonid fishes. In *Protein Polymorphism: Adaptive and Taxonomic Significance* (Oxford, G.S. & Rollinson, D., eds), pp. 85–99. Academic Press, London.

Ferguson, A. & Mason, F.M. (1981) Allozyme evidence for reproductively isolated sympatric populations of brown trout *Salmo trutta* L. in Loch Melvin, Ireland. *Journal of Fish Biology* **18**, 629–42.

Ferguson, A., Hynes, R.A., Prodohl, P.A. & Taggart, J.B. (1995) Molecular approaches to the study of genetic variation in salmonid fishes. *Nordic Journal of Freshwater Research* **71**, 13–32.

Ferguson, M.M. (1992) Enzyme heterozygosity and growth in rainbow trout: genetic and physiological explanations. *Heredity* **68**, 115–22.

Ferguson, M.M. (1994) Molecular genetics and the impact of aquaculture on fisheries. *Reviews of Fish Biology and Fisheries* **4**, 351–73.

Ferguson, M.M. (1995) The role of molecular genetic markers in the management of cultured fishes. In *Molecular Genetics in Fisheries* (Carvalho, G.R. & Pitcher, T.J., eds), pp. 81–103. Chapman & Hall, London.

Ferguson, M.M., Danzman, R.G. & Allendorf, F.W. (1987) Adaptive significance of developmental rate in rainbow trout: an experimental test. *Biological Journal of the Linnean Society* **33**, 205–16.

Ferris, S.D. & Berg, W.J. (1987) The utility of mitochondrial DNA in fish genetics and management. In *Population Genetics and Fishery Management* (Ryman, N. & Utter, F., eds), pp. 277–301. University of Washington Press, Seattle.

Ferris, S.D., Sage, R.D., Prager, E.M., Ritte, U. & Wilson, A.C. (1983) Mitochondrial DNA evolution in mice. *Genetics* **105**, 681–721.

Fisher, R.A. (1930) *The Genetical Theory of Natural Selection*. Clarendon Press, Oxford.

Flint, R.F. (1971) *Glacial and Quaternary Geology*. Wiley, New York.

Foerster, R.E. (1954a) On the relation of adult sockeye salmon (*Oncorhynchus nerka*) returns to known smolt seaward migrations. *Journal of the Fisheries Research Board of Canada* **11**, 339–50.

Foerster, R.E. (1954b) Sex ratio in sockeye salmon (*Oncorhynchus nerka* Walb.). *Journal of the Fisheries Research Board of Canada* **11**, 988–97.

Foerster, R.E. (1968) The sockeye salmon, *Oncorhynchus nerka*. *Bulletin of the Fisheries Research Board of Canada* **162**, 1–422.

Fontaine, P.M., Dodson, J.J., Bernatchez, L. & Slettan, A. (1997) A genetic test of metapopulation structure in Atlantic salmon (*Salmo salar*) using microsatellites. *Canadian Journal of Fisheries and Aquatic Sciences* **54**, 2434–42.

Fontdevila, A. (1992) Genetic instability and rapid speciation: are they coupled? *Genetica* (Netherlands) **86** (1–3), 247–58.

Foote, C.J. (1988) Male mate choice dependent on male size in salmon. *Behaviour* **106**, 63–80.

Foote, C.J., Wood, C.C. & Withler, R.E. (1989) Biochemical genetic comparison of sockeye salmon and kokanee, the anadromous and non-anadromous forms of *Oncorhynchus nerka*. *Canadian Journal of Fisheries and Aquatic Sciences* **46**, 149–58.

Forbes, S.H. & Allendorf, F.W. (1991) Associations between mitochondrial and nuclear genotypes in cutthroat hybrid swarms. *Evolution* **45**, 1332–49.

Ford, E.B. (1940) Polymorphism and taxonomy. In *The New Systematics* (Huxley, J.S., ed.), pp. 493–513. Clarendon Press, Oxford.

Foster, R.W., Bagatell, C. & Fuss, H.J. (1980) Return of one-year-old pink salmon to stream in Puget Sound. *Progress in Fish Culture* **43**, 31.

Fournier, D.A., Beacham, T.D., Riddell, B.E. & Busack, C.A. (1984) Estimating stock composition in mixed-stock fisheries using morphometric, meristic, and electrophoretic characteristics. *Canadian Journal of Fisheries and Aquatic Sciences* **41**, 400–8.

Frankham, R. (1995) Effective populations size/adult population size ratios in widlife: a review. *Genetic Research* **66**, 95–107.

French, R.H., Bilton, M., Osako, M. & Hartt, A. (1976) Distribution and origin of sockeye salmon (*Oncorhynchus nerka*) in offshore water of the North Pacific Ocean. *Bulletin of the International North Pacific Fisheries Commission* **34**, 1–113.

Frolov, S.V. (1981) Investigation of karyotype of *Salmo mykiss* by the differential chromosome staining method. *Marine Biology (Russia)* **6**, 73–5.

Gagalchy, N.G. (1985) Biochemical polymorphism of the Kamchatka pink salmon *Oncorhynchus gorbuscha* (Walb.). 1. The allelic frequencies of polymorphic loci in the odd-year generation. *Genetika* **21**, 854–60.

Gagalchy, N.G. (1986) Biochemical polymorphism of the Kamchatka salmon *Oncorhynchus gorbuscha* (Walb.). 2. The allelic frequencies of polymorphic loci in the even-year generation. *Genetika* **22**, 2839–46.

Gagalchy, N.G. (1988) Inheritance of liver isocitrate dehydrogenase in the pink salmon, *Oncorhynchus gorbuscha* (Walb.). *Genetika* **24**, 324–30.

Galvin, P., Cross, T. & Ferguson, A. (1994) Genetic differentiation and gene flow in Atlantic salmon, *Salmo salar* L: a case study of the River Shannon system in Ireland. *Aquaculture and Fishery Management* **25** (Suppl. 2), 131–45.

Galvin, P., Taggart, J., Ferguson, A., O'Farrell, M. & Cross, T. (1996) Population genetics of Atlantic salmon (*Salmo salar*) in the River Shannon system in Ireland. *Canadian Journal of Fisheries and Aquatic Sciences* **53**, 1933–42.

Garcia-Marin, J.L. & Pla, C. (1996) Origins and relationships of native populations of *Salmo trutta* (brown trout) in Spain. *Heredity* **77**, 313–23.

Garcia-Marin, J.L., Jorde, P.E., Ryman, N., Utter, F. & Pla, C. (1991) Management implications of genetic differentiation between native and hatchery populations of brown trout (*Salmo trutta*) in Spain. *Aquaculture* **95**, 235–49.

Garcia-Marin, J.L., Utter, F.M. & Pla, C. (1999) Postglacial colonization of brown trout in Europe based on distribution of allozyme variants. *Heredity* **82**, 46–56.

Gardner, M.L. (1976) A review of factors which may influences the sea – age and maturation of Atlantic salmon (*Salmo salar* L.). *Journal of Fish Biology* **4**, 289–327.

Gavrilets, S. & Hastings, A. (1996) Founder effect speciation: a theoretical reassessment. *American Nature* **147**, 466–91.

Gharrett, A.J. & Smoker, W.W. (1991) Two generations of hybrids between even-and odd-year pink salmon (*Oncorhynchus gorbuscha*): a test for outbreeding depression? *Canadian Journal of Fisheries and Aquatic Sciences* **48**, 1744–9.

Gharrett, A.J., Shirley, S.M. & Tromble, G.R. (1987) Genetic relationships among populations of Alaskan chinook salmon (*Oncorhynchus tshawytscha*). *Canadian Journal of Fisheries and Aquatic Sciences* **44**, 765–74.

Gharrett, A.J., Smoot, C., McGregor, A.J. & Holmes, P.B. (1988) Genetic relationships of even-year north-western Alaskan pink salmon. *Transactions of the American Fisheries Society* **117**, 536–45.

Gilpin, M.E. (1991) The genetic effective size of a metapopulation. In *Metapopulation Dynamics: Empirical and Theoretical Investigations* (Gilpin, M. & Hanski, I., eds), pp. 165–75. Academic Press, London.

Ginatulina, L.K. (1992) Genetic differentiation among chum salmon, *Oncorhynchus keta* (Walbaum), from Primorye and Sakhalin. *Journal of Fish Biology* **40**, 33–8.

Ginatulina, L.K. & Maksimovich, A.A. (1994) Inheritance of mitochondrial DNA in the artificial hybrids of pink salmon *Oncorhynchus gorbuscha* (Walb.) and masu salmon *Oncorhynchus masou* (Brev.). *Genetika* **30**, 1248–50.

Ginatulina, L.K. & Mashkin, S.A. (1990) Intraspecific mitochondrial DNA polymorphism of chum salmon (*Oncorhynchus keta* Walb.) from Primorye and Sakhalin. *Genetika* **26**, 729–38.

Gjedrem, T. (1983) Genetic variation in quantitative traits and selective breeding in fish and shellfish. *Aquaculture* **33**, 51–72.

Gjerde, B., Gunnes, K. & Gjedrem, T. (1983) Effect of inbreeding on survival and growth in rainbow trout. *Aquaculture* **34**, 327–32.

Gliddon, C. & Goudet, J. (1994) The genetic structure of metapopulations and conservation biology. In *Conservation Genetics* (Loeschcke, V.L., Tomiuk, J. & Jain, S.K., eds), pp. 107–14. Birkhauser Verlag, Basel.

Glubokovsky, M.K. (1995) *Evolutionary Biology of Salmonid Fishes*. Nauka, Moscow.

Glubokovsky, M.K. & Glubokovskaya, E.V. (1981) The paths of evolution of Pacific salmon of the genus *Oncorhynchus* Suckley. In *Fishes in Ecosystems of Far East Salmon Rivers* (Levanidov, V.Ya., ed.), pp. 5–66. DVNTS, Vladivostok.

Glubokovsky, M.K. & Zhivotovsky, L.A. (1986) Population structure in pink salmon: a system of fluctuating stocks. *Marine Biology (Russia)* **2**, 39–44.

Glubokovsky, M.K. & Zhivotovsky, L.A. (1989) Population structure of pink salmon: facts and models. In *Genetics in Aquaculture* (Kirpichnikov, V.S., ed.), pp. 47–67. Nauka, Leningrad.

Glubokovsky, M.K., Zhivotovsky, L.A., Viktorovsky, R.M. *et al.* (1989) Population structure of pink salmon. *Genetika* **25**, 1275–85.

Glubokovsky, M.K., Efremov, V.V., Frolov, S.V., Ribnikova, I.G. & Katugin, O.N. (1993) Phylogenetic relationships and systematics of charrs in Elgigitgin Lake. In *Nature of Elgigitgin Lake Depression* (Beliy, V.F. & Chereshnev, I.A., eds), pp. 149–77. IBPS, Magadan.

Glushkova, O.Yu. (1984) Morphology and palaeogeography of late-Pleistocene glaciations in the north-east of the USSR. In *Pleistocene Glaciations of Eastern Asia* (Bespaliy, V.G., ed.), pp. 28–43. DVNTS, Magadan.

Godfrey, H., Henry, K.A. & Machidori, S. (1975) Distribution and abundance of coho salmon in the offshore waters of the North Pacific Oceans. *Bulletin of the International North Pacific Fisheries Commission* **31**, 1–80.

Goldstein, D.B., Linares, A.R. & Cavalli-Sforza, L.L. (1995) An evaluation of genetic distances for use with microsatellite loci. *Genetics* **139**, 463–71.

Gorshkov, S.A. (1988) Some features of intraspecies structure of pink salmon, *Oncorhynchus gorbuscha* (Salmoniformes, *Salmonidae*). 2. Morphological diversity and spatial isolation of spawning groups. *Journal of Zoology (Russia)* **67**, 576–88.

Gorshkov, S.A., Dorofeeva, E.A., Klyukanov, V.A. & Kulikova, N.I. (1979) Osteological features of Pacific salmon of the genus *Oncorhynchus*. *Journal of Ichthyology (Russia)* **19**, 963–82.

Gorshkova, G.V. & Gorshkov, S.A. (1983) Chromosome polymorphism of pink salmon in odd and even years. *Proceedings of the USSR Academy of Sciences* **273**, 1023–4.

Gorshkova, G.V. & Gorshkov, S.A. (1985) Karyotypes of Kamchatkan Pacific trout of the genus *Salmo*. *Journal of Ichthyology (Russia)* **25**, 553–60.

Gorshkova, G.V., Gorshkov, S.A. & Kinas, N.M. (1988) Chromosome polymorphism in pink salmon *Oncorhynchus gorbuscha* (Walb.) from the populations of even- and odd-year broodlines. *Genetika* **24**, 1873–81.

Grant, W.S., Milner, G.B., Krasnowski, P. & Utter, F.M. (1980) Use of biochemical genetic variants for identification of sockeye salmon (*Oncorhynchus nerka*) stocks in Cook Inlet, Alaska. *Canadian Journal of Fisheries and Aquatic Sciences* **37**, 1326–47.

Greene, B.A. & Seeb, J.E. (1997) SINE and transposon sequences generate high-resolution DNA fingerprints, SINE prints, that exhibit faithful Mendelian inheritance in pink salmon (*Oncorhynchus gorbuscha*). *Molecular Marine Biology and Biotechnology* **6**, 328–38.

Grewe, P., Krueger, C.C., Aquadro, C.F., Bermingham, E., Kincaid, H.L. & May, B. (1993) Mitochondrial DNA variation among lake trout (*Salvelinus namaycush*) strains stocked into Lake Ontario. *Canadian Journal of Fisheries and Aquatic Sciences* **50**, 2397–403.

Gritsenko, O.F. (1981) On population structure of pink salmon *Oncorhynchus gorbuscha* (Walbaum). *Journal of Ichthyology (Russia)* **21**, 787–99.

Gritsenko, O.F. (1990) Population structure of Sakhalin's pink salmon *Oncorhynchus gorbuscha*. *Journal of Ichthyology (Russia)* **30**, 825–35.

Groot, C. & Margolis, L. (eds) (1991) *Pacific Salmon Life Histories*. University of British Columbia Press, Vancouver.

de Groot, S.J. & Roberts, R.J. (eds) (1994) Homing and straying in salmon. *Aquaculture and Fisheries Management* **5** (Suppl. 2), 1–190.

Gross, M.R. (1985) Disruptive selection for alternative life histories in salmon. *Nature* **313**, 47–8.

Gross, R. & Nilsson, J. (1995) Application of heteroduplex analysis for detecting variation within the growth hormone 2 gene in *Salmo trutta* L. (brown trout). *Heredity* **74**, 286–95.

Guyomard, R. & Krieg, F. (1983) Electrophoretic variation in six populations of brown trout (*Salmo trutta* L.). *Canadian Genetics and Cytology* **25**, 403–13.

Gyllensten, U. & Wilson, A.C. (1987) Mitochondrial DNA of salmonids: inter- and intraspecific variability detected with restriction enzymes. In *Population Genetics and Fishery Management* (Ryman, N. & Utter, F., eds), pp. 301–17. University of Washington Press, Seattle.

Gyllensten, U., Leary, R.F., Allendorf, F.W. & Wilson, A.C. (1985) Introgression between two cutthroat trout subspecies with substantial karyotypic, nuclear and mitochondrial genomic divergence. *Genetics* **111**, 905–15.

Haldane, J.B.S. (1957) The cost of natural selection. *Journal of Genetics* **55**, 511–24.

Haldane, J.B.S. (1960) More precise expression for the cost of natural selection. *Journal of Genetics* **57**, 351–60.

Hamilton, K.A., Ferguson, A., Taggart, J.B., Tomasson, T., Walker, A. & Fahy, E. (1989) Postglacial colonization of brown trout, *Salmo trutta* L.: LDH-5 as a phylogeographic marker locus. *Journal of Fish Biology* **35**, 651–64.

Hansen, L.P., Jonsson, J. & Andersen, R. (1989) Salmon ranching experiments in the River Imsa: is homing dependent on sequential imprinting of the smolts? In *Salmon Migration and Distribution Symposium* (Brannon, E. & Jonsson, B., eds.), pp. 19–29. University of Washington, Seattle.

Hansen, M.M. & Loeschcke, V. (1996) Temporal variations in mitochondrial DNA haplotype frequencies in a brown trout (*Salmo trutta* L.) population that shows stability in nuclear allele frequencies. *Evolution* **50**, 454–7.

Hansen, M.M. & Mensberg, R.L.D. (1998) Genetic differentiation and relationship between genetic and geographical distance in Danish sea trout (*Salmo trutta* L.) populations. *Heredity* **81**, 493–504.

Hansen, M.M., Loeschcke, V., Rasmussen, G. & Simonsen, V. (1993) Genetic differentiation among Danish brown trout (*Salmo trutta*) populations. *Hereditas* **118**, 177–85.

Hansen, M.M., Hynes, R.A., Loeschcke, V. & Rasmussen, G. (1995) Assessment of the stocked or wild origin of anadromous brown trout (*Salmo trutta*) in the Danish river system, using mitochondrial DNA RFLP analysis. *Molecular Ecology* **4**, 189–98.

Hansen, M.M., Mensberg, K.L.D., Rasmussen, G. & Simonsen, V. (1997) Genetic variation within and among Danish brown trout (*Salmo trutta* L) hatchery strains, assessed by PCR-RFLP analysis of mitochondrial DNA segments. *Aquaculture* **153**, 15–29.

Hanski, I. (1991) Single-species metapopulation dynamics. In *Metapopulation Dynamics: Empirical and Theoretical Investigations* (Gilpin, M. & Hanski, I., eds), pp. 17–38. Academic Press, London.

Hanski, I. & Simberloff, D. (1997) The metapopulation approach, its history, conceptual domain, and application to conservation. In *Metapopulation Biology. Ecology, Genetics and Evolution* (Hanski, I. & Gilpin M.E., eds), pp. 5–26. Academic Press, San Diego.

Hanson, A.J. & Smith, H.D. (1967) Male selection in a population of sockeye salmon (*Oncorhynchus nerka*) of mixed age group. *Journal of the Fisheries Research Board of Canada* **24**, 1955–77.

Harris, H. (1966) Enzyme polymorphism in man. *Proceedings of the Royal Society B* **164**, 298–310.

Harris, H. (1977) *The Principles of Human Biochemical Genetics*. North-Holland, Amsterdam.

Harris, H., Hopkinson, D.A. & Edwards Y.H. (1977) Polymorphism and the subunit structure of isozymes: contribution to the neutralist-selectionist controversy. *Proceedings of the National Academy of Sciences of the USA* **74**, 698–701.

Hart, J.L. (1973) Pacific fishes of Canada. *Bulletin of the Fisheries Research Board of Canada* **180**, 1–740.

Hartley, S.E. (1996) High incidence of Atlantic salmon × brown trout hybrids in a Lake District stream. *Journal of Fish Biology* **48**, 151–4.

Hartman, W.L. & Raleigh, R.F. (1964) Tributary homing of sockeye salmon at Brook and Karluk lakes, Alaska. *Journal of the Fisheries Research Board of Canada* **21**, 485–504.

Hasler, A.D. (1966) *Underwater Guidepost: Homing of Salmon*. University of Wisconsin Press, Madison.

Hasler, A.D. & Scholz, A.T. (1983) Olfactory imprinting and homing in salmon. Investigation into the mechanism of the imprinting process. *Zoophysiology* **14**, 1–134.

Hasler, A.D. & Wisby, W.J. (1951) Discrimination of stream odors by fishes and relation to parent stream behavior. *American Nature* **85**, 223–38.

Hayashi, K. (1985) PCR-SSCP: a simple and sensitive method for detection of mutations in the genomic DNA. *PCR Methods and Applications* **1**, 34–8.

Healey, M.C. (1986) Optimum size and age at maturity in Pacific salmon and effects of size-selective fisheries. Salmonid age at maturity. *Canadian Special Publications in Fisheries and Aquatic Sciences* **89**, 39–52.

Healey, M.C. (1991) Life history of chinook salmon (*Oncorhynchus tshawytscha*). In *Pacific Salmon Life Histories* (Groot, C. & Margolis, L., eds), pp. 311–93. University of British Columbia Press, Vancouver.

Heard, W.R. (1991) Life history of pink salmon (*Oncorhynchus gorbuscha*). In *Pacific Salmon Life Histories* (Groot, C. & Margolis, L., eds), pp. 121–30. University of British Columbia Press, Vancouver.

Heard, W.R. (1996) Ocean ranching: an assessment. In *Principles of Salmonid Culture* Development in Aquaculture and Fisheries Science, Vol. 29 (Pennell, W. & Barton, B.A., eds), pp. 833–69. Elsevier, Amsterdam.

Heath, D.D., Iwama, G.K. & Delvin, R.H. (1994) DNA fingerprinting used to test for family effects on precocious sexual maturation in two populations of *Oncorhynchus tshawytscha* (Chinook salmon). *Heredity* **73**, 616–24.

Heggberget, T.G., Johnsen, B.O., Hindar, K. *et al.* (1993) Interactions between wild and cultured Atlantic salmon: a review of the Norwegian experience. *Fisheries Research* **18**, 123–46.

Heincke, F. (1898) *Naturgeschichte des Herings. (Wanderungen des hering in den europeischen meeren.)* O. Salle, Berlin.

Helle, J.H. (1981) Significance of the stock concept in artificial propagation of salmonids in Alaska. *Canadian Journal of Fisheries and Aquatic Sciences* **38**, 1665–71.

Helle, J.H. & Hoffman, M.S. (1995) Size decline and older age at maturity of two chum salmon (*Oncorhynchus keta*) stocks in western North America, 1972–1992. *Canadian Special Publications on Fisheries and Aquatic Sciences* **121**, 245–60.

Henry, T. & Ferguson, A. (1985) Kinetic studies of the lactate dehydrogenase (LDH-5) isozymes of brown trout, *Salmo trutta* L. *Comparative Biochemistry and Physiology* B **82**, 95–8.

Hesthagen, T. & Hansen, L.P. (1991) Estimates of the annual loss of Atlantic salmon, *Salmo salar* L., in Norway due to acidification. *Aquaculture and Fisheries Management* **22**, 85–91.

Hill, W.G. (1981) Estimation of effective population size from data of linkage disequilibrium. *Genetical Research* (Cambridge) **38**, 209–16.

Hindar, K., Ryman, N. & Utter, F. (1991) Genetic effects of cultured fish on natural fish populations. *Canadian Journal of Fisheries and Aquatic Science* **48**, 945–57.

Hiroi, O. (1998) Historical trends of salmon fisheries and stock conditions in Japan. In *Assessment and Status of Pacific Rim Salmonid Stocks. NPAFC Bulletin No. 1* (Welch, D.W., Eggers, D.M., Wakabayashi, K. & Karpenko, V., eds), pp. 23–7. North Pacific Anadromous Fish Commission, Vancouver.

Hoar, W.S. (1976) Smolt transformation: evolution, behaviour and physiology. *Journal of the Fisheries Research Board of Canada* **33**, 1234–52.

Hodgins, H. (1972) Serological and biochemical studies in racial identification of fishes. In *Stock Concept in Pacific Salmon*. H.R. MacMillan lectures in fishes (Simon, R. & Larkin P., eds), pp. 199–208. University of British Columbia Press, Vancouver.

Hodgins, H., Ames, W.E. & Utter, F.M. (1969) Variants of lactate dehydrogenase isozymes in sera of sockeye salmon (*Oncorhynchus nerka*). *Journal of the Fisheries Research Board of Canada* **26**, 15–19.

Hopkins, D.M. (1967) *The Bering Land Bridge*. Stanford University Press, Paulo Alto.

Huang, T., Cottingham, R., Ledbetter, D. & Zoghbi, H. (1992) Genetic mapping of four dinucleotide repeat loci, *DXS453*, *DXS458*, *DXS454* and *DXS424* on the X-chromosome using multiplex polymerase chain reaction. *Genomics* **13**, 375–80.

Hunter, J.G. (1959) Survival and production of pink and chum salmon in a coastal stream. *Journal of the Fisheries Research Board of Canada* **16**, 835–86.

Hunter, R.L. & Markert, C.L. (1957) Histochemical demonstration of enzymes separated by zone electrophoresis in starch gels. *Science* **125**, 1294–5.

Hurst, L.D. (1992) Intragenomic conflict as an evolutionary force. *Proceedings of the Royal Society* **B 248**, 135–40.

Hutchings, J.A. & Myers, R.A. (1985) Mating between anadromous and non-anadromous Atlantic salmon, *Salmo salar*. *Canadian Journal of Zoology* **63**, 2219–21.

Hutchinson, W.F., Carvalho, G.R. & Rogers, F.I. (1999) A non-destructive technique for the recovery of DNA from dried fish otoliths for subsequent molecular genetic analysis. *Molecular Ecology* **8**, 893–4.

Hynes, R.A., Duke, E.J. & Joyce, P. (1989) Mitochondrial DNA as a genetic marker for brown trout, *Salmo trutta* L., populations. *Journal of Fish Biology* **35**, 687–701.

Hynes, R.A., Ferguson, A. & McCann, M.A. (1996) Variation in mitochondrial DNA and postglacial colonization of north-western Europe by brown trout. *Journal of Fish Biology* **48**, 54–67.

Il'in, V.E., Konovalov, S.M. & Shevlyakov, A.G. (1983) Migration coefficient and spatial structure of Pacific salmon. In *Biological Bases for the Development of Salmon Farming in USSR Reservoirs* (Skarlato, O.A., ed.), pp. 9–31. Nauka, Moscow.

Innis, M.A., Gelfand, D.H., Sninsky, J.J. & White, T.J. (1990) *PCR Protocols: A Guide to Methods and Applications.* Academic Press, San Diego, CA.

International Union of Biochemistry Nomenclature Committee (1984) *Enzyme Nomenclature.* Academic Press, Orlando, FL.

Isaksson, A. (1994) Ocean ranching: its role and contribution to Pacific and Atlantic salmon fisheries. The state of the world's fisheries resources. In *Proceedings of the World Fisheries Congress, Plenary Sessions* (Voigtlander, C.W., ed.), pp. 133–49. Oxford & IBH Publishers, New Delhi.

Ishida, T. (1966) Salmon of the North Pacific Ocean. Part III. A review of the life history of North Pacific salmon. 2. Pink salmon in the Far East. *Bulletin of the International North Pacific Fisheries Commission* **18**, 29–39.

Ishida, Y., Ito, S., Kaeriyama, M., McKinnell, S. & Nagasawa, K. (1993) Recent changes in age and size of chum salmon (*Oncorhynchus keta*) in the North Pacific Ocean and possible causes. *Canadian Journal of Fisheries and Aquatic Sciences* **50**, 290–5.

IUCN (1978) *A World Conservation Strategy*, 2nd draft. International Union for Conservation of Nature and Natural Resources, Morges, Switzerland.

IUCN (1980) *World Conservation Strategy. Living Resource Conservation for Sustainable Development.* International Union for Conservation of Nature and Natural Resources, Gland.

Ivankov, V.N. (1967) On season races in pink salmon. *Bulletin of the Pacific Research Institute of Fisheries and Oceanography* **61**, 143–51.

Ivankov, V.N. (1968) Pacific salmon of Iturup Island (Kuril Islands). *Bulletin of the Pacific Research Institute of Fisheries and Oceanography* **65**, 49–74.

Ivankov, V.N. (1971) Season races in pink salmon of the Kuril Islands. *Scientific Papers of Far East University* **15** (3), 34–43.

Ivankov, V.N. (1985) Ecotypes of salmonid fishes. In *Morphology and Systematics of Salmonid Fishes* (Korovina, V.M. & Dorofeeva, E.A., eds), pp. 85–91. ZIN AN SSSR, Leningrad.

Ivankov, V.N. (1986) Peculiarities of population structure in pink salmon and rational utilization of this salmon. *Marine Biology (Russia)* **2**, 44–51.

Ivankov, V.N., Mitrofanov, Yu.A. & Bushuev, V.P. (1975) Maturation of pink salmon, *Oncorhynchus gorbuscha* (Walbaum), in the age smaller than one year. *Journal of Ichthyology (Russia)* **15**, 556–7.

Ivankov, V.N., Mitrofanov, Yu.A. & Omelchenko, V.T. (1987) Maturation of pink salmon on the first year of life. *Marine Biology (Russia)* **1**, 28–31.

Ivankov, V.N., Dobritsky, O.Yu., Skuba, N.S. & Karpenko, A.I. (1996) Differentiation of the population of pink salmon, *Oncorhynchus gorbuscha*, in southern Sakhalin. *Marine Biology (Russia)* **22**, 167–73.

Jackson, T.R., Ferguson, M.M., Danzmann, R.G. *et al.* (1998) Identification of two QTL influencing upper temperature tolerance in three rainbow trout (*Oncorhynchus mykiss*) half-sib families. *Heredity* **80**, 143–51.

Jarvi, T.H. (1948) On the periodicity of salmon reproduction in the northern Baltic area and its causes. *Rapp. Proces-Verbaux Conc. Perm. L'explor. Mer.* **69**, 1–131.

Jeffreys, A.J., Wilson, V. & Thein, S.L. (1985a) Hypervariable 'minisatellite' regions in human DNA. *Nature* **316**, 67–73.

Jeffreys, A.J., Wilson, V. & Thein, S.L. (1985b) Individual specific 'fingerprints' of human DNA. *Nature* **316**, 76–9.

Jeffreys, A.J., Royle, N.J., Wilson, V. & Wong, Z. (1988) Spontaneous mutation rates to new length alleles at tandem repetitive hypervariable loci in human DNA. *Nature* **332**, 278–81.

Jin, L. & Chakraborty, R. (1994) Estimation of genetic distance and coefficient of gene diversity from single-probe multi-locus DNA fingerprinting data. *Molecular Biology and Evolution* **11**, 12–127.

Johnson, K.R. (1979) *Genetic variation in populations of pink salmon* (Oncorhynchus gorbuscha) *from Kodiak Island, Alaska*. MSc Thesis, University of Washington, Seattle.

Johnson, K.R., Wright, J.E. & May, B. (1987) Linkage relationships reflecting ancestral tetraploidy in salmonid fish. *Genetics* **116**, 579–91.

Jordan, W.C. & Youngson, A.F. (1991) Genetic protein variation and natural selection in Atlantic salmon (*Salmo salar* L.) parr. *Journal of Fish Biology* **39**, (Suppl. A), 185–92.

Jordan, W.C., Youngson, A.F. & Webb, J.H. (1990) Genetic variation at the malic enzyme-2 locus and age at maturity in sea-run Atlantic salmon (*Salmo salar*). *Canadian Journal of Fisheries and Aquatic Sciences* **47**, 1672–7.

Jordan, W.C., Youngson, A.F., Hay, D.W. & Ferguson, A. (1992) Genetic protein variation in natural populations of Atlantic salmon (*Salmo salar*) in Scotland: temporal and spatial variation. *Canadian Journal of Fisheries and Aquatic Sciences* **49**, 1863–72.

Jordan, W.C., Verspoor, E & Youngson, A.F. (1997) The effect of natural selection on estimates of genetic divergence amongst populations of the Atlantic Salmon. *Journal of Fish Biology* **51**, 546–60.

Jorde, P.E. (1994) Allozymes in Scandinavian brown trout (*Salmo trutta* L). In *Report of Division of Population Genetics*, pp. 1–54. Stockholm University, Stockholm.

Jorde, P.E. & Ryman, N. (1995) Temporal allele frequency change and estimation of effective size in populations with overlapping generations. *Genetics* **139**, 1077–90.

Jorde, P.E. & Ryman, N. (1996) Demographic genetics of brown trout populations and estimation of effective size from temporal change of allele frequencies. *Genetics* **143**, 1369–81.

Jorde, P.E., Gitt, A. & Ryman, N. (1991) New biochemical genetic markers in the brown trout (*Salmo trutta* L.). *Journal of Fish Biology* **39**, 451–4.

Jorstad, K.E. & Naevdal, G. (1996) Breeding and genetics. In *Principles of Salmonid Culture. Development in Aquaculture and Fisheries Sciences*, No. 29 (Pennell, W. & Barton, B.A., eds), pp. 655–725. Elsevier, Amsterdam.

Jukes, T.H. (1971) Comparisons of polypeptide chains of globins. *Journal of Molecular Evolution* **1**, 46–62.

Kaeriyama, M. (1998) Dynamics of a chum salmon, *Oncorhynchus keta*, population released from Hokkaido in Japan. In *Assessment and Status of Pacific Rim Salmonid Stocks*. NPAFC Bulletin No. 1 (Welch, D.W., Eggers, D.M., Wakabayashi, K. & Karpenko, V., eds) pp. 90–102. North Pacific Anadromous Fish Commission, Vancouver.

Kalabushkin, B.A. & Salmenkova, E.A. (1988) Investigation of population genetic structure in pink salmon of the Sakhalin–Kuril region by means of multidimensional analysis. In *The*

Modern State of Research of Salmonid Fishes. Abstracts of the All-Union Meeting on Salmonid Fishes (Konovalov, S.M., ed.), pp. 144–5. Institute of Ecology of the Volga River Basin, Tol'yatti.

Kalabushkin, B.A., Salmenkova, E.A., Omelchenko, V.T. *et al.* (1998) Population structure and gene migration in pink salmon, *Oncorhynchus gorbuscha*, from the Sakhalin–Kuril region. *Genetika* **34**, 1675–85.

Kaplan, N.O. (1965) Evolution of the dehydrogenases. In *Evolving Genes and Proteins* (Bryson, V. & Vogel, J.H., eds), pp. 245–79. Academic Press, New York.

Karakousis, Y. & Triataphyllidis, C. (1990) Genetic structure and differentiation among Greek brown trout (*Salmo trutta* L.) populations. *Heredity* **64**, 297–307.

Karl, S.K. & Avise, J.C. (1992) Balancing selection at allozyme loci in oysters: implications from nuclear RFLPs. *Science* **256**, 100–102.

Karl, S.K., Bowen, B.W. & Avise, J.C. (1992) Global population genetic structure and male-mediated gene flow in the green turtle (*Chelonia mydas*): RFLP analysis of anonymous nuclear loci. *Genetics* **131**, 163–73.

Karn, M.N. & Penrose, L.S. (1951) Birth weight and gestation time in relation to maternal age, parity, and infant survival. *Annals of Eugenics* **16**, 147–64.

Karpevich, A.F. (1975) *Theory and Practice of Acclimatization of Hydrobionts*. Pishchevaya Promishlennost, Moscow.

Karpevich, A.F., Agapov, V.S. & Magomedov, G.M. (1991) *Acclimatization and Culture of Introduced Fishes*. VNIRO, Moscow.

Kartavtsev, Yu.F. (1991) The variability of allele frequencies in space and time in populations of pink salmon, *Oncorhynchus gorbuscha*. *Journal of Ichthyology (Russia)* **31**, 487–95.

Kartavtsev, Yu.F., Polyakova, N.E. & Karpenko, A.I. (1985) Inheritance, qualitative and quantitative variability of allozymes of malate dehydrogenase, malic enzyme and lactate dehydrogenase in chum salmon, *Oncorhynchus keta*. *Genetika* **21**, 845–53.

Kartavtsev, Yu.F., Salmenkova, E.A., Rubtzova, G.A. & Afanasyev, K.I. (1990) The family analysis of allozyme variability and its relationship with body length and viability of progeny in pink salmon, *Oncorhynchus gorbuscha* (Walb.). *Genetika* **26**, 1610–19.

Kato, F. (1991) Life history of masu and amago salmon (*Oncorhynchus masou* and *Oncorhynchus rhodurus*). In *Pacific Salmon Life Histories* (Groot, C. & Margolis, L., eds), pp. 447–520. UBC Press, Vancouver.

Kazakov, R.V. (1990) *Artificial Forming of Populations of Anadromous Salmonid Fishes*. Agropromizdat, Moscow.

Kazakov, R.V. (1993) On pathway and time of colonization by Atlantic salmon into freshwater basins of Europe. In *Problems of Salmonid Fishes in the North of Europe* (Kitaev, S.P., eds), pp. 35–57. Karel'sky Nauchnyi Tsentr, Petrozavodsk.

Kazakov, R.V. (ed.) (1998) *Atlantic Salmon*. Nauka, St Petersburg.

Kazakov, R.V. & Titov, S.F. (1991) Geographical patterns in the population genetics of Atlantic salmon, *Salmo salar* L., on USSR territory, as evidence for colonization routes. *Journal of Fish Biology* **39**, 1–6.

Kazakov, R.V. & Titov, S.F. (1992) Population genetic aspect of fish breeding in the European North. *Annual Proceedings of Scientific Papers of the State Research Institute on Lake and River Fisheries* **304**, 109–24.

Kazakov, R.V. & Titov, S.F. (1993) Population genetics of salmon, *Salmo salar* L., in northern Russia. *Aquaculture and Fishery Management* **24**, 495–506.

Kazakov, R.V. & Titov, S.F. (1995) *Population Genetic Structure of Atlantic Salmon.* Nauchniye Tetradi, N1. GOSNIORKH, St Petersburg.

Kazarnovsky, M.Ya. (1987) *Japanese Salmon Fishing in the North-west Pacific.* VNIRO, Moscow.

Kazarnovsky, M.Ya. (1989a) *Fishing of Pacific Salmon by the USSR in the North-west Pacific.* VNIRO, Moscow.

Kazarnovsky, M.Ya. (1989b) *Salmon Fishing by the USA in the North-west Pacific.* VNIRO, Moscow.

Kazarnovsky, M.Ya. (1989c) *Fishing of Pacific Salmon by Canada in the North Pacific.* VNIRO, Moscow.

Kerster, H.W. (1964) Neighbourhood size in the rustry lizard, *Sceloporus olivacens. Evolution* **18**, 445–57.

Kerster, H.W. & Levin, D.A. (1968) Neighbourhood size in *Litosphermum carolinence. Genetics* **60**, 577–83.

Khalturin, D.K. (1978) The age of smolts and sea life period of salmon (Salmonidae). *Journal of Ichthyology (Russia)* **18**, 983–98.

Khorevin, L.D. (1986) Pacific salmon culture in the Sakhalin region. *Marine Biology (Russia)* **2**, 17–27.

Khorevin, L.D. (1994) The importance of salmon culture for supporting their resources in the Sakhalin region. In *Systematics, Biology and Breeding Biotechnics of Salmonid Fishes* (Reshetnikov, Yu.S., ed.), pp. 204–6. GosNIORKH, St Petersburg.

Khudyakov, G.I., Denisov, E.P., Korotkiy, A.M., Kulakov, A.P., Nikonova, I.I. & Chernobrovkina, E.I. (1972) *The South of the Far East. History of Relief Development of Siberia and the Far East.* Nauka, Moscow.

Kijima, A. & Fujio, Y. (1977) Genetic control of isocitrate dehydrogenase isozymes in chum salmon. *Tohoku Journal of Agricultural Research* **28**, 96–102.

Kijima, A. & Fujio, Y. (1979) Geographical distribution of IDH and LDH isozymes in chum salmon populations. *Bulletin of the Japanese Society of Scientific Fisheries* **45**, 287–95.

Kijima, A. & Fujio, Y. (1980) Duplicated isozyme loci in chum salmon. *Tohoku Journal of Agricultural Research* **31**, 159–72.

Kijima, A. & Fujio, Y. (1982) Correlation between geographical distance and genetic distance in populations of chum salmon, *Oncorhynchus keta. Bulletin of the Japanese Society of Scientific Fisheries* **48**, 1703–9.

Kijima, A. & Fujio, Y. (1984) Relationship between average heterozygosity and river population size in chum salmon. *Bulletin of the Japanese Society of Scientific Fisheries* **50**, 603–8.

Killick, S.R. & Clemens, W.A. (1963) The age, sex ratio and size of Fraser River sockeye salmon, 1915–1960. *International North Pacific Salmon Fisheries Commission Bulletin* **14**, 1–140.

Kimmel, P. (1975) Fishes of the Miocene–Pliocene Berr Butte formation, Southern Oregon. *Papers on Palaeontology* **14**, 62–87.

Kimura, M. (1953) 'Stepping stone' model of population. *Annual Report of the National Institute of Genetics Mishima* **3**, 63–5.

Kimura, M. (1960a) Optimum mutation rate and degree of dominance as determined by the principle of minimum genetic load. *Journal of Genetics* **57**, 21–34.

Kimura, M. (1960b) Genetic load of a population and its significance evolution. *Japanese Journal of Genetics* **35**, 7–33.

Kimura, M. (1968) Genetic variability maintained in a finite population due to mutational production of neutral and nearly neutral isoalleles. *Genetical Research* **11**, 247–69.

Kimura, M. (1983) *Neutral Theory of Molecular Evolution*. University Press, Cambridge.

Kimura, M. & Crow, J.F. (1963) The measurement of effective population number. *Evolution* **17**, 279–88.

Kimura, M. & Maruyama, T. (1971) Pattern of neutral polymorphism in a geographically structured population. *Genetical Research* **18**, 125–31.

Kimura, M. & Ohta, T. (1971) *Theoretical Aspects of Population Genetics*. Princeton University Press, Princeton, NJ.

Kimura, M. & Ohta, T. (1973) Mutation and evolution at the molecular level. *Genetics* **73** (Suppl.), 19–35.

Kimura, M. & Weiss, G.H. (1964) The stepping-stone model of population structures and the decrease of genetic correlation with distance. *Genetics* **49**, 561–7.

Kincaid, H.L. (1976) Inbreeding in rainbow trout (*Salmo gairdneri*). *Journal of the Fisheries Research Board of Canada* **33**, 2420–26.

Kincaid, H.L. (1983) Inbreeding in fish populations used for aquaculture. *Aquaculture* **33**, 215–27.

Kirpichnikov, V.S. (1933) Problem of races in fish. *Biological Journal (Russia)* **2**, 609–27.

Kirpichnikov, V.S. (1935) Biology–systematic essay of smelt from the White Sea, Chesky Bay and the Pechora River. *Proceedings of the All-Union Research Institute of Fisheries and Oceanography* **2**, 103–87.

Kirpichnikov, V.S. (1943) Experimental systematics of carp. *Bulletin of USSR Academy of Sciencies, Biological Seria* **4**, 189–218.

Kirpichnikov, V.S. (1981) *Genetic Bases of Fish Selection*. Springer, Berlin.

Kirpichnikov, V.S. (1987) *Genetics and Selection of Fish*, 2nd edn. Nauka, Leningrad.

Kirpichnikov, V.S. & Ivanova, I.M. (1977) Space, temporal and age-dependent variations in Pacific sockeye salmon for *LDH-B1* and *PGM-1* loci. *Genetika* **13**, 1183–93.

Kirpichnikov, V.S. & Muske, G.A. (1981) Population genetics of Kamchatka sockeye salmon, *Oncorhynchus nerka* Walb. In *Genetics and Reproduction of Marine Organisms* (Kasyanov, V.L. & Pudovkin, A.I., eds), pp. 59–71. DVNTS AN SSSR, Vladivostok.

Knox, D. & Verspoor, E. (1991) A mitochondrial DNA restriction fragment length polymorphism of potential use for discrimination of farmed Norwegian and wild Atlantic salmon populations in Scotland. *Aquaculture* **98**, 249–57.

Kobayashi, T., Milner, G.B., Teel, D.J. & Utter, F. (1984) Genetic basis for electrophoretic variation of adenosine deaminase in chinook salmon. *Transactions of the American Fisheries Society* **113**, 86–9.

Koch, H.L.A., Bergstrom, E. & Evans, J.C. (1964) The microelectrophoretic separation on starch gel of the haemoglobins of *Salmo salar* L. *Mededelingen Vlaamse Academie Klasse der Wetenschappen (Brussels)* **26** (9), 1–33.

Koljonen, M. (1989) Electrophoretically detectable genetic variation in natural and hatchery stocks of Atlantic salmon in Finland. *Hereditas* **110**, 23–35.

Koljonen, M.L. & Pella, J.J. (1997) The advantage of using smolt age with allozymes for assessing wild stock contributions to Atlantic salmon catches in the Baltic Sea. *International Council for the Exploration of the Sea, Journal of Marine Science* **54**, 1015–30.

Kolmogorov, A.N. (1935) Deviations from Hardy's formulae in the case of partial isolation. *Proceedings of the USSR Academy of Sciences* **3**, 129–32.

Kondzela, C.M., Guthrie, C.M., Hawkins, S.L., Russell, C.D., Helle, J.H. & Gharrett, A.J. (1994) Genetic relationships among chum salmon populations in south-east Alaska and northern British Columbia. *Canadian Journal of Fisheries and Aquatic Sciences* **51** (Suppl. 1), 50–64.

Konovalov, S.M. (1971) *Differentiation of Local Stocks of Sockeye Salmon*. Nauka, Leningrad.

Konovalov, S.M. (1980) *Population Biology of Pacific Salmon*. Nauka, Leningrad.

Kontrimavichus, V.L. (1973) *Beringia in the Cenozoic Era*. Oxonian Press, New Delhi.

Korochkin, L.I., Serov, O.L., Pudovkin, A.I. *et al.* (1977) *Genetics of Isozymes*. Nauka, Moscow.

Krieg, F. & Guyomard, R. (1985) Population genetics of French brown trout (*Salmo trutta* L.): large geographical differentiation of wild populations and high similarity of domesticated stocks. *Genetics, Selection and Evolution* **17**, 225–42.

Krimbas, C.B. & Tsakas, S. (1971) The genetics of *Dacus oleae*. V. Changes of esterase polymorphism in a natural population following insecticide control – selection or drift? *Evolution* **25**, 454–60.

Krogius, F.V. (1960) Growth rate and the age groupings of sockeye salmon, *Oncorhynchus nerka* (Walb.) in the sea. *Journal of Ichthyology (Russia)* **16**, 67–88.

Krogius, F.V. (1972) The linear growth of *Oncorhynchus nerka* fry in Lake Dal'nee. *Bulletin of the Pacific Research Institute of Fisheries and Oceanography* **82**, 19–31.

Krogius, F.V. (1975) Population dynamics and growth of the sockeye salmon fry in Lake Dal'nee (Kamchatka). *Journal of Ichthyology (Russia)* **93**, 612–29.

Krogius, F.V. (1978) On the meaning of genetic and ecologic factors in the dynamics of a population of sockeye salmon *Oncorhynchus nerka* (Walb.) of Dal'nee Lake. *Journal of Ichthyology (Russia)* **18**, 211–21.

Krogius, F.V. (1979) On the relationship between the freshwater and marine life spans of sockeye salmon from Lake Dal'nee. *Marine Biology (Russia)* **3**, 24–9.

Krogius, F.V. (1981) The role of dwarf forms in reproduction of anadromous sockeye salmon, *Oncorhynchus nerka* (Walb.). *Journal of Ichthyology (Russia)* **12**, 976–84.

Krogius, F.V. & Krokhin, E.M. (1956) Results of investigations in sockeye salmon biology, reserves and fluctuation of number in Kamchatka waters. *Journal of Ichthyology (Russia)* **7**, 3–20.

Krogius, F.V., Krokhin, Ye.M. & Menshutkin, V.V. (1969) *A Community of Pelagic Fish in Lake Dal'nee*. Nauka, Leningrad.

Krueger, C.C. & May, B. (1987a) Stock identification of naturalized brown trout in Lake Superior tributaries: differentiation based on allozyme data. *Transactions of the American Fisheries Society* **116**, 785–94.

Krueger, C.C. & May, B. (1987b) Genetic comparison of naturalized rainbow trout populations among Lake Superior tributaries: differentiation based on allozyme data. *Transactions of the American Fisheries Society* **116**, 795–806.

Krueger, C.C. & May, B. (1991) Ecological and genetic effects of salmonid introductions in North America. *Canadian Journal of Fisheries and Aquatic Sciences* **48** (Suppl. 1), 66–77.

Kulida, S.V. (1985) Data on biology of the North Dvina Atlantic salmon. In *Studies of*

Population Biology and Ecology of the Salmonid Fishes in the North Water Bodies (Skarlato, O.A., ed.), pp. 11–6. ZIN AN SSSR, Leningrad.

Kulida, S.V. & Derets, I.P. (1985) Biological characteristics of the Soyana population of Atlantic salmon. In *Ecology and Reproduction of Anadromous Salmonid Fishes in the Basin of the White and Barentz Seas* (Nizovtsev, G.P., ed.), pp. 6–24. PINRO, Murmansk.

Kulikova, N.I. (1972) Variability and evolution of chum salmon *Oncorhynchus keta* (Walb.) *Journal of Ichthyology (Russia)* **12**, 211–25.

Kuprina, N.P. (1970) *Stratigraphy and History of Deposit Accumulating in the Pleistocene Sedimentation of Central Kamchatka*. Nauka, Moscow.

Kwain, W. (1981) Population dynamics and exploitation of rainbow trout in Stockely Creek, eastern Lake Superior. *Transactions of the American Fisheries Society* **110**, 210–15.

Kwain, W. & Chappel, J.A. (1978) First evidence for even-year spawning pink salmon, *Oncorhynchus gorbuscha*, in Lake Superior. *Journal of the Fisheries Research Board of Canada* **35**, 1373–6.

L'Abée-Lund, J.H. (1989) Significance of mature male parr in a small populations of Atlantic salmon (*Salmo salar*). *Canadian Journal of Fisheries and Aquatic Science* **46**, 928–31.

Labelle, M. (1992) Straying patterns of coho salmon (*Oncorhynchus kisutch*) stocks from south-east Vancouver Island, British Columbia. *Canadian Journal of Fisheries and Aquatic Sciences* **49**, 1843–55.

Laikre, L., Prodohl, P.A., Jorde, P.E. & Ryman, N. (1995) Genetic variability at minisatellite and allozyme loci in brown trout (*Salmo trutta* L.) – a comparison. *Hereditas* **123**, 191–5.

Laird, L.M. (1996) History and applications of salmonid culture. In: *Principles of Salmonid Culture* pp. 1–28. *Development in Aquaculture and Fisheries Science*, no. 29 (Pennell, W. & Barton, B.A., eds), Elsevier, Amsterdam.

Lande, R. (1979) Effective deme sizes during long-term evolution estimated from rates of chromosomal rearrangement. *Evolution* **33**, 234–51.

Lane, S., McGregor, A.J., Taylor, S.G. & Gharrett, A.J. (1990) Genetic marking of an Alaskan pink salmon population, with an evaluation of the mark and marking process. *American Fisheries Society Symposium* **7**, 397–406.

Lapin, Yu.E. (1964) Peculiarities of number dynamics in the Pacific pink salmon, *Oncorhynchus gorbuscha* (Walb.). *Journal of Ichthyology (Russia)* **4**, 233–42.

Lapin, Yu.E. (1971) *Regularity of Dynamics of Fish Populations in Connection with the Longevity of Life Cycle*. Nauka, Moscow.

Larson, A., Wake, D.B. & Yanev, K.P. (1984) Measuring of gene flow among populations having high levels of genetic fragmentation. *Genetics* **106**, 293–308.

Leary, R.F., Allendorf, F.W. & Knudsen, K.L. (1985) Developmental instability as an indicator of the loss of genetic variation in hatchery trout. *Transactions of the American Fisheries Society* **114**, 230–35.

Lebedev, N.V. (1967) *Elementary Populations of Fishes*. Pishchevaya Promyshlennost, Moscow.

Lee, C.-Y., Charles, D., Brouson, D., Griffin, M. & Bennett, L. (1979) Analysis of mouse and *Drosophila* proteins by two-dimensional gel electrophoresis. *Molecular and General Genetics* **176**, 303–11.

Lem, A. & Shehadeh, Z.N. (1997) International trade in aquaculture products. *FAO Aquaculture Newsletters* **17**, 3–6.

Lever, C. (1996) *Naturalized Fishes of the World*. Academic Press, San Diego.

Levins, R. (1965) The genetic consequences of natural selection. In *Theoretical and Mathematical Biology* (Waterman, T.H. & Morowitz, H.J., eds), pp. 388–97. Blaisdell Publishing Company, New York.

Levins, R. (1970) Extinction. In *Some Mathematical Problems in Biology* (Gerstenhaber, M., ed.), pp. 75–107. American Mathematical Society, Providence, RI.

Lewin, B. (1997) *Genes*. Oxford University Press, Oxford.

Lewontin, P. (1993) *Human Individuality: Heredity and Environment*. Univers, Moscow.

Lewontin, R.S. (1978a) *The Genetic Bases of Evolution*. Mir, Moscow.

Lewontin, R.S. (1978b) The genetic heterogeneity of electrophoretic alleles. In *Abstracts of 14th International Congress of Genetics*, Part 1 (Belyaev, D.K., ed.), pp. 465. Mir, Moscow.

Leyzerovich, Kh.A. & Mel'nikova, M.N. (1979) Seaward migration and return into river of dwarf males of Atlantic salmon. *Journal of Ichthyology (Russia)* **19**, 178–81.

Li, C.C. (1955) *Population Genetics*. University of Chicago Press, Chicago.

Li, C.C. (1967) Genetic equilibrium under selection. *Biometrics* **23**, 397–484.

Li, C.C. (1976) *A First Course in Population Genetics*. Boxwood Press, Pacific Grove, CA.

Li, C.C. (1978) *A First Course in Population Genetics*. Mir, Moscow.

Lidicker, W.Z. & Patton, J.L. (1987) Patterns of dispersal and genetic structure in populations of small rodents. In *Mammalian Dispersal Patterns: The Effects of Social Structure on Population Genetics* (Chepko-Sade, B.D. & Halpin, Z.T., eds), pp. 144–61. University of Chicago Press, Chicago.

Lie, O., Slettan, A., Lingaas, F., Olsaker, L., Hordvic, I. & Refstie, T. (1994) Haploid gynogenesis: a powerful strategy for linkage analysis in fish. *Animal Biotechnology* **5**, 33–45.

de Ligny, W. (1969) Serological and biochemical studies on fish populations. *Oceanography and Marine Biology Annual Review* **7**, 411–513.

Lim, S.T. & Bailey, G.S. (1977) Gene duplication in salmonid fish: evidence for duplicated but catalytically equivalent A(4) lactate dehydrogenases. *Biochemical Genetics* **15**, 707–21.

Lindberg, G.U. (1972) *Great Fluctuations of Ocean Level in the Quarternary Period*. Nauka, Leningrad.

Lindsey, C.C. & McPhail, J.D. (1986) Zoogeography of fishes of the Yukon and Mackenzie basins. In *Zoogeography of North American Freshwater Fishes* (Hocutt, C.H. & Wiley, E.O., eds), pp. 639–73. Wiley & Sons, New York.

Lindsey, C.C., Northcote, T.G. & Hartman, G.F. (1959) Homing of rainbow trout to inlet and outlet spawning streams at Loon Lake, British Columbia. *Journal of the Fisheries Research Board of Canada* **16**, 695–719.

Liskauskas, A.P. & Ferguson, M.M. (1991) Genetic variation and fitness: a test in a naturalized population of brook trout (*Salvelinus fontinalis*). *Canadian Journal of Fisheries and Aquatic Sciences* **48**, 2152–62.

Livshits, G. & Kobyliansky, E. (1985) Lerner's concept of developmental homeostasis and problem of heterozygosity level in natural populations. *Heredity* **55**, 341–53.

Low, L.-L. (1998) Symposium review. In *Assessment and Status of Pacific Rim Salmonid Stocks, NPAFC Bulletin No. 1* (Welch, D.W., Eggers, D.M., Wakabayashi, K. & Karpenko, V., eds), p. 511. North Pacific Anadromous Fish Commission, Vancouver.

Lozano, R., Ruiz Rejon, C. & Ruiz Rejon, M. (1992) A comparative analysis of NORs in diploid and triploid salmonids: implications with respect to the diploidization process occurring in this fish group. *Heredity* **69**, 450–57.

Lynch, M. (1990) The similarity index and DNA fingerprinting. *Molecular Biology and Evolution* **7**, 478–84.

Lynch, M. & Walsh, B. (1998) *Genetics and Analysis of Quantitative Traits*. Sinauer Associates, Sunderland, MA.

McAtee, W.L. (1937) Survival of the ordinary. *Quarterly Review of Biology* **12**, 47–64.

McCart, P. (1969) Digging behaviour of *Oncorhynchus nerka* spawning in streams at Babine Lake, British Columbia. In *Proceedings of the Symposium on Salmon and Trout in Streams*. H.R. McMillan lectures in fisheries (Northcote, T.G., ed.), pp. 39–51. UBC Press, Vancouver.

McConkey, E.H., Taylor, B.I. & Phan, D. (1979) Human heterozygosity: a new estimate. *Proceedings of the National Academy of Sciences of the USA* **76**, 6500–504.

McConnell, S.K., O'Reilly, P., Hamilton, L., Wright, L.N. & Bentzen, P. (1995) Polymorphic microsatellite loci from Atlantic salmon (*Salmo salar*): genetic differentiation of North American and European populations. *Canadian Journal of Fisheries and Aquatic Sciences* **52**, 1863–72.

McConnell, S.K.J., Ruzzante, D.E., O'Reilly, P.T., Hamilton, L. & Wright, J.M. (1997) Microsatellite loci reveal highly significant genetic differentiation among Atlantic salmon (*Salmo salar* L.) stocks from the east coast of Canada. *Molecular Ecology* **6**, 1075–89.

MacCrimmon, H.R. (1971) World distribution of rainbow trout (*Salmo gairdneri*). *Journal of the Fisheries Research Board of Canada* **28**, 663–704.

MacCrimmon, H.R. & Marshall, T.L. (1968) World distribution of brown trout, *Salmo trutta*. *Journal of the Fisheries Research Board of Canada* **25**, 2527–48.

MacCrimmon, H.R., Marshall, T.L. & Gots, B.L. (1970) World distribution of brown trout, *Salmo trutta*: further observations. *Journal of the Fisheries Research Board of Canada* **27**, 811–7.

McDowall, R.M. (1990) *New Zealand Freshwater Fishes*. Heinemann Reed, Auckland.

McElligot, E.A. & Cross, T.F. (1991) Protein variation in wild Atlantic salmon, with particular reference to southern Ireland. *Journal of Fish Biology* **39** (Suppl. A), 35–42.

McElroy, D., Moran, P., Bermingham, E. & Kornfield, J. (1991) *RELP: The Restriction Enzyme Analysis Package, Version 4.0*. Maine University, Orono, ME.

McGowan, C. & Davidson, W.S. (1992) Unidirectional natural hybridization between brown trout (*Salmo trutta*) and Atlantic salmon (*S. salar*) in Newfoundland. *Canadian Journal of Fisheries and Aquatic Sciences* **49**, 1953–8.

McGregor, A.J., Lane, S., Thomason, M.A., Zhivotovsky, L.A., Smoker, W.W. & Gharrett, A.J. (1998) Migration timing, a life history trait important in the genetic structure of pink salmon. In *Assessment and Status of Pacific Rim Salmonid Stocks. NPAFC Bulletin No. 1* (Welch, D.W., Eggers, D.M., Wakabayashi, K. & Karpenko, V., eds), pp. 262–73. North Pacific Anadromous Fish Commission, Vancouver.

McIntyre, J.D. & Johnson, A.K. (1977) Relative yield of two transferrin phenotypes in coho salmon. *Progress in Fish Culture* **39**, 175–7.

McKean, M. (ed.) (1992) *FRED (Fisheries, rehabilitation, enhancement and development)*. 1990 Annual report to the Alaska State Legislature. FRED Technical Report Series No. 109. Alaska Department of Fish and Game, Juneau, AK.

McPhail, J.D. & Lindsey, C.C. (1970) Freshwater fishes of north-western Canada and Alaska. *Bulletin of the Fisheries Research Board of Canada* **173**, 1–375.

McPhail, J.D. & Lindsey, C.C. (1986) Zoogeography of the freshwater fishes of Cascadia (the

Columbia system and rivers north to the Stikine). In *Zoogeography of North American Freshwater Fishes* (Hocutt, C.H. & Wiley, E.O., eds), pp. 615–37. Wiley & Sons, New York.

McVeigh, H.P., Hynes, R.A. & Ferguson, A. (1995) Mitochondrial DNA differentiation of sympatric populations of brown trout, *Salmo trutta* L., Lough Melvin, Ireland. *Canadian Journal of Fisheries and Aquatic Sciences* **52**, 1617–22.

Madison, D.M. (1969) Homing behaviour of the red-cheeked salamander *Plethodon jordani*. *Animal Behaviour* **17**, 25–39.

Mahnken, C., Ruggerone, G., Waknitz, W. & Flagg, T. (1998) A historical perspective on salmonid production from Pacific Rim hatcheries. In *Assessment and Status of Pacific Rim Salmonid Stocks, NPAFC Bulletin No. 1* (Welch, D.W., Eggers, D.M., Wakabayashi, K. & Karpenko, V., eds), pp. 38–53. North Pacific Anadromous Fish Commission Vancouver.

Major, R.L., Ito, J., Ito, S. & Godfrey, H. (1978) Distribution and origin of chinook salmon (*Oncorhynchus tshawytscha*) in offshore waters of the North Pacific ocean. *Bulletin of the International North Pacific Fisheries Commission* **38**, 1–54.

Makhrov, A.A. (1999) Genetic differentiation of brown trout (*Salmo trutta* L.) populations in the White sea basin. PhD thesis, Vavilov Institute of General Genetics, Moscow.

Makhrov, A.A., Salmenkova, E.A., Zelinsky, Yu.P. & Gruzdev, A.I. (1994) Distribution of *LDH-5* genotypes in age groups of three populations of brown trout (*Salmo trutta* L.). *Genetika* **30**, 92–4.

Makhrov, A.A., Kuzishchin, K.V. & Altukhov, Yu.P. (1997) Association of allozyme heterozygosity with growth rate and ecological differentiation in brown trout *Salmo trutta* L. *Genetika* **33**, 681–6.

Makhrov, A.A., Skaala, O., Altukhov, Yu.P. & Saunders, R.L. (1998) *ESTD** allozyme locus as a marker of genetic differentiation for the Atlantic salmon population of Europe and North America. *Proceedings of the USSR Academy of Sciences* **360**, 850–52.

Makoedov, A.N., Pustovoit, S.P., Ermolenko, L.N., Kovalev, V.G. & Viktorovsky, R.M. (1993). Population genetic research on pink salmon breeding in the rivers of north-eastern Russia. *Genetika* **29**, 1366–74.

Malecot, G. (1948) *Les Mathematiques de l'heredité*. Masson, Paris.

Malecot, G. (1955) Decrease of relationship with distance. *Cold Spring Harbor Symposium on Quantitative Biology* **20**, 52–3.

Malecot, G. (1967) Identical loci and relationship. *Proceedings of the 5th Berkley Symposium on Mathematical and Statistical Problems* **4**, 317–32.

Manzer, J.I., Ishida, T., Peterson, A.E. & Hanavan, M.G. (1965) Salmon of the North Pacific Ocean. V. Offshore distribution of salmon. *Bulletin of the International North Pacific Fisheries Commission* **15**, 1–452.

Markert, C.L. & Möller, F. (1959) Multiple forms of enzymes: tissue, ontogenetic and specific patters. *Proceedings of the National Academy of Sciences of the USA* **45**, 753–63.

Marriott, R.A. (1964) Stream catalogue of the Wood River lake system, Bristol Bay, Alaska. *Special Scientific Report on Fisheries* **494**, 1–210.

Martynov, V.G. & Kuznetsov, G.M. (1985) Atlantic salmon of the Umba River. In *Ecology and Reproduction of Anadromous Salmonid Fishes in Basins of the White and Barentz Seas* (Nizovtsev, G.P., ed.), pp. 3–15. PINRO, Murmansk.

Maruyama, T. (1970) On the rate of decrease of heterozygosity in circular stepping stone models of populations. *Theoretical Population Biology* **1**, 101–19.

Maruyama, T. (1971) The rate of decrease of heterozygosity in a population occupying a circular or a linear habitat. *Genetics* **67**, 437–54.

Maruyama, T. (1972) Rate of decrease of genetic variability in a two-dimensional continuous population of finite size. *Genetics* **70**, 639–51.

Maruyama, T. & Kimura, M. (1980) Genetic variability and effective population size when local extinction and recolonization of subpopulations are frequent. *Proceedings of the National Academy of Sciences of the USA* **77**, 6710–14.

Massaro, E.I. & Markert, C.L. (1968) Isozyme patterns of salmonid fishes: evidence for multiple cistrons for lactate dehydrogenase polypeptides. *Journal of Experimental Zoology* **168**, 223–38.

May, B. (1980) *The salmonid genome: evolutionary restructuring following a tetraploid event.* DPhil thesis, Pennsylvania State University.

May, B. & Johnson, K.R. (1993) Composite linkage map of salmonid fishes (*Salvelinus, Salmo, Oncorhynchus*). In *Genetic Maps: Locus Maps of Complex Genomes*, 6th edn, Book 4 (O'Brien, S.J., ed.), pp. 309–17.

May, B., Utter, F.M. & Allendorf, F.W. (1975) Biochemical genetic variation in pink salmon and chum salmon. *Journal of Heredity* **66**, 227–32.

May, B., Stoneking, M. & Wright, J.E. (1979a) Joint segregation of malate dehydrogenase and diaphorase loci in brown trout (*Salmo trutta*). *Transactions of the American Fisheries Society* **108**, 373–7.

May, B., Wright, J.E. & Stoneking, M. (1979b) Joint segregation of biochemical loci in Salmonidae: results from experiments with *Salvelinus* and review of the literature on other species. *Journal of the Fisheries Research Board of Canada* **36**, 1114–28.

May, B., Stoneking, M. & Wright, J.E. (1980) Joint segregation of biochemical loci in Salmonidae: II. Linkage associations from a hybridized *Salvelinus* genome (*S. namaycush* × *S. fontinalis*). *Genetics* **95**, 707–26.

May, B., Wright, J.E. & Johnson, K.R. (1982) Joint segregation of biochemical loci in Salmonidae. III. Linkage associations in Salmonidae including data from rainbow trout (*Salmo gairdneri*). *Biochemical Genetics* **20**, 29–39.

May, B., Johnson, K.R. & Wright, J.E. (1989) Sex-linkage in salmonids: evidence from a hybridized genome of brook trout and Arctic charr. *Biochemical Genetics* **27**, 291–301.

Mayr, E. (1968) *Animal Species and Evolution.* Mir, Moscow.

Mayr, E. (1974) *Populations, Species and Evolution.* Mir, Moscow.

Mednikov, B.M., Bannikova, A.A., Lomov, A.A., Melnikova, M.N. & Shubina, E.A. (1995) Restrictase analysis of repeating nuclear DNA, species criterion and mechanism of evolution. *Molecular Biology* (Russia) **29**, 1308–19.

Melekestsev, I.V., Braytseva, O.A., Erlikh, E.N. *et al.* (1974) *Kamchatka Peninsula, Kuriles and Commander Islands. The History of Relief Development of Siberia and the Far East.* Nauka, Moscow.

Meylan, A.B. (1982) Sea turtle migration – evidence from tag returns. In *Biology and Conservation of Sea Turtles* (Bjorndal, K.A., ed.), pp. 91–100. Smithsonian Institute Press, Washington DC.

Millar, R.B. (1990) Comparison of methods for estimating mixed stock fishery composition. *Canadian Journal of Fisheries and Aquatic Sciences* **47**, 2235–41.

Miller, K.M., Withler, R.E. & Beacham, T.D. (1996) Stock identification of coho salmon

(*Oncorhynchus kisutch*) using minisatellite DNA variation. *Canadian Journal of Fisheries and Aquatic Sciences* **53**, 181–95.

Miller, R.B. (1957) Have the genetic patterns of fishes been altered by introductions or by selective fishing? *Journal of the Fisheries Research Board of Canada* **14**, 797–806.

Milner, G.B., Teel, D.J., Utter, F.M. & Winans, G.A. (1985) A genetic method of stock identification in mixed populations of Pacific salmon, *Oncorhynchus* spp. *US National Marine Fisheries Service, Marine Fisheries Review* **47**, 1–8.

Mina, M.V. (1978) On the population structure of a species in fish: evaluating some hypotheses. *Journal of General Biology (Russia)* **39**, 453–60.

Mitton, J.B. & Grant, M.C. (1984) Associations among protein heterozygosity, growth rate, and developmental homeostasis. *Annual Review of Ecological Systems* **15**, 479–99.

Moller, D. (1970) Transferrin polymorphism in atlantic salmon (*Salmo salar*). *Journal of the Fisheries Research Board of Canada* **27**, 1617–25.

Moran, P.A.P. (1959) The theory of some genetical effects of population subdivision. *Australian Journal of Biology* **12**, 109–18.

Moran, P., Pendas, A.M., Garcia-Vazquez, E. & Izquierdo, J.I. (1994) Genetic variation among Atlantic salmon in six Spanish rivers. *Journal of Fish Biology* **45**, 831–7.

Moran, P., Pendas, A.M., Garcia-Vazquez, E., Izquierdo, J.I. & Lobon-Cervia, J. (1995) Estimates of gene flow among neighbouring populations of brown trout. *Journal of Fish Biology* **46**, 593–602.

Moran, P., Pendas, A.M., Beall, E. & Garcia-Vazquez, E. (1996) Genetic assessment of the reproductive success of Atlantic salmon precocious parr by means of VNTR loci. *Heredity* **77**, 655–60.

Moritz, C., Dowling, T.E. & Brown, W.M. (1987) Evolution of animal mitochondrial DNA: relevance for population biology and systematics. *Annual Review of Ecological Systems* **18**, 269–92.

Mork, J. (1991) One-generation effects of farmed fish immigration on the genetic differentiation of wild Atlantic salmon in Norway. *Aquaculture* **98**, 267–76.

Morris, D.B., Richard, K.R. & Wright, J.M. (1996) Microsatellites from rainbow trout (*Oncorhynchus mykiss*) and their use for genetic study in salmonids. *Canadian Journal of Fisheries and Aquatic Science* **53**, 120–26.

Morrison, W.J. & Wright, J.E. (1966) Genetic analysis of three lactate dehydrogenase isozyme systems in trout: evidence for linkage of genes coding subunits A and B. *Journal of Experimental Zoology* **163**, 259–70.

Morton, N.E., Crow, J.F. & Muller, H.J. (1956) An estimation of mutational damage in men from data on consanguineous marriages. *Proceedings of the National Academy of Sciences of the USA* **42**, 855–63.

Mousseau, T.A., Ritland, K. & Heath, D.D. (1998) A novel method for estimating heritability using molecular markers. *Heredity* **80**, 218–24.

Muller, H.J. (1950) Our load mutations. *American Journal of Human Genetics* **2**, 111–76.

Mullis, K., Faloona, F., Scharf, S., Saiki, R., Horn, G. & Erlich, H. (1986) Specific enzymatic amplificaftion of DNA *in vitro*: the polymerase chain reaction. *Cold Spring Harbor Symposium on Quantitative Biology* **51**, 263–73.

Murata, S., Takasaki, N., Saitoh, M. & Okada, N. (1993) Determination of the phylogenetic relationships among Pacific salmonids by using short interspersed elements (SINEs) as temporal landmarks of evolution. *Proceedings of the National Academy of Sciences of the USA* **90**, 6995–9.

Murata, S., Takasaki, N., Saitoh, M., Tachida, H. & Okada, N. (1996) Details of retropositional genome dynamics that provide a rationale for a genetic division: the distinct branching of the Pacific salmon and trout (*Oncorhynchus*) from the Atlantic salmon and trout (*Salmo*). *Genetics* **142**, 915–26.

Muske, G.A. (1983) Investigation of genetic structure of sockeye salmon, *Oncorhynchus nerka* (Walb.) populations. In *Biological Basis of Fish Breeding: Genetics and Selection Problems* (Kirpichnikov, V.S., ed.), pp. 186–94. Nauka, Leningrad.

Muske, G.A. (1988) Polymorphism of liver esterase in sockeye salmon (*Oncorhynchus nerka* Walb.) In *Modern State of Investigations of Salmonid Fishes* (Konovalov, S.M., ed.), pp. 212–13. IEVB AN SSSR, Tol'yatti.

Nagylaki, T. (1985) Homozygosity, effective number of alleles, and interdeme differentiation in subdivided populations. *Proceedings of the National Academy of Sciences of the USA* **82**, 8611–13.

Nagylaki, T. (1989) Gustave Malecot and the transition from classical to modern population genetics. *Genetics* **122**, 253–68.

Nakajima, M., Kita, A. & Fujio, Y. (1986) Genetic features of natural and cultured populations in masu salmon (*Oncorhynchus masou*). *Tohoku Journal of Agricultural Research* **37**, 31–42.

Neave, F. (1952) 'Even-year' and 'odd-year' pink salmon populations. *Transactions of the Royal Society of Canada, Series 3* **46**, (5), 55–70.

Neave, F. (1958) The origin and speciation of *Oncorhynchus*. *Transactions of the Royal Society of Canada, Series 3* **52**, (5), 25–49.

Neave, F. (1966a) Salmon of the North Pacific Ocean. Part III. A review of the life history of North Pacific salmon. 5. Pink salmon in British Columbia. *Bulletin of the International North Pacific Fisheries Commission* **18**, 71–9.

Neave, F. (1966b) Salmon of the North Pacific Ocean. Part III. A review of the life history of North Pacific salmon. 6. Chum salmon in British Columbia. *Bulletin of the International North Pacific Fisheries Commission* **18**, 81–5.

Neave, F., Yonemori, T. & Bakkala, R.G. (1976) Distribution and origin of chum salmon in offshore waters of the North Pacific Ocean. *Bulletin of the International North Pacific Fisheries Commission* **35**, 1–79.

Neel, J.V. & Schull, W. (1958) *Human Heredity*. Inostrannaya Literatura, Moscow.

Nehlsen, W., Williams, J.E. & Lichatowich, J.A. (1991) Pacific salmon at the crossroads: stocks at risk from California, Oregon, Idaho and Washington. *Fisheries* **16**, 4–21.

Nei, M. (1972) Genetic distance between populations. *American Nature* **106**, 283–92.

Nei, M. (1975) *Molecular Population Genetics and Evolution*. North Holland, Amsterdam.

Nei, M. (1978) Estimation of average heterozygosity and genetic distance from a small number of individuals. *Genetics* **89**, 583–90.

Nei, M. (1987) *Molecular Evolutionary Genetics*. Columbia University Press, New York.

Nei, M. & Imaizumi, Y. (1966a) Genetic structure of human population. 1. Local differentiation of blood group gene frequencies in Japan. *Heredity* **21**, 9–35.

Nei, M. & Imaizumi, Y. (1966b) Genetic structure of human population. 2. Differentiation of blood group gene frequencies among isolated human populations. *Heredity* **21**, 183–90.

Nei, M. & Roychoudhury, A.K. (1974) Genetic variation within and between the three major races in man, Caucasoides, Negroids and Mongoloids. *American Journal of Human Genetics* **26**, 421–43.

Nei, M. & Tajima, F. (1981) DNA polymorphism detectable by restriction endonucleases. *Genetics* **105**, 207–17.

Nekludov, M.N. (1985) Age and growth of Atlantic salmon from the Ponoy River. In *Ecology and Reproduction of Anadromous Salmonid Fishes in Basins of the White and Barentz Seas* (Nizovtsev, G.P., ed.), pp. 42–58. PINRO, Murmansk.

Nelson, J.S. (1984) *Fishes of the World*. John Wiley & Sons, New York.

Nelson, K. & Soulé, M. (1987) Genetical conservation of exploited fishes. In *Population Genetics and Fishery Management* (Ryman, N. & Utter, F., eds), pp. 345–68. University of Washington Press, Seattle.

Nevo, E., Beilis, A. & Ben-Shlomo, R. (1984) The evolutionary significance of genetic diversity: ecological, demographic and life history correlates. In *Evolutionary Dynamics of Genetic Diversity* (Mani, G.S., ed.), pp. 13–213. Springer, Berlin.

Nielsen, J.L., Can, C. & Thomas, W.K. (1994) Differences in genetic diversity for mitochondrial DNA between hatchery and wild populations of *Oncorhynchus*. *Canadian Journal of Fisheries and Aquatic Sciences* **51** (Suppl. 1), 290–97.

Nielsen, E.E., Hansen, M.M. & Loeschcke, V. (1996) Genetic structure of European populations of *Salmo salar* L. (Atlantic salmon) inferred from mitochondrial DNA. *Heredity* **77**, 351–8.

Nielsen, E.E., Hansen, M.M. & Loeschcke, V. (1997) Analysis of microsatellite DNA from old scale samples of Atlantic salmon, *Salmo salar*: a comparison of genetic composition over 60 years. *Molecular Ecology* **6**, 487–92.

Nikol'sky, G.V. (1971) *Special Ichthyology*. Vysshaya Shkola, Moscow.

Nilsson, J. (1997) mtDNA microsatellite variation in Baltic Atlantic salmon. *International Council for the Exploration of the Sea, Journal of Marine Science* **54**, 1173–6.

Noakes, D.J., Beamish, R.J., Klyashtorin, L. & McFarlane, G.A. (1998) On the coherence of salmon abundance trends and environmental factors. In *Assessment and Status of Pacific Rim Salmonid Stocks. NPAFC Bulletin No. 1* (Welch, D.W., Eggers, D.M., Wakabayashi, K. & Karpenko, V., eds), pp. 454–63. North Pacific Anadromous Fish Commission, Vancouver.

Norden, C.R. (1961) Comparative osteology of representative salmonid fishes, with particular reference to the grayling (*Thymallus arcticus*) and its phylogeny. *Journal of the Fisheries Research Board of Canada* **18**, 679–791.

Nordeng, H. (1977) A pheromone hypothesis for homeward migration in anadromous salmonids. *Oikos* **28**, 155–9.

Novosel'skaya, A.Yu., Novosel'sky, Yu.I. & Altukhov, Yu.P. (1982) Physicochemical characteristics of the spawning grounds and hereditary heterogeneity of Azabachye Lake sockeye salmon stocks. *Genetika* **18**, 1004–11.

Nygren, A., Nilsson, B. & Jahnke, M. (1971) Cytological studies in *Salmo trutta* and *Salmo alpinus*. *Hereditas* **67**, 259–67.

Nyman, L. (1970) Electrophoretic analysis of hybrids between salmon (*S. salar* L.) and trout (*S. trutta* L.). *Transactions of the American Fisheries Society* **99**, 229–36.

Nyman, L. (1972) A new approach to the taxonomy of the '*Salvelinus alpinus* species complex'. *Report of the Institute of Freshwater Research, Drottningholm* **52**, 103–31.

O'Connell, M. & Wright, J.M. (1997) Microsatellite DNA in fishes. *Reviews in Fish Biology and Fisheries* **7**, 331–64.

O'Connell, M., Skibinski, D.O.F. & Beardmore, J.A. (1995) Mitochondrial DNA and allozyme variation in Atlantic salmon (*Salmo salar*) populations in Wales. *Canadian Journal of Fisheries and Aquatic Sciences* **52**, 171–8.

Ohno, S. (1970a) *Evolution by Gene Duplication*. Springer Verlag, Berlin.

Ohno, S. (1970b) The enormous diversity in genome size of fish as a reflection of nature's extensive experiments with gene duplication. *Journal of the Fisheries Research Board of Canada* **99**, 120–30.

Ohno, S. (1974) Chordata 1: Protochordata, Cyclostomata, and Pisces. In *Animal Cytogenetics*, Vol. 4 (John B., ed.), pp. 1–92. Gebruder-Borntrage, Berlin.

Ohno, S., Wolf, U. & Atkin, N.B. (1968) Evolution from fish to mammals by gene duplication. *Hereditas* **59**, 169–87.

Ohno, S., Stenius, C., Christian, L.C. & Schipmann, G. (1969) *De novo* mutation-like events observed at the *6-Pgd* locus of Japanese quail and the principle of polymorphism breeding more polymorphism. *Biochemical Genetics* **3**, 417–28.

Okazaki, T. (1978) Genetic differences of two chum salmon (*Oncorhynchus keta*) populations returning to the Tokachi River. *Bulletin of the Far Seas Fishery Research Laboratory* **16**, 121–8.

Okazaki, T. (1982a) Geographical distribution of allelic variations of enzymes in chum salmon, *Oncorhynchus keta*, river populations of Japan and the effects of transplantation. *Bulletin of the Japanese Society of Scientific Fisheries* **48**, 1525–35.

Okazaki, T. (1982b) Genetic study on population structure in chum salmon (*Oncorhynchus keta*). *Bulletin of the Far Seas Fishery Research Laboratory* **19**, 25–116.

Okazaki, T. (1983) Genetic structure of chum salmon *Oncorhynchus keta* river populations. *Bulletin of the Japanese Society of Scientific Fisheries* **49**, 189–96.

Okazaki, T. (1986a) Studies on closely related species *Salmo gairdneri* and *Salmo mykiss*; their distribution and migration in the North Pacific and systematics. *Bulletin of the Far Seas Fishery Research Laboratory* **23**, 1–68.

Okazaki, T. (1986b) Genetic variation and population structure in masu salmon *Oncorhynchus masou* of Japan. *Bulletin of the Japanese Society of Scientific Fisheries* **52**, 1365–76.

Oleinik, A.G. (1994) *Molecular evolution of salmonid fishes*. PhD thesis, Institute of Marine Biology, Vladivostok.

Olsen, J.B., Wenburg, J.K. & Bentzen, P. (1996) Semi-automated multi-locus genotyping of Pacific salmon (*Oncorhynchus* spp.) using microsatellites. *Molecular Marine Biology and Biotechnology* **5**, 259–72.

Omelchenko, V.T. (1974) Electrophoretic analysis of haemoglobins in fish of the Far East. *Genetika* **10**, 35–43.

Omelchenko, V.T. (1975) Application of the protein electrophoregrams in species taxonomy of the genus *Salvelinus*. *Marine Biology (Russia)* **4**, 76–9.

Omelchenko, V.T. (1994) Population organization of pink salmon, *Oncorhynchus gorbuscha*: analysis of concepts. *Journal of Ichthyology* **34**, 83–91.

Omelchenko, V.T. & Gerasimenko, T.P. (1981) Molecular organization of the genomes and diploid/tetraploid ratios of coho salmon and herring. *Genetika* **17**, 338–47.

Omelchenko, V.T. & Salmenkova, E.A. (1998) Genetic differences between Arctic charr group (*Salvelinus alpinus* L. and *Salvelinus taranetzi* Kaganovsky) and the Pacific Dolly Varden trout (*Salvelinus malma* Walbaum). *Genetika* **34**, 1518–22.

Omelchenko, V.T. & Vyalova, G.P. (1990) Population structure of pink salmon. *Marine Biology (Russia)* **1**, 3–13.

Omelchenko, V.T., Malinina, T.V. & Tsigir, V.V. (1991) Population structure of masu salmon, *Oncorhynchus masou* (Brev.): biochemical–genetic investigation of populations in south Primorye. *Genetika* **27**, 290–98.

Omelchenko, V.T., Salmenkova, E.A. & Afanasyev, K.I. (1992) Genetic structure of chum salmon populations of the Primorye region. *Genetika* **28**, 101–13.

Omelchenko, V.T., Salmenkova, E.A., Ivanov, A.N., Afanas'ev, K.I., Savvushkina, N.E. & Roslyi, Yu.S. (1994) Genetic identification of the origin of the chum salmon, *Oncorhynchus keta*, taken by marine fishery along north-west Sakhalin. *Journal of Ichthyology (Russia)* **34**, 820–26.

Omelchenko, V.T., Politov, D.V., Salmenkova, E.A., Malinina, T.V. & Frolov, S.V. (1996) Genetic differentiation of sympatric charrs of the genus *Salvelinus* from the Yama River. *Genetika* **32**, 1562–8.

Omelchenko, V.T., Salmenkova, E.A., Malinina, T.V. & Frolov, S.V. (1998) Genetic differentiation of sympatric populations of charrs of the genus *Salvelinus* from the Achchen lake (Chukotka). *Genetika* **34**, 399–405.

O'Reilly, P. & Wright, J.M. (1995) The evolving technology of DNA fingerprinting and its application to fisheries and aquaculture. *Journal of Fish Biology* **47** (Suppl. A), 29–55.

Orr, H.A. (1995) The population genetics of speciation: the evolution of hybrid incompatibilities. *Genetics* **139**, 1805–13.

Osinov, A.G. (1984) On the origin of the modern range of brown trout, *Salmo trutta* L. (Salmonidae): data on biochemical gene markers. *Journal of Ichthyology (Russia)* **24**, 11–22.

Osinov, A.G. (1988) Brown trout, *Salmo trutta* (Salmonidae), of the Black and Caspian Sea basins: population genetic analysis. *Genetika* **24**, 2173–86.

Osinov, A.G. (1990) Some data on level of genetic diversity and differentiation of trout (*Salmo trutta*) from Tajikistan. *Bulletin of Moscow State University, Seria 16, Biology* **1**, 37–41.

Osinov, A.G. & Bernatchez, L. (1996) 'Atlantic' and 'Danube' phylogenetic groups of the brown trout *Salmo trutta* complex: genetic divergence, evolution and protection. *Journal of Ichthyology (Russia)* **36**, 762–86.

Osinov, A.G. & Pavlov, S.D. (1993) Genetic similarity of Kamchatka's trout and rainbow trout. *Journal of Ichthyology (Russia)* **33**, 626–30.

Osinov, A.G., Pavlov, S.D. & Maximov, V.A. (1996) Allozyme variation and genetic differentiation in populations of Arctic charr *Salvelinus alpinus* from the Baltic Sea–Taimyr part of the range. *Genetika* **32**, 547–59.

Osterdahl, L. (1969) The smolt run of a small Swedish river. *Report of the Swedish Salmon Research Institute* **8**, 205–15.

Paabo, S. (1989) Ancient DNA: extraction, characterization, molecular cloning and enzymatic amplification. *Proceedings of the National Academy of Sciences of the USA* **86**, 1939–43.

Palva, T.K. & Palva, E.T. (1987) Restriction site polymorphism in mitochondrial DNA of rainbow trout, *Salmo gairdneri* (Richardson), stocks in Finland. *Aquaculture* **67**, 283–9.

Palva, T.K., Lehvaslaiho, H. & Palva, E.T. (1989) Identification of anadromous and non-anadromous salmon stocks in Finland by mitochondrial DNA analysis. *Aquaculture* **81**, 237–44.

Parensky, V.A. (1992) *Aetiology of spawning of sockeye salmon* (Oncorhynchus nerka *Walbaum*). Dalnauka, Vladivostok.

Parensky, V.A. & Kovalev, M.Yu. (1983) Factors of spawning behaviour of sockeye salmon (*Oncorhynchus nerka* Walbaum) males. In *Morphology, Population Structure and Problems of Rational Utilization of Salmonid Fishes* (Skarlato, O.A., ed.), p. 161. Nauka, Leningrad.

Park, L.K. & Moran, P. (1994) Developments in molecular genetic techniques in fisheries. *Review of Fish Biology and Fisheries* **4**, 272–99.

Park, L.K. & Moran, P. (1995) Development in molecular genetic techniques in fisheries. In *Molecular Genetics in Fisheries* (Carvalho, G.R. & Pitcher, T.J., eds), pp. 1–28. Chapman & Hall, London.

Park, L.K., Brainard, M.A., Dingman, D.A. & Winans, G.A. (1993) Low levels of infraspecific variation in the mitochondrial DNA of chum salmon (*Oncorhynchus keta*). *Molecular Marine Biology and Biotechnology* **2**, 362–70.

Paterson, A.H., Lander, E.S., Hewitt, J.D., Peterson, S., Lincoln, S.E. & Tanksley, S.D. (1988) Resolution of quantitative traits into Mendelian factors by using a complete linkage map of restriction fragment length polymorphism. *Nature* **335**, 721–6.

Pavlov, D.A. (1989) *Salmonid Fishes: Biology of Development and Reproduction*. MGU, Moscow.

Peacock, A.C., Bunting, S.L. & Quenn, K.G. (1965) Serum protein electrophoresis in acrylamide gel. *Science* **147**, 1451–2.

Pella, J.J. & Milner, G.B. (1987) Use of genetic markers in stock composition analysis. In *Population Genetics and Fishery Management* (Ryman, N. & Utter, F., eds), pp. 247–76. University of Washington Press, Seattle.

Pennell, W. & Barton, B.A. (eds) (1996) Principles of salmonid culture. In *Development in Aquaculture and Fisheries Science*, Vol. 29. Elsevier, Amsterdam.

Petras, M.L. (1967) Studies of natural populations of *Mus*. 1. Biochemical polymorphism and their bearing on breeding structure. *Evolution* **21**, 250–74.

Pezzack, D.S. & Duggan, D.R. (1986) Evidence of migration and homing of lobster (*Homarus americanus*) on the Scotian shelf. *Canadian Journal of Fisheries and Aquatic Sciences* **43**, 2206–11.

Phelps, S.R., Baker, B.M., Hulett, P.L. & Leider, S.A. (1994a) *Genetic analysis of Washington steelhead: initial electrophoretic analysis of wild and hatchery steelhead and rainbow trout.* Seattle (Wash.) Report No. 94–9, pp. 1–63. Washington Department of Fish and Wildlife. Fishery Management Program; WA 98501-1091.

Phelps, S.R., LeClair, L.L., Yong, S. & Blankenship, H.L. (1994b) Genetic diversity of chum salmon in the Pacific north-west. *Canadian Journal of Fisheries and Aquatic Sciences* **51** (Suppl. 1), 65–83.

Philipp, D.P., Childers, W.F. & Whitt, G.S. (1985) Correlations of allele frequences with physical and environmental variables for populations of largemouth bass, *Micropterus salmoides* (Lacepede). *Journal of Fish Biology* **27**, 347–65.

Phillips, R.B. & Oakley, T.H. (1997) Phylogenetic relationships among the Salmoninae based on nuclear and mitochondrial DNA sequences. In *Molecular Systematics in Fishes* (Kocher, T.D. & Stepien, C.A., eds), pp. 145–62. Academic Press, San Diego.

Phillips, R.B., Pleyte, K.A. & Brown, M.R. (1992) Salmonid phylogeny inferred from ribosomal DNA restriction maps. *Canadian Journal of Fisheries and Aquatic Sciences* **49**, 2345–53.

Pogson, G.H., Mesa, K.A. & Bautilier, R.G. (1995) Genetic population structure and gene flow in the Atlantic cod, *Gadus morhua*: a comparison of allozyme and nuclear RFLP loci. *Genetics* **139**, 375–85.

Poinar, G.O. (1993) Still life in amber. *The Sciences*, March//April, 34–8.

Politov, D.V., Omelchenko, V.T., Salmenkova, E.A., & Malinina, T.V. (1998) Genetic differentiation of Arctic and Far Eastern charrs of the genus *Salvelinus*. *Genetika* **34**, 83–92.

Pollak, E. (1983) A new method for estimating the effective population size from allele frequency changes. *Genetics* **104**, 531–48.

Pollard, S.M., Danzmann, R.G. & Clautor, R.R. (1994) Association between the regulatory locus *PGM-1r** and life-history types of juvenile Atlantic salmon (*Salmo salar*). *Canadian Journal of Fisheries and Aquatic Sciences* **51**, 1322–9.

Polyakova, N.E., Malinina, T.V., Galeev A.V., Zaikin, D.V., Brykov, V.A. & Altukhov, Yu.P. (1992) Variability in the mitochondrial DNA of pink salmon *Oncorhynchus gorbuscha* (Walbaum). *Genetika* **28**, 122–29.

Polyakova, N.E., Skurikhina, L.A., Kukhlevsky, A.D. *et al.* (1996) Population genetic structure of pink salmon *Oncorhynchus gorbuscha* (Walbaum) according to restriction analysis of mitochondrial DNA: II. Comparison of non-overlapping generations of even and odd years. *Genetika* **32**, 1256–62.

Pope, T.R. (1992) The influence of dispersal patterns and mating system on genetic differentiation within and between populations of the red howler monkey (*Alouata seniculus*). *Evolution* **46**, 1112–28.

Poulik, M.D. (1957) Starch gel electrophoresis in a discontinuous system of buffers. *Nature* **180**, 1477.

Prakash, S., Lewontin, R.C. & Hubby, T.L. (1969) A molecular approach to the study of genetic heterozygosity in natural populations. IV. Patterns of genetic variation in central, marginal and isolated populations of *Drosophila pseudoobscura*. *Genetics* **61**, 841–58.

Presa, P., Krieg, F., Estoup, A. & Guyomard R. (1994) Diversité et gestion génétique de la truite commune: apport de l'étude du polymorphisme des locus protéiques et microsatellites. *Genetics, Selection and Evolution* **26** (Suppl. 1), 183–202.

Pritchard, A.L. (1940) Studies on the age of the coho salmon (*Oncorhynchus kisutch*) and spring salmon (*Oncorhynchus tshawytscha*) in British Columbia. *Proceedings and Transactions of the Royal Society of Canada, Series 3* **34**, 99–120.

Prodohl, P.A., Taggart, J.B. & Ferguson, A. (1994) Single-locus minisatellite variation in brown trout, *Salmo trutta* L. populations. In *Genetics and Evolution of Aquatic Organisms* (Beaumont, A.R., ed.), pp. 263–70. Chapman & Hall, London.

Prodohl, P.A., Walker, A.F., Hynes, R., Taggart, J.B. & Ferguson, A. (1997) Genetically monomorphic brown trout (*Salmo trutta* L.) populations, as revealed by mitochondrial DNA, multi-locus and single-locus minisatellite (VNTR) analysis. *Heredity* **79**, 208–13.

Pustovoit, S.P. (1993) Genetic heterogeneity of sockeye salmon, *Oncorhynchus nerka* Walbaum, populations of Kamchatka rivers. *Genetika* **29**, 808–19.

Pustovoit, S.P. (1994) Within-population genetic variation and among-population differentiation in Asiatic sockeye salmon, *Oncorhynchus nerka* (Walbaum). *Genetika* **30**, 101–6.

Putivkin, S.V. (1989) On forming of hydrologic conditions of spawning ground for Anadir's chum salmon. *Journal of Ichthyology (Russia)* **29**, 96–103.

Quatro, J.M. & Vrijenhoek, R.C. (1989) Fitness differences among remnant populations of the endangered sonoran topminnow. *Science* **245**, 976–8.

Quinn, T.P. (1984) Homing and straying in Pacific salmon. In *Mechanisms of Migration in Fishes* (McCleave, J.D., Arnold, G.P., Dodson, J.J. & Neil, W.H., eds), pp. 357–62. Plenum Press, New York.

Quinn, T.P. (1990) Current controversies in the study of salmon homing. *Ethology, Ecology and Evolution* **2**, 49–63.

Quinn, T.P. (1993) A review of homing and straying of wild and hatchery-produced salmon. *Fisheries Research* **18**, 29–44.

Quinn, T.P. & Fresh, K. (1984) Homing and straying in chinook salmon (*Oncorhynchus tshawytscha*) from Cowlitz River Hatchery, Washington. *Canadian Journal of Fisheries and Aquatic Sciences* **41**, 1078–82.

Quinn, T.P., Wood, C.C., Margolis, L., Riddell, B.E. & Hyatt, K.D. (1987) Homing in wild sockeye salmon (*Oncorhynchus nerka*) populations as inferred from differences in parasite prevalance and allozyme allele frequencies. *Canadian Journal of Fisheries and Aquatic Sciences* **44**, 1963–71.

Racine, R.R. & Langley, C.H. (1980) Genetic heterozygosity in a natural population of *Mus musculus* assessed using two-dimensional electrophoresis. *Nature* **283**, 855–7.

Radchenko, V.I. (1998) Historical trends of fisheries and stock condition of Pacific salmon in Russia. In *Assessment and Status of Pacific Rim Salmonid Stocks. NPAFC Bulletin No. 1* (Welch, D.W., Eggers, D.M., Wakabayashi, K. & Karpenko, V., eds), pp. 28–37. North Pacific Anadromous Fish Commission, Vancouver.

Radchenko, O.A., Malyarchuk, B.A. & Solovenchuk, L.L. (1997) Comparative restriction enzyme analysis of a fragment of the cytochrome *b* gene in pink, chum and coho salmon. *Genetika* **33**, 471–4.

Rannala, B. & Hartigan, J.A. (1995) Identity by descent in island–mainland populations. *Genetics* **139**, 429–37.

Raven, P.H. (1979) Future directions in plant population biology. In *Topics in Plant Population Biology* (Solbrig, O.T., Jain, S., Jonson, J.B. & Raven, P.H., eds), pp. 461–81. Columbia University Press, New York.

Reinitz, G.L. (1977) Inheritance of muscle and liver types of supernatant NADP-dependent isocitrate dehydrogenase in rainbow trout (*Salmo gairdneri*). *Biochemical Genetics*, **15**, 445–54.

Reisenbichler, R.R. (1988) Relation between distance transferred from natal stream and recovery rate for hatchery coho salmon. *North American Fisheries Management* **8**, 172–4.

Reisenbichler, R.R. & McIntyre, J.D. (1977) Genetic differences in growth and survival of juvenile hatchery and wild steelhead trout, *Salmo gairdneri*. *Canadian Journal of Fisheries and Aquatic Sciences* **34**, 123–8.

Reisenbichler, R.R. & Phelps, S.R. (1987) Genetic variation in chinook, *Oncorhynchus tshawytscha*, and coho, *O. kisutch*, salmon from the north coast of Washington. *Fisheries Bulletin* **85**, 681–701.

Reisenbichler, R.R. & Phelps, S.R. (1989) Genetic variation in steelhead (*Salmo gairdneri*) from the north coast of Washington. *Canadian Journal of Fisheries and Aquatic Sciences* **46**, 66–73.

Reshetnikov, Yu.S. (1980) *Ecology and Systematics of Coregonid Fishes*. Nauka, Moscow.

Rhymer, J.M. & Simberloff, D. (1996) Extinction by hybridization and introgression. *Annual Review of Ecological Systems* **27**, 83–109.

Ricker, W.E. (1954) Stock and recruitment. *Journal of the Fisheries Research Board of Canada* **11**, 559–623.

Ricker, W.E. (1966) Sockeye salmon in British Columbia. *Bulletin of the International North Pacific Fisheries Commission* **18**, 59–70.

Ricker, W.E. (1972) Hereditary and environmental factors affecting certain salmonid populations. In *The Stock Concept in Pacific Salmon, H.R. MacMillan Lectures in Fisheries*, pp. 19–160. University of British Columbia, Vancouver.

Ricker, W.E. (1981) Changes in the average age of Pacific salmon. *Canadian Journal of Fisheries and Aquatic Sciences* **38**, 1636–56.

Riddell, B.E. (1993) Spatial organization of Pacific salmon: what to conserve? In *Genetic Conservation of Salmonid Fishes* (Cloud, J.G. & Thorgaard, G.H., eds), pp. 23–41. Plenum Press, New York.

Riffel, M., Storch, V. & Schreiber, A. (1995) Allozyme variability of brown trout (*Salmo trutta* L.) populations across the Rhenanian–Danubian watershed in south-west Germany. *Heredity* **74**, 241–9.

Roberts, F.L., Wohnus, I.F. & Ohno, S. (1969) Phosphoglucomutase polymorphism in the rainbow trout. *Experientia* **25**, 1109–10.

Romashov, D.D. (1931) Conditions of equilibrium in a population. *Journal of Experimental Biology (Russia)* **A7**, 442–54.

Ropers, H.H., Engel, W. & Wolf, U. (1973) Inheritance of the S-form of NADP-dependent isocitrate dehydrogenase polymorphism in rainbow trout. In *Genetics and Mutagenesis of Fish* (Schroder, J.H., ed.), pp. 319–27. Springer, Berlin.

Rukhlov, F.N. (1982) *Life of the Pacific Salmon*. Dal'nevostochnoye Knizhnoye Izdatel'stvo, Uzhno-Sakhalinsk.

Ryabova, G.C. & Goncharova, A.A. (1981) Comparative analysis of allozyme diversity of *Salmo mykiss* (Walb.) and *Salmo gairdneri* (Rich.). In *Genetics, Selection, and Hybridization of Fishes* (Kirpichnikov, V.S., ed.), pp. 52–3. AzCherNIRO, Postov on Don.

Ryabova, G.C., Goncharova, A.A., Titova, A.Yu. & Altukhov, Yu.P. (1978) On the factors of stationary distribution of lactate dehydrogenase and phosphoglucomutase allele frequencies in sockeye salmon populations of the Azabachye Lake (Kamchatka). In *Abstracts of the International Symposium on Biology of Pacific Salmon* (Konovalov, S.M., ed.), pp. 88–90. TINRO, Vladivostok.

Rychkov, Yu.G. (1968) Response of populations to isolation. In *Problems of Evolution*, Vol. 1 (Vorontsov, N.N., ed.), pp. 212–36. Nauka, Novosibirsk.

Rychkov, Yu.G. (1969) Some population genetic approaches to anthropology of Siberia. *Journal of Anthropology (Russia)* **33**, 16–33.

Rychkov, Yu.G. (1973) A system of ancient human isolates in Northern Asia in view of the stability and evolution of the populations. *Journal of Anthropology (Russia)* **44**, 3–22.

Rychkov, Yu.G. (1975) Definition of natural selection on ABO group genotypes by data of genetic population structure for humans from North Asia. In *Statistic Methods in Population Genetics*, Vol. 49 (Malyutov, M.B., ed.), pp. 19–44. MGU, Moscow.

Rychkov, Yu.G. (1984) Space and time in genogeography. *Bulletin of the USSR Academy of Medical Sciences* **7**, 11–16.

Rychkov, Yu.G. (1986) Genochronology of historical events. *Journal of Anthropology (Russia)* **77**, 3–18.

Rychkov, Yu.G. & Balanovskaya, E.V. (1990) Ethnogenetics: on alignment of adaptive and selective-neutral genetic differentiation in ethnoses. *Genetika* **26**, 541–9.

Rychkov, Yu.G. & Movsesyan, A.A. (1972) Genetic-anthropologic analysis of distribution of skull anomaly for Siberian mongoloids with problems of their origin. *Bulletin of the Moscow Society of Naturalists, Biological Section* **43**, 114–32.

Rychkov, Yu.G. & Sheremetyeva, V.A. (1976) Genetics of the circumpolar populations of Eurasia in connection with the problem of human adaptation. In *Resources of the Biosphere*, Vol. 3 (Rodin, L.Ye. & Smirnov, N.N., eds), pp. 10–41. Nauka, Leningrad.

Rychkov, Yu.G. & Sheremetyeva, V.A. (1977) The genetic process in the system of ancient human isolates in North Asia. In: *Population Structure and Human Variation*, Part 2 (Harrison, G.A., ed.), pp. 47–108. University Press, Cambridge.

Rychkov, Yu.G. & Sheremetyeva, V.A. (1979) The genetics of circumpolar populations of Eurasia related to the problem of human adaptation. In *The Human Biology of Circumpolar Populations*, International Biological Programme, Vol. 21 (Milan, F.A., ed.), pp. 37–80. Cambridge University Press, Cambridge.

Rychkov, Yu.G. & Yashchuk, Ye.V. (1980) Genetics and ethnogenesis. *Journal of Anthropology (Russia)* **64**, 23–39.

Rychkov, Yu.G. & Yashchuk, Ye.V. (1983) Genetics and ethnogenesis: trends of genetic processes in connection with the features of development of the European population. *Journal of Anthropology (Russia)* **72**, 3–17.

Rychkov, Yu.G. & Yashchuk, Ye.V. (1985) Genetics and ethnogenesis. Historical order of genetic differentiation of the human population: model and reality. *Journal of Anthropology (Russia)* **75**, 97–116.

Rychkov, Yu.G. & Yashchuk, Ye.V. (1986) Genetic memory about ethnogenesis. In *Ethnic Relationship of North Asian and American People by Data of Anthropology* (Velikanova, M.S. & Zolotareva, I.M., eds), pp. 149–66. Nauka, Moscow.

Rychkov, Yu.G., Rusakova, O.L. & Rapoport, M.P. (1973) Factors of genetic differentiation of the population systems of the Northern Asia native populations. *Genetika* **9**, 136–45.

Rychkov, Yu.G., Yashchuk, Ye.V. & Veselovskaya, E.V. (1982) Genetics and ethnogenesis. *Journal of Anthropology (Russia)* **69**, 3–18.

Ryman, N. (1970) A genetic analysis of recapture frequencies of released young of salmon (*Salmo salar* L.). *Hereditas* **65**, 159–60.

Ryman, N. (1981) Conservation of genetic resources: experiences from brown trout (*Salmo trutta*). *Ecological Bulletin* **34**, 61–74.

Ryman, N. (1983) Patterns of distribution of biochemical genetic variation in salmonids: differences between species. *Aquaculture* **33**, 1–21.

Ryman, N. & Ståhl, G. (1980) Genetic changes in hatchery stocks of brown trout (*Salmo trutta*). *Canadian Journal of Fisheries and Aquatic Sciences* **37**, 95–105.

Ryman, N. & Ståhl, G. (1981) Genetic perspectives of the identification and conservation of Scandinavian stocks of fish. *Canadian Journal of Fisheries and Aquatic Sciences* **38**, 1562–75.

Ryman, N. & Utter, F (eds) (1987) *Population Genetics and Fishery Management*. University of Washington Press, Seattle.

Ryman, N., Allendorf, F.W. & Ståhl, G. (1979) Reproductive isolation with little genetic divergence in sympatric populations of brown trout (*Salmo trutta*). *Genetics* **92**, 247–62.

Sachko, G.D. (1973) Genetics of lactate dehydrogenase isozymes in Pacific salmon. In *Fish Biochemical Genetics* (Troshin, A.S., ed.), pp. 155–60. Nauka, Leningrad.

Saiki, R.K., Gelfand, D.H., Stoffel, S. *et al.* (1988) Premier-directed enzymatic amplification of DNA with thermostable DNA polymerase. *Science* **239**, 487–91.

Sakamoto, T., Okamoto, N., Ikeda, Y., Nakamura, Y. & Sato, T. (1994) Dinucleotide-repeat polymorphism in DNA in rainbow trout and its application in fisheries science. *Journal of Fish Biology* **44**, 1093–6.

Salmenkova, E.A. (1973) Genetics of fish isozymes. *Advances in Modern Biology (Russia)* **75**, 217–35.

Salmenkova E.A. (1989) General results and aims of population genetic studies of salmonid fishes. In *Genetics in Aquaculture* (Kirpichnikov, V.S., ed.), pp. 7–29. Nauka, Leningrad.

Salmenkova, E.A. & Omelchenko, V.T. (1978) Polymorphism of proteins in the populations of diploid and tetraploid fishes. *Marine Biology (Russia)* **4**, 67–74.

Salmenkova, E.A. & Omelchenko, V.T. (1982) Genetic polymorphism of 6-phosphogluconate dehydrogenase in the pink salmon, *Oncorhynchus gorbuscha* (Walb.). *Marine Biology (Russia)* **4**, 55–8.

Salmenkova, E.A. & Omelchenko, V.T. (1983a) Polymorphism of NADP-dependent isocitrate dehydrogenase and malate dehydrogenase in populations of chum and pink salmon. *Marine Biology (Russia)* **3**, 24–8.

Salmenkova, E.A. & Omelchenko, V.T. (1983b) Genetic analysis of some polymorphic enzyme systems of pink salmon. In *Genetics of Commercial and Cultivated Species* (Burtsev, I.A., ed.), pp. 8–15. Lyogkaya i Pishchevaya Promishlennost, Moscow.

Salmenkova, E.A. & Volokhonskaya, L.G. (1973) Biochemical polymorphism in the populations of diploid and tetraploid fishes. In *Fish Biochemical Genetics* (Troshin, A.S., ed.), pp. 54–61. Nauka, Leningrad.

Salmenkova, E.A., Omelchenko, V.T., Malinina, T.V., Afanasyev, K.I. & Altukhov, Yu.P. (1981) Population genetic differences between successive generations of pink salmon spawning in rivers of the Asian coasts of the North Pacific. In *Genetics and Reproduction of Marine Organisms* (Kasyanov, V.L. & Pudovkin, A.I., eds), pp. 95–104. DVNTS AN SSSR, Vladivostok.

Salmenkova, E.A., Omelchenko, V.T., Pobedonostseva, E.Yu., Milishnikov, A.N. Nikonorov, S.I. & Altukhov, Yu.P. (1983) Population genetic analysis of the effectiveness of transferring Kuril chum salmon eggs to south-western Sakhalin. *Genetika* **19**, 1660–67.

Salmenkova, E.A., Altukhov, Yu.P., Viktorovsky, R.M. *et al.* (1986) The genetic structure of chum salmon (*Oncorhynchus keta*) populations spawning in the rivers of the Far East and north-east of the USSR. *Journal of General Biology (Russia)* **47**, 529–49.

Salmenkova, E.A., Omelchenko, V.T. & Roslyi, Yu.S. (1989) Differences and genetic structure of summer and autumn races of Amur chum salmon. In *Genetics in Aquaculture* (Kirpichnikov, V.S., ed.), pp. 80–86. Nauka, Leningrad.

Salmenkova, E.A., Omelchenko, V.T., Khorevin, L.D. & Semenov, B.A. (1990) Population origin of immature chum salmon from south-western Sakhalin. *Marine Biology (Russia)* **6**, 51–8.

Salmenkova, E.A., Omelchenko, V.T. & Altukhov, Yu.P. (1992) Genogeographic analysis of chum salmon, *Oncorhynchus keta* (Walbaum), populations of the Asian part of species range. *Genetika* **28**, 76–92.

Salmenkova, E.A., Omelchenko, V.T., Roslyi, Yu.S., Kalabushkin, B.A. & Probatov, N.S. (1994) Genetic differentiation of chum salmon of the Amur River. *Genetika* **30**, 518–28.

Salo, E.O. (1991) Life history of chum salmon (*Oncorhynchus keta*). In *Pacific Salmon Life Histories* (Groot, C. & Margolis, L., eds), pp. 231–309. UBC Press, Vancouver.

Sanchez, J.A., Clabby, C., Ramos, D. *et al.* (1996) Protein and microsatellite single locus variability in *Salmo salar* L. (Atlantic salmon). *Heredity* **77**, 423–32.

Sandercock, F.K. (1991) Life history of coho salmon (*Oncorhynchus kisutch*). In *Pacific Salmon Life Histories* (Groot, C. & Margolis, L., eds), pp. 397–445. UBC Press, Vancouver.

Sanger, F., Nicklen, S. & Coulson A.R. (1977) DNA sequencing with chain-terminating inhibitors. *Proceedings of the National Academy of Sciences of the USA* **74**, 5463–7.

Sano, S. (1966) Salmon of the North Pacific ocean – part III. A review of the life history of North Pacific salmon. 3. Chum salmon in the Far East. *Bulletin of the International North Pacific Fisheries Commission* **18**, 41–56.

Saunders, R.L., Henderson, E.B. & Glebe, B.D. (1982) Precocious sexual maturation and smoltification in male Atlantic salmon (*Salmo salar*). *Aquaculture* **28**, 211–29.

Savvaitova, K.A. (1973) Ecology and systematics of freshwater charr of the genus *Salvelinus* (Nilson) Richardson of some Kamchatka basins. *Journal of Ichthyology (Russia)* **13**, 67–78.

Savvaitova, K.A. (1975a) The population structure of *Salmo mykiss* in Kamchatka. *Journal of Ichthyology* **15**, 876–88.

Savvaitova, K.A. (1975b) Population structure of *Salmo mykiss* Walbaum on the natural range. *Journal of Ichthyology (Russia)* **15**, 984–97.

Savvaitova, K.A. (1995) Patterns of diversity and processes of speciation in Arctic charrs. *Nordic Journal of Freshwater Research* **71**, 81–91.

Savvaitova, K.A. & Maksimov, V.A. (1969) Age and growth of the Kamchatkan anadromous trout (*Salmo penshinensis* Pallas) and the Kamchatkan river trout (*Salmo mykiss* Walbaum). *Journal of Ichthyology* **9**, 536–46.

Savvaitova, K.A. & Movchan, Yu.V. (1973) On the problem of taxony of charrs of the genus *Salvelinus* (Salmonidae, Salmoniformes) from the South Kuril Islands. *Journal of Hydrobiology (Russia)* **4**, 94–8.

Schaffer, W.M. & Elson, P.F. (1975) The adaptive significance of variations in life history among local populations of Atlantic salmon in North America. *Ecology* **56**, 577–90.

Schmidt, J. (1917) *Zoarces viviparus* and local races of the same. *Compte rendue traveau laboratory, Carlsberg, Copenhague* **13**, 279–396.

Schmidt, J. (1930) The atlantic cod (*Gadus calarias* L.) and local races of the same. *Compte rendue traveau laboratory, Carlsberg, Copenhague* **18** (6), 1–72.

Schmidt, D.Yu. (1947) *Migrations of Fishes*. USSR Academy of Sciences, Leningrad.

Scott, W.B. & Crossman, E.J. (1973) Freshwater fishes of Canada. *Journal of the Fisheries Research Board of Canada* **184**, 1–966.

Scribner, K.T., Gust, J.R. & Fields, R.L. (1996) Isolation and characterization of novel microsatellite loci: cross-species amplification and population genetic applications. *Canadian Journal of Fisheries and Aquatic Science* **53**, 685–93.

Seeb, J.E. & Seeb, L. (1986) Gene mapping of isozyme loci in chum salmon. *Journal of Heredity* **77**, 399–402.

Seeb, J.E., Habicht, C., Olsen, J.B., Bentzen, P., Shaklee, J.B. & Seeb, L.W. (1998) Allozyme, mtDNA, and microsatellite variants describe structure of populations of pink and sockeye salmon in Alaska. In *Assessment and Status of Pacific Rim Salmonid Stocks. NPAFC Bulletin No. 1* (Welch, D.W., Eggers, D.M., Wakabayashi, K. & Karpenko, V., eds), pp. 300–18. North Pacific Anadromous Fish Commission, Vancouver.

Seeley, C.M. & McCammon, G.W. (1966) Kokanee. In *Inland Fisheries Management* (Calhoun, A.C., ed.), pp. 274–94. California Department of Fish and Game, Sacramento, CA.

Selander, R.K. (1970) Behaviour and genetic variation in natural populations. *American Zoology* **10** (10), 53–66.

Semenova, S.K. & Slynko, V.I. (1988) Polymorphism of proteins of Atlantic salmon (*Salmo salar* L.), brown trout (*S. trutta* L.) and their hybrids. *Genetika* **24**, 548–55.

Seppa, P. & Pamilo, P. (1995) Gene flow and population viscosity in *Myrmica* ants. *Heredity* **74**, 200–209.

Serebrovsky, A.S. (1927) Genetic analysis of hens breeding on the farms of Daghestan highlanders. *Journal of Experimental Biology (Russia) Series A* **3**, 62–146.

Serebrovsky, A.S. (1935) The genegeography of hens in Armenia. *Advances in Zootechnic Sciences (Russia)* **1**, 317–48.

Shaklee, J.B. & Phelps, S.R. (1990) Operation of a large-scale, multi-agency programme for genetic stock identification. *American Fisheries Society Symposium* **7**, 817–30.

Shaklee, J.B. & Phelps, S.R. (1992) Chinook salmon $NADP^+$-dependent cytosolic isocitrate dehydrogenase: electrophoretic and genetic dissection of a complex isozyme system and geographic patterns of variation. *Biochemical Genetics* **30**, 455–89.

Shaklee, J.B. & Salini, J. (1993) Electrophoretic characterizations of multiple genetic stocks of barramundi perch in Queensland, Australia. *Transactions of the American Fisheries Society* **122**, 685–701.

Shaklee, J.B. & Varnavskaya, N.V. (1994) Electrophoretic characterization of odd-year pink salmon (*Oncorhynchus gorbuscha*) populations from the Pacific coast of Russia, and comparison with selected North American populations. *Canadian Journal of Fisheries and Aquatic Sciences* **51** (Suppl. 1), 158–70.

Shaklee, J.B., Allendorf, F.W., Morizot, D.C. & Whitt, G.S. (1990a) Gene nomenclature for protein-coding loci in fish. *Transactions of the American Fisheries Society* **199**, 2–15.

Shaklee, J.B., Busack, C., Marshall, A., Miller, M. & Phelps, S. (1990b) The electrophoretic analysis of mixed-stock fisheries of Pacific salmon. In *Isozymes: Structure, Function, and Use in Biology and Medicine. Progress in Clinical and Biological Research* (Ogita, Z.I. & Markert, C.L., eds), pp. 235–65. Wiley-Liss, New York.

Shaklee, J.B., Klaybor, D.C., Houng, S. & White, B.A. (1991) Genetic stock structure of odd-year pink salmon, *Oncorhynchus gorbuscha* (Walbaum), from British Columbia and potential mixed-stock fisheries applications. *Journal of Fish Biology* **39** (Suppl. A), 21–34.

Shapovalov, L. & Taft, A.C. (1954) The life histories of the steelhead rainbow trout (*Salmo gairdneri gairdneri*) and silver salmon (*Oncorhynchus kisutch*) with special reference to Waddel Creek, California, and recommendations regarding their management. *Fisheries Bulletin of the California Department of Fish and Game* **98**, 1–375.

Shed'ko, S.V. (1991a) On systematics of *Salmo mykiss* (Salmoniformes, Salmonidae). Restrictase analysis of mitochondrial DNA. *Journal of Zoology (Russia)* **70**, 151–5.

Shed'ko, S.V. (1991b) Evolutionary rate of mitochondrial DNA in salmonid fishes. *Journal of Evolutionary Biochemistry and Physiology (Russia)* **27**, 249–54.

Sheffield, V.C., Beck, J.S., Kwitek, A.E. *et al.* (1993) The sensitivity of single-strand conformation polymorphism analysis for the detection of single base substitutions. *Genomics* **16**, 325.

Sheppard, D. (1972) The present status of the steelhead trout stocks along the Pacific coast. In

A Review of the Oceanography and Renewable Resources of the Northern Gulf of Alaska (Rosenberg, D.H., ed.), pp. 519–36. University of Alaska, Juneau.

Sheremetyeva, V.A. & Rychkov, Yu.G. (1978) *Population Genetics of Peoples of North-East Asia.* Moscow University Press, Moscow.

Sheridan, W.L. (1962) Relation of stream temperatures to timing of pink salmon escapements in south-east Alaska. In *Symposium on Pink Salmon H.R. MacMillan Lectures in Fisheries*, pp. 87–102. University of British Columbia, Vancouver.

Shilov, I.A. (1977) *Ecologic and Physiologic Bases of Population Relations Among Animals.* MGU, Moscow.

Sholes, W.H. & Hallock, R.J. (1979) An evaluation of rearing fall-run chinook salmon (*Oncorhynchus tshawytscha*) to yearlings at Feather River Hatchery, with a comparison of returns from hatchery and downstream releases. *Californian Fish and Game* **65**, 239–55.

Simon, R.C. (1963) Chromosome morphology and species evolution in the five North American species of Pacific salmon (*Oncorhynchus*). *Journal of Morphology* **112**, 77–97.

Simon, R.C., McIntyre, J.D. & Hemmingsen, A.R. (1986) Family size and effective population size in a hatchery stock of coho salmon (*Oncorhynchus kisutch*). *Canadian Journal of Fisheries and Aquatic Sciences* **43**, 2434–42.

Simonova, N.A. (1975) Observation on spawning of sockeye salmon. *Bulletin of the Pacific Research Institute of Fisheries and Oceanography* **97**, 145–6.

Singer, M.F. (1982) SINEs and LINEs: highly repeated short and long interspersed sequences in mammalian genomes. *Cell* **28**, 433–4.

Singh, R.S. (1979) Genetic heterogeneity within electrophoretic 'alleles' and the pattern of variation among loci in *Drosophila pseudoobscura. Genetics* **93**, 997–1018.

Skaala, O. (1992) Genetic population structure of Norwegian brown trout. *Journal of Fish Biology* **41**, 631–46.

Skaala, O. & Jorstad, K.E. (1994) Genetic markers in Atlantic salmon, *Salmo salar* L., their potential for experimental studies of gene flow and adaptation. *Aquaculture and Fisheries Management* **25** (Suppl. 2), 121–9.

Skrochowska, S. (1969) Migration of the sea trout (*Salmo trutta* L.), brown trout (*S. t. m. fario* L.) and their crosses. Part I. Problem, methods and results of taggings. *Polish Archives of Hydrobiology* **16** (29), 125–40.

Slatkin, M. (1993) Isolation by distance in equilibrium and non-equilibrium populations. *Evolution* **47**, 264–79.

Slatkin, M. (1995) A measure of population subdivision based on microsatellite allele frequencies. *Genetics* **139**, 457–62.

Slatkin, M. (1996) In defence of founder-flush theories of speciation. *American Nature* **147**, 493–505.

Slettan, A., Olsaker, I. & Lie, O. (1993) Isolation and characterization of variable $(GT)_n$ repetitive sequences from Atlantic salmon, *Salmo salar* L. *Animal Genetics* **24**, 195–7.

Slettan, A., Olsaker, I. & Lie, O. (1997) Segregation studies and linkage analysis of Atlantic salmon microsatellites using haploid genetics. *Heredity* **78**, 620–27.

Slynko, V.I. (1971) Polymorphism of malate dehydrogenase isozymes in Pacific salmon. In *Science Information of the Institute of Marine Biology*, Vol. 2 (Zhirmunsky, A.V., ed.), pp. 207–11. DVNTS, Vladivostok.

Slynko, V.I. (1976) Electrophoretic analysis of malate dehydrogenase isozymes in fishes of Salmonidae. *Proceedings of the USSR Academy of Sciences* **226**, 448–51.

Slynko, V.I. (1978) Biochemical polymorphism in salmonid fishes and its relationship with ecology of chum salmon *Oncorhynchus keta* and pink salmon *O. gorbuscha*. PhD thesis, Valvilov Institute of General Genetics, Moscow.

Small, M.P., Beacham, T.D., Withler, R.E. & Nelson, R.J. (1998) Discriminating coho salmon (*Oncorhynchus kisutch*) populations within the Fraser River. *Molecular Ecology* **7**, 141–55.

Smirnov, A.I. (1975) *Biology, Reproduction and Development of Pacific Salmon*. MGU, Moscow.

Smith, F.A. (1895) *A History of Scandinavian Fishes*, Part 2, pp. 567–1240. Stockholm.

Smith, G.R. (1975) Fishes of the Pliocene Glenns Ferry formation, South-west Idaho. *Papers on Palaeontology* **14**, 1–68.

Smith, G.R. & Stearley, R.F. (1989) The classification and scientific names of rainbow and cutthroat trout. *Fisheries* **14**, 4–10.

Smith, G.R., Swirydczuk, K., Kimmel, P.G. & Wilkinson B.H. (1982) Fish biostratiography of late Miocene to Pleistocene sediments of the western Snake River plain, Idaho. In *Cenizoic Geology of Idaho. IBMG Bulletin 26* (Bonnichsen, B. & Breckenridge, R.M., eds), pp. 519–41. Idaho Bureau of Mines and Geology, Moscow, ID.

Soin, S.G. (1980) Types of development of salmonid fishes and their taxonomic significance. *Journal of Ichthyology (Russia)* **20**, 65–72.

Sokolov, L.V. (1991) *Philopatry and Dispersion of Birds. Proceedings of the Zoological Institute of the USSR Academy of Sciences* **230**, 1–200.

Sokolov, V.E. & Shatunovsky, M.I. (1995) Is it possible to conserve biologic diversity? *Herald of the Russian Academy of Sciences* **66**, 422–4.

Soulé, M. (ed.) (1987) *Viable Populations for Conservation*. Cambridge University Press, Cambridge.

Southern, E.M. (1975) Detection of specific DNA fragments separated by gel electrophoresis. *Journal of Molecular Biology* **98**, 503–17.

Spruell, P. & Thorgaard, G.H. (1996) SINE sequences detect DNA fingerprints in salmonid fishes. *Heredity* **76**, 317–24.

Stabell, O.B. (1984) Homing and olfaction in salmonids, a critical review with special reference to the Atlantic salmon. *Biological Review of the Cambridge Philosophical Society* **59**, 334–88.

Stacey, P.B., Johnson, V.A. & Toper, M.L. (1997) Migration within metapopulation: the impact upon local population dynamics. In *Metapopulation Biology. Ecology, Genetics, and Evolution* (Hanski, I. & Gilpin, M.E., eds), pp. 267–91. Academic Press, San Diego.

Ståhl, G. (1983) Differences in the amount and distribution of genetic variation between natural populations and hatchery stocks of Atlantic salmon. *Aquaculture* **33**, 23–32.

Ståhl, G. (1987) Genetic population structure of Atlantic salmon. In *Population Genetics and Fishery Management* (Ryman, N. & Utter, F.M., eds), pp. 121–40. University of Washington Press, Seattle.

Ståhl, G. & Ryman, N. (1982) Simple Mendelian inheritance at a locus coding for L-glycerophosphate dehydrogenase in brown trout (*Salmo trutta*). *Hereditas* **96**, 313–15.

Stone, C.E., Taggart, J.B. & Ferguson, A. (1997) Single-locus minisatellite DNA variation in European populations of Atlantic salmon (*Salmon salar* L.). *Hereditas* **126**, 269–75.

Stoneking, M., May, B. & Wright, J.E. (1979) Genetic variations, inheritance, and quaternary structure of malic enzyme in brook trout. *Biochemical Genetics* **17**, 599–619.

Suzumoto, B.K., Schreck, C.B. & McIntyre, J.D. (1977) Relative resistance of three transferrin genotypes of coho salmon *Oncorhynchus kisutch* and their haematological responses to bacterial kidney disease. *Journal of the Fisheries Research Board of Canada* **34**, 1–8.

Svärdson, G. (1945) Chromosome studies on Salmonidae. *Report of the Institute of Freshwater Fisheries Research, Drottningholm* **23**, 1–151.

Svärdson, G. & Fagerstrom, A. (1982) Adaptive differences in the long-distance migration of some trout (*Salmo trutta* L.) stocks. *Report of the Institute of Freshwater Fisheries Research, Drottningholm*, **60**, 51–80.

Taggart, J.B. & Ferguson, A. (1984a) An electrophoretically detectable genetic tag for hatchery-reared brown trout (*Salmo trutta* L.). *Aquaculture* **41**, 119–30.

Taggart, J.B. & Ferguson, A. (1984b) Allozyme variation in the brown trout (*Salmo trutto* L.): single-locus and joint segregation inheritance studies. *Heredity* **53**, 339–59.

Taggart, J.B. & Ferguson, A. (1986) Electrophoretic evaluation of a supplemental stocking programme for brown trout, *Salmo trutta* L. *Aquaculture and Fishery Management* **17**, 155–62.

Taggart, J.B. & Ferguson, A. (1990) Hypervariable minisatellite DNA single-locus probes for the Atlantic salmon, *Salmo salar* L. *Journal of Fish Biology* **37**, 991–3.

Taggart, J.B., Ferguson, A. & Mason, G.M. (1981) Genetic variation in Irish populations of brown trout (*Salmo trutta* L.): electrophoretic analysis of isozymes. *Comparative Biochemistry and Physiology* **69B**, 393–412.

Taggart, J.B., Prodohl, P.A. & Ferguson, A. (1995a) Genetic markers for Atlantic salmon (*Salmo salar* L.): single-locus inheritance and joint segregation analysis of minisatellite (VNTR) DNA loci. *Animal Genetics* **26**, 13–20.

Taggart, J.B., Verspoor, E., Galvin, P.T., Moran, P. & Ferguson, A. (1995b) A minisatellite DNA marker for discrimination between European and North American Atlantic salmon (*Salmo salar*). *Canadian Journal of Fisheries and Aquatic Sciences* **52**, 2305–11.

Tajima, F. (1992) Statistical method for estimating the effective population size in Pacific salmon. *Journal of Heredity* **83**, 309–11.

Takagi, K., Aro, K.V., Hartt, A.C. & Dell, M.B. (1981) Distribution and origin of pink salmon (*Oncorhynchus gorbuscha*) in offshore waters of the North Pacific Ocean. *Bulletin of the International North Pacific Fisheries Commission* **40**, 1–196.

Takezaki, N. & Nei, M. (1996) Genetic distances and reconstruction of phylogenetic trees from microsatellite DNA. *Genetics* **144**, 389–99.

Tallman, R.F. & Healey, M.C. (1994) Homing, straying and gene flow among seasonally separated populations of chum salmon (*Oncorhynchus keta*). *Canadian Journal of Fisheries and Aquatic Sciences* **51**, 577–88.

Tautz, D. (1989) Hypervariability of simple sequences as a general source for polymorphic DNA markers. *Nucleic Acids Research* **17**, 6463–71.

Taylor, G. (ed.) (1997) *Laboratory Methods for the Detection of Mutations and Polymorphisms in DNA*. CRC Press, Boca Raton, FL.

Taylor, E.B., Beacham, T.D. & Kaeriyama, M. (1994) Population structure and identification of North Pacific Ocean chum salmon (*Oncorhynchus keta*) revealed by analysis of minisatellite DNA variation. *Canadian Journal of Fisheries and Aquatic Sciences* **51**, 1430–42.

Taylor, E.B., Foote, C.J. & Wood, C.C. (1996) Molecular genetic evidence for parallel life-history evolution within a Pacific salmon (sockeye salmon and kokanee, *Oncorhynchus nerka*). *Evolution* **50**, 401–16.

Taylor, E.B., Harvey, S., Pollard, S. & Volpe, J. (1997) Postglacial genetic differentiation of reproductive ecotypes of kokanee *Oncorhynchus nerka* in Okanagan Lake, British Columbia. *Molecular Ecology* **6**, 503–17.

Tchernavin, V.V. (1921) Origin of breeding dress in salmon. *Journal of the Petrograd Agricultural Institute* **3** (4), 177–200.

Templeton, A.R. (1986) Coadaptation and outbreeding depression. In *Conservation Biology: the Sciences of Scarcity and Diversity* (Soule, M.E., ed.), pp. 105–16. Sinauer, Sunderland, MA.

Tessier, N., Bernatchez, L., Presa, P. & Angers, B. (1995) Gene diversity analysis of mitochondrial DNA, microsatellites and allozymes in landlocked Atlantic salmon. *Journal of Fish Biology* **47** (Suppl. A), 156–63.

Tessier, N., Bernatchez, L. & Wright, J.M. (1997) Population structure and impact of supportive breeding inferred from mitochondrial and microsatellite DNA analysis in landlocked Atlantic salmon, *Salmo salar* L. *Molecular Ecology* **6**, 735–50.

Thomas, W.K. & Beckenbach, A.T. (1989) Variation in salmonid mitochondrial DNA: evolutionary constraints and mechanisms of substitution. *Journal of Molecular Evolution* **29**, 233–45.

Thomas, W.K. & Withler, R.E. & Beckenbach, A.T. (1986) Mitochondrial DNA analysis of Pacific salmonid evolution. *Canadian Journal of Zoology* **64**, 1058–64.

Thorpe, J.E. (1975) Early maturity in male Atlantic salmon. *Scottish Fish Bulletin* **42**, 15–17.

Thorpe, J.E. (1986) Age and first maturity in Atlantic salmon, *Salmo salar*: freshwater period influences and conflicts with smolting. *Canadian Special Publications on Fisheries and Aquatic Sciences* **89**, 7–14.

Thorpe, J.E. (1993) Impacts of fishing on genetic structure of salmonid populations. In *Genetic Conservation of Salmonid Fishes* (Cloud, J.G. & Thorgaard, G.H., eds), pp. 67–80. Plenum Press, New York.

Thorpe, J.E. (1994) Significance of straying in salmonids and implications for ranching. *Aquaculture and Fishery Management* **25** (Suppl. 2), 183–90.

Thorpe, J.E. & Mitchell, K.A. (1981) Stocks of Atlantic salmon (*Salmo salar*) in Britain and Ireland: discreteness, and current management. *Canadian Journal of Fisheries and Aquatic Sciences* **38**, 1576–90.

Timofeev-Ressovsky, N.W. & Timofeev-Ressovsky, H.A. (1927) Genetische Analyse einer freilebenden *Drosophila melanogaster* Population. *Wilhelm Roux Archives Entwicklungsmech Organism (Germany)* **109**, 70–109.

Tinkle, D.W. (1965) Population structure and effective size of a lizard populations. *Evolution* **19**, 569–73.

Torrisen, K.R., Male, R. & Naevdal, G. (1993) Trypsin isozymes in Atlantic salmon, *Salmo salar* L.: studies of heredity, egg quality and effect on growth of three different populations. *Aquaculture and Fishery Management* **24**, 407–15.

Torroni, A., Sucernik, R.I., Schurr, T.G. *et al.* (1993) mtDNA variation of aboriginal Siberians reveals distinct genetic affinities with native Americans. *American Journal of Human Genetics* **53**, 591–608.

Tsuyuki, H. & Roberts, E. (1966) Interspecies relationships within the genus *Oncorhynchus* based on biochemical systematics. *Journal of the Fisheries Research Board of Canada* **23**, 101–7.

Tsuyuki, H. & Ronald, A.P. (1970) Existence in salmonid haemoglobins of molecular species with three and four different polypeptides. *Journal of the Fisheries Research Board of Canada* **27**, 1326–8.

Tsuyuki, H., Roberts, E., Vanstone, W.E. & Markert, J.R. (1965) The species specificity and constancy of muscle myogen and haemoglobin electrophoregrams of *Oncorhynchus*. *Journal of the Fisheries Research Board of Canada* **22**, 215–17.

Turner, C.E. & Bilton, H.T. (1968) Another pink salmon (*Oncorhynchus gorbuscha*) in its third year. *Journal of the Fisheries Research Board of Canada* **25**, 1993–6.

Utter, F.M. (1991) Biochemical genetics and fishery management: an historical perspective. *Journal of Fish Biology* **39**(Suppl. A), 1–20.

Utter, F.M. & Hodgins, H.O. (1970) Phosphoglucomutase polymorphism in sockeye salmon. *Comparative Biochemistry and Physiology* **36**, 195–9.

Utter, F.M., Ames, W.E. & Hodgins, H.O. (1970) Transferrin polymorphism in coho salmon (*Oncorhynchus kisutch*). *Journal of the Fisheries Research Board of Canada* **27**, 2371–3.

Utter, F.M., Hodgins, H.O. & Johnson, A.G. (1972) Biochemical studies of genetic differences among species and stocks of fish. *International North Pacific Fishery Commission, Annual Report*, 98–101.

Utter, F.M., Allendorf, F.W. & Hodgins, H.O. (1973) Genetic variability and relationships in Pacific salmon and related trout based on protein variations. *Systematic Zoology* **22**, 257–70.

Utter, F.M., Hodgins, H.O. & Allendorf, F.M. (1974) Biochemical genetic studies of fishes: potentialities and limitations. In *Biochemical and Biophysical Perspectives in Marine Biology*, Vol. 1 (Malines, D.C. & Sargents, J.R., eds), pp. 213–38. Academic Press, New York.

Utter, F.M., Allendorf, F.W. & May, B. (1976) The use of protein variation in the management of salmonid populations. *Transactions of the 41st North American Wildlife and Nature Research Conference* **41**, 373–84.

Utter, F.M., Campton, D., Grant, S., Milner, G., Seeb, J. & Wishard, L. (1978) Population structures of indigenous salmonid species of the Pacific north-west: I. A within- and between-species examination of natural populations based on genetic variations of proteins. In *Salmonid Ecosystems of the North Pacific* (McNeil, W.J. & Himsworth D.C., eds), pp. 285–304. Oregon State University Press, Corvallis.

Utter, F.M., Allendorf, F.W. & May, B. (1979) Genetic basis of creatine kinase isozymes in skeletal muscle of salmonid fishes. *Biochemical Genetics* **17**, 1079–91.

Utter, F.M., Aebersold, P.B., Helle, J. & Winans, G. (1984) Genetic characterization of a population in the south-eastern range of sockeye salmon. In *Proceedings of the Olympic Wild Fish Conference* (Walton, J.M. & Houston, D.B., eds), pp. 17–32. Peninsula College, Port Angeles, WA.

Utter, F.M., Aebersold, P.B. & Winans, G.A. (1987) Interpreting genetic variation detected by electrophoresis. In *Population Genetics and Fishery Management* (Ryman, N., & Utter F.M., eds), pp. 21–45. University of Washington Press, Seattle.

Utter, F.M., Milner, G., Ståhl, G. & Teel, D. (1989) Genetic population structure of chinook salmon, *Oncorhynchus tshawytscha*, in the Pacific north-west. *Fisheries Bulletin* **87**, 239–64.

Varnavskaya, N.V. (1984) Adaptive genetic structure and its relationship with intrapopulation differentiation by sex, age and growth rate in Pacific sockeye salmon, *Oncorhynchus nerka* (Walb.). PhD thesis, Vavilov Institute of General Genetics, Moscow.

Varnavskaya, N.V. (1992) Genetic variation at 19 polymorphic loci in pink salmon (*Oncorhynchus gorbuscha* Walbaum) populations of odd-year broodline of east Kamchatka and North America. *Genetika* **28** (9), 127–40.

Varnavskaya, N.V. & Beacham, T.D. (1992) Biochemical genetic variation in odd-year pink salmon (*Oncorhynchus gorbuscha*) from Kamchatka. *Canadian Journal of Zoology* **70**, 2115–20.

Varnavskaya, N.V., Varnavsky, V.S., Vetsler, I.L. & Nepomnyashchiy, K.Yu. (1988) Peculiarities of genetic spatial differentiation of sockeye salmon (*Oncorhynchus nerka* Walbaum) populations of lakes Nachiki, Dalneye and Dvuyurtochnoye (Kamchatka). *Genetika* **24**, 723–31.

Varnavskaya, N.V., Wood, C.C. & Everett, R.J. (1994a) Genetic variation in sockeye salmon (*Oncorhynchus nerka*) populations of Asia and North America. *Canadian Journal of Fisheries and Aquatic Sciences* **51** (Suppl. 1), 132–46.

Varnavskaya, N.V., Wood, C.C., Everett, R.J. *et al.* (1994b) Genetic differentiation of subpopulations of sockeye salmon (*Oncorhynchus nerka*) within lakes of Alaska, British Columbia, and Kamchatka, Russia. *Canadian Journal of Fisheries and Aquatic Sciences* **51** (Suppl. 1), 147–57.

Vasiliev, V.P. (1977) Polyploidy in fish and some aspects of salmon karyotype evolution. *Journal of General Biology (Russia)* **38**, 380–92.

Vasiliev, V.P. (1985) *Evolutionary Karyology of Fish*. Nauka, Moscow.

Vaughan, E. (1947) Time of appearance of pink salmon runs in southern Alaska. *Copeia* **1**, 40–50.

Vernon, E.H. (1957) Morphometric comparison of three races of kokanee (*Oncorhynchus nerka*) within a large British Columbia lake. *Journal of the Fisheries Research Board of Canada* **14**, 573–98.

Verspoor, E. (1986) Spatial correlations of transferrin allele frequencies in Atlantic salmon (*Salmo salar*) populations from North America. *Canadian Journal of Fisheries and Aquatic Science* **43**, 1074–8.

Verspoor, E. (1988) Reduced genetic variability in first-generation hatchery populations of Atlantic salmon (*Salmo salar*). *Canadian Journal of Fisheries and Aquatic Sciences* **45**, 1686–90.

Verspoor, E. (1994) The evolution of genetic divergence at protein-coding loci among anadromous and non-anadromous populations of Atlantic salmon *Salmo salar*. In *Genetics and Evolution of Aquatic Organisms* (Beaumont, A.R., ed.), pp. 52–67. Chapman & Hall, London.

Verspoor, E. & Jordan, W.C. (1989) Genetic variation at the *Me-2* locus in the Atlantic salmon within and between rivers: evidence for its selective maintenance. *Journal of Fish Biology* **35**, (Suppl. A), 205–13.

Verspoor, E., Fraser, N.H.C. & Youngson, A.F. (1991) Protein polymorphism in Atlantic salmon within a Scottish river: evidence for selection and estimates of gene flow between tributaries. *Aquaculture* **98**, 217–30.

Viktorovsky, R.M. (1975) Karyotypes of white-spotted charr (*Salvelinus leucomaenis*) and Dolly Varden (*S. malma*). *Journal of Zoology (Russia)* **54**, 787–9.

Viktorovsky, R.M. (1978) *Speciation Mechanisms in Charrs of Lake Kronotsky*. Nauka, Moscow.

Viktorovsky, R.M. & Glubokovsky, M.K. (1977) Mechanisms and rates of speciation in charrs of the genus *Salvelinus* (Salmonidae, Pisces). *Proceedings of the USSR Academy of Sciences* **235**, 946–9.

Viktorovsky, R.M., Bachevskaya, L.T., Ermolenko, L.N. *et al.* (1986) Genetic population

structure of chum salmon from the north-east USSR and problems of rational utilization of their resources. *Marine Biology (Russia)* **2**, 51–9.
Vladimirov, V.I. (1946) Trout from diatomic deposits of Pliocene in Armenia. *Proceedings of the Armenian Academy of Sciences* **4**, 123–8.
Vladimirov, V.I. (1948) Brook trout of Armenia and its relationship with other species of the genus *Salmo*. *Proceedings of Lake Sevan Hydrobiological Station* **10**, 87–178.
Vladykov, V.D. (1963) A review of salmonid genera and their broad geographical distribution. *Transactions of the Royal Society of Canada, Series IV, Section III* **1**, 459–504.
Vogel, F. & Motulsky, A.G. (1986) *Human Genetics*, 2nd edn. Springer-Verlag, Berlin.
Vosberg, H.P. (1977) Molecular cloning of DNA. An introduction into techniques and problems. *Human Genetics* **40**, 1–72.
Vronsky, B.B. (1972) Materials on the reproduction of chinook salmon *Oncorhynchus tshawytscha* (Walbaum) of the Kamchatka River. *Journal of Ichthyology (Russia)* **12**, 293–308.
Vuorinen, J. (1984) Reduction of genetic variability in a hatchery stock of brown trout, *Salmo trutta* L. *Journal of Fish Biology* **24**, 339–48.
Vuorinen, J. & Berg, O.K. (1989) Genetic divergence of anadromous and non-anadromous Atlantic salmon (*Salmo salar*) in the River Namsen, Norway. *Canadian Journal of Fisheries and Aquatic Sciences* **46**, 406–9.
Vuorinen, J. & Piironen, J. (1984) Electrophoretic identification of Atlantic salmon (*Salmo salar*), brown trout (*S. trutta*), and their hybrids. *Canadian Journal of Fisheries and Aquatic Sciences* **41**, 1834–7.
Wahlund, S. (1928) Zusammensetzung von populationen und korrelationserscheinungen von standpunkt der vererbungslehre aus betrachtet. *Hereditas* **11**, 65–106.
Wallace, B. (1966) On the dispersal of *Drosophila*. *American Nature* **100**, 551–62.
Wallace, B. (1979) The migration of a mutant gene into isolated populations. *Genetica* (Netherlands) **50**, 67–72.
Walton, K.E., Stuer, D. & Gruenstein, E.A. (1979) Genetic polymorphism in normal human fibroblasts as analysed by two-dimensional polyacrylamide gel electrophoresis. *Journal of Biological Chemistry* **254**, 7951–60.
Waples, R.S. (1990a) Conservation genetics of Pacific salmon. II. Effective population size and the rate of loss of genetic variability. *Journal of Heredity* **81**, 267–76.
Waples, R.S. (1990b) Conservation genetics of Pacific salmon. III. Estimating effective population size. *Journal of Heredity* **81**, 277–89.
Waples, R.S. (1991) Genetic interaction between hatchery and wild salmonids: lessons from the Pacific north-west. *Canadian Journal of Fisheries and Aquatic Sciences* **48** (Suppl. 1), 124–33.
Waples, R.S. & Teel, D.J. (1990) Conservation genetics of Pacific salmon. I. Temporal changes in allele frequency. *Conservation Biology* **4**, 144–56.
Ward, R.D. & Grewe, P. (1994) Appraisal of molecular genetic techniques in fisheries. *Review of Fish Biology and Fisheries* **4**, 300–25.
Ward, R.D., Woodwark, M. & Skibinski, D.O.F. (1994) A comparison of genetic diversity levels in marine, freshwater and anadromous fish. *Journal of Fish Biology* **44**, 213–42.
Waser, N.M. & Price, M.V. (1989) Optimal outcrossing in *Ipomopsis aggregata*: seed set and offspring fitness. *Evolution* **43**, 1097–109.
Washington, P.M. & Koziol, A.M. (1993) Overview of the interactions and environmental

impacts of hatchery practices on natural and artificial stocks of salmonids. *Fisheries Research* **18**, 105–22.

Weaver, J.E. & Stolte, L.W. (1992) Atlantic salmon restoration and management in the United States. *Proceedings of Scientific Papers of the State Research Institute on Lake and River Fisheries* **304**, 61–84.

Wehrhahn, C.F. & Powell, R. (1987) Electrophoretic variation, regional differences, and gene flow in coho salmon (*Oncorhynchus kisutch*) of southern British Columbia. *Canadian Journal of Fisheries and Aquatic Sciences* **44**, 822–31.

Wenburg J.K., Bentzen P. & Foote C.J. (1998) Microsatellite analysis of genetic population structure in an endangered salmonid – the coastal cutthroat trout (*Oncorhynchus clarki clarki*). *Molecular Ecology* **7**, 733–49.

Whitlock, M.C. & Barton, N.H. (1997) The effective size of a subdivided population. *Genetics* **146**, 427–41.

Wilding, C.S., Beaumont, A.R. & Latchford, J.W. (1997) Mitochondrial DNA variation in the scallop *Pecten maximus* (L) assessed by a PCR-RFLP method. *Heredity* **79**, 178–89.

Wilkins, N.P. (1968) Multiple haemoglobins of the Atlantic salmon (*Salmo salar*). *Journal of the Fisheries Research Board of Canada* **25**, 2651–63.

Wilkins, N.P. (1970) The subunit composition of the haemoglobins of the Atlantic salmon (*Salmo salar* L.). *Biochemica et Biophysica Acta* **214**, 52–63.

Wilkins, N.P. (1989) *Ponds, Passes and Parks. Aquaculture in Victorian Ireland*. Glendale, Dublin.

Williamson, R.B. (1988) Status of exploitation of Atlantic salmon in Scotland. In *Atlantic Salmon: Planning for the Future. Proceedings of the 3rd International Atlantic Salmon Symposium*, Biarritz, October, 1986 (Mills, D. & Piggins, D., eds), pp. 91–116. Timber Press, Portland, ME.

Williscroft, S.N. & Tsuyuki, H. (1970) LDH systems of rainbow trout: evidence for polymorphism in liver and additional subunits in gills. *Journal of the Fisheries Research Board of Canada* **27**, 1563–7.

Wilmot, R.L. & Burger, C.V. (1985) Genetic differences among population of Alaskan sockeye salmon. *Transactions of the American Fisheries Society* **114**, 236–43.

Wilmot, R.L., Everett, R., Spearman, W.J. & Baccus, R. (1992) *Genetic stock identification of Yukon River chum and chinook salmon, 1987 to 1990*. Progress Report. US Fish and Wildlife Service, Anchorage, AK.

Wilmot, R.L., Everett, R.J., Spearman, W.J., Baccus, R., Varnavskaya, N.V. & Putivkin, S.V. (1994) Genetic stock structure of western Alaska chum salmon and comparison with Russian Far East stock. *Canadian Journal of Fisheries and Aquatic Sciences* **51**, (Suppl. 1), 84–94.

Wilmot, R.L., Kondzela, C.M., Guthrie, C.M. & Masuda, M.S. (1998) Genetic stock identification of chum salmon harvested incidentally in the 1994 and 1995 Bering Sea trawl fishery. In *Assessment and status of Pacific rim salmonid stocks. NPAFC Bulletin No. 1* (Welch, D.W., Eggers, D.M., Wakabayashi, K. & Karpenko, V., eds), pp. 285–9. North Pacific Anadromous Fish Commission, Vancouver.

Wilson, A.C. (ed.) (1988) *Biodiversity*. National Academy Press, Washington DC.

Wilson, G.M., Thomas, W.K. & Beckenbach, A.T. (1987) Mitochondrial DNA analysis of Pacific north-west populations of *Oncorhynchus tshawytscha*. *Canadian Journal of Fisheries and Aquatic Sciences* **44**, 1301–5.

Wilson, I.E., Bourke, E.A. & Cross, T.F. (1995) The triose phosphate isomerase polymorphism in the Atlantic salmon, *Salmo salar* L. *Biochemical Genetics* **33**, 25–33.

Winans, G.A., Aebersold, P.B., Urawa, S. & Varnavskaya, N.V. (1994) Determining continent of origin of chum salmon (*Oncorhynchus keta*) using genetic stock identification techniques: status of allozyme baseline in Asia. *Canadian Journal of Fisheries and Aquatic Sciences* **51**, (Suppl. 1), 95–113.

Winter, G.W., Schreck, C.B. & McIntyre, J.B. (1980) Resistance of different stocks and transferrin genotypes of coho salmon *Oncorhynchus kisutch* and steelhead trout *Salmo gairdneri* to bacterial kidney disease and vibriosis. *Fisheries Bulletin* **77**, 795–802.

Withler, F.C. (1982) Transplanting Pacific salmon. *Canadian Technical Report on Fisheries and Aquatic Sciences* **1079**, 1–27.

Withler, R.E. (1985) *LDH-4* allozyme variability in North American sockeye salmon, (*Oncorhynchus nerka*) populations. *Canadian Journal of Zoology* **63**, 2924–32.

Withler, R.E., Beacham, T.D., Watkins, R.F. & Stevens, T.A. (1994) Identification of farm-reared and native chinook salmon (*Oncorhynchus tshawytscha*) on the west coast of Vancouver Island, British Columbia, using the nuclear DNA probe B2-2. *Canadian Journal of Fisheries and Aquatic Sciences* **51**, (Suppl. 1), 267–76.

Wolf, P.G. & Soltis, P.S. (1992) Estimates of gene flow among populations, geographic races, and species in the *Ipomopsis aggregata* complex. *Genetics* **130**, 639–47.

Wood, C.C., McKinnell, S., Mulligan, T.J. & Fournier, D.A. (1987) Stock identification with the maximum-likelihood mixture model: sensitivity analysis and application to complex problems. *Canadian Journal of Fisheries and Aquatic Sciences* **44**, 866–81.

Wood, C.C., Rutherford, T.D., & McKinnell, S. (1989) Identification of sockeye salmon (*Oncorhynchus nerka*) stocks in mixed-stock fisheries in British Columbia and south-east Alaska using biological markers. *Canadian Journal of Fisheries and Aquatic Sciences* **46**, 2108–10.

Wood, C.C., Riddel, B.E., Rutherford, D.T. & Withler, R.E. (1994) Biochemical genetic survey of sockeye salmon (*Oncorhynchus nerka*) in Canada. *Canadian Journal of Fisheries and Aquatic Sciences* **51**, (Suppl. 1), 114–31.

Wright, J. (1993) DNA fingerprinting in fishes. In *Biochemistry and Molecular Biology of Fishes*, Vol. 2 (Hochachka, P. & Mommsen, T., eds), pp. 57–91. Elsevier, Amsterdam.

Wright, J.M. (1994) Mutation at VNTRs: are minisatellites the evolutionary progeny of microsatellites? *Genome* **37**, 345–7.

Wright, J.M. & Atherton, L.M. (1970) Polymorphism for LDH and transferrin loci in brook trout populations. *Transactions of the American Fisheries Society* **99**, 179–92.

Wright, J.M. & Bentzen, P. (1994) Microsatellites: genetic markers for the future. *Review of Fish Biology and Fisheries* **4**, 384–8.

Wright, J.E., Heckman, J.R. & Atherton, L.M. (1975) Genetic and developmental analysis of LDH isozymes in trout. In *Isozymes III: Developmental Biology* (Markert, C.L. ed.), pp. 375–401. Academic Press, New York.

Wright, J.E., May, B., Stoneking, M. & Lee, G.M. (1980) Pseudolinkage of the duplicate loci for supernatant aspartate aminotransferase in brook trout *Salvelinus fontinalis*. *Journal of Heredity* **71**, 223–8.

Wright, J.E., Johnson, K., Hollister, A. & May, B. (1983) Meiotic models to explain classical linkage, pseudolinkage, and chromosome pairing in tetraploid derivative salmonids. In

Isozymes: Current Topics in Biological and Medical Research, Vol. 10 (Rattazzi, M.C., Scandalios, J.G. & Whitt, G.S., eds) pp. 239–60. Liss, New York.

Wright, S. (1931) Evolution in Mendelian populations. *Genetics* **16**, 97–159.

Wright, S. (1932) The role of mutation, inbreeding, crossbreeding, and selection in evolution. *Proceedings of the 6th International Congress on Genetics* **1**, 356–66.

Wright, S. (1938) Size of populations and breeding structure in relation to evolution. *Science* **87**, 430–31.

Wright, S. (1943a) Isolation by distance. *Genetics* **28**, 114–38.

Wright, S. (1943b) An analysis of local variability of flower colour in *Linanthus parryae*. *Genetics* **28**, 139–56.

Wright, S. (1951) The genetic structure of populations. *Annals of Eugenics* **15**, 323–54.

Wright, S. (1969) *Evolution and the Genetics of Populations. The Theory of Gene Frequencies*, Vol. 2. University of Chicago Press, Chicago.

Wright, S. (1970) Random drift and shifting balance theory of evolution. In *Mathematical Topics in Population Genetics* (Kojima, K., ed.), pp. 1–31. Springer-Verlag, Berlin.

Wright, S. (1977) *Evolution and the Genetics of Populations. Experimental Results and Evolutionary Deductions*, Vol. 3. University of Chicago Press, Chicago.

Wright, S. (1978) *Evolution and the Genetics of Populations. Variability Within and Among Natural Populations* Vol. 4. University of Chicago Press, Chicago.

Yablokov, A.V. (1987) *Population Biology*. Visshaya Shkola, Moscow.

Yakovenko, M.Ya. (1998) Acclimatization of pink salmon as a source for replenishment of fishery bioresources of the White Sea. In *Problems of Investigation, Rational Exploitation and Protection of Natural Resources of the White Sea* (Alimov, A.F., ed.), pp. 41–7. Ministry of Science and Technologies of Russia, St Petersburg.

Yakovlev, V.N. (1961) Distribution of freshwater Neogene fishes in holarctic and zoogeographic districts. *Journal of Ichthyology (Russia)* **1**, 209–20.

Yakovlev, V.N. (1964) History of formation of freshwater faunistic complexes. *Journal of Ichthyology (Russia)* **4**, 10–22.

Yamanaka, H., Yamaguchi, K., Hashimoto, K. & Matsuura, F. (1967) Starch gel electrophoresis of fish haemoglobins. III. Salmonid fishes. *Bulletin of the Japanese Society of Scientific Fisheries* **33** (3), 195–203.

Yanulov, K.P. (1962) About groups of deepwater redfish (*Sebastes mentella* Travin) in the Labrador–Newfoundland region. In *Soviet Fisheries Research in the North-West Atlantic* (Marti, Yu.Yu., ed.), pp. 285–96. Rybnoye Khozyaistvo, Moscow.

Young, W.P., Wheeler, P.A., Coryell, V.H., Keim, P. & Thorgaard, G.H. (1998) A detailed linkage map of rainbow trout using doubled haploids. *Genetics* **148**, 839–50.

Yves, H. (1992) Pacific salmon in Atlantic water. *International Council for the Exploration of the Sea, Marine Science Symposium* **194**, 31–55.

Zardoya, R., Garrido-Pertierra, A. & Bautista, J.M. (1995) The complete nucleotide sequence of the mitochondrial DNA of the rainbow trout, *Oncorhynchus mykiss*. *Journal of Molecular Evolution* **41**, 942–51.

Zelinsky, Yu.P. (1985) *Structure and Differentiation of Populations and Forms in Atlantic salmon*. Nauka, Leningrad.

Zelinsky, Yu.P. & Gruzdev, A.I. (1992) On the value of variation of the *LDH-5* locus in an analysis of population sample of *Salmo trutta* adapted to gumification. *Journal of Ichthyology (Russia)* **32**, 57–62.

Zenkevich, L.A. (1933) Some aspects of zoogeography of the North Polar basin in connection with problems of its palaeogeographic past. *Journal of Zoology (Russia)* **12**, 17–34.

Zhivotovsky, L.A. (1979) Similarity index of populations for polymorphic traits. *Journal of General Biology (Russia)* **40**, 587–602.

Zhivotovsky, L.A. (1981) On Fisher's theorem. *Genetika* **17**, 324–31.

Zhivotovsky, L.A. (1984) *Integration of Polygenic Systems in Populations*. Nauka, Moscow.

Zhivotovsky, L.A. & Feldman, M.W. (1995) Microsatellite variability and genetic distance. *Proceedings of the National Academy of Sciences of the USA* **92**, 11549–52.

Zhivotovsky, L.A., Afanas'yev, K.I. & Rubtsova, G.A. (1987) Selection for enzymatic loci in pink salmon, *Oncorhynchus gorbuscha* (Walb.). *Genetika* **23**, 1876–83.

Zhivotovsky, L.A., Glubokovsky, M.K., Viktorovsky, R.M. *et al.* (1989) Genetic differentiation in pink salmon. *Genetika* **25**, 1261–74.

Zhivotovsky, L.A., Gharrett, A.J., McGregor, A.J. & Glubokovsky, M.K. (1994) Gene differentiation in Pacific salmon (*Oncorhynchus* spp.): facts and models with reference to pink salmon (*O. gorbuscha*). *Canadian Journal of Fisheries and Aquatic Sciences* **51** (Suppl. 1), 223–32.

Zhivotovsky, L.A., Khramtsov, V.V. & Glubokovsky, M.K. (1996) The model of abundance dynamics in pink salmon. *Journal of Ichthyology (Russia)* **36**, 369–85.

Zolotareva, I.M. (1980) Estimation of breeding effectivity of pink salmon in Sakhalin salmon hatcheries. In *Salmonid Fishes* (Skarlato, O.A., ed.), pp. 259–61. Nauka, Leningrad.

Zouros, E. & Foltz, D.W. (1987) The use of allelic isozyme variation for the study of heterosis. In *Isozymes: Current Topics in Biological and Medical Research*, Vol. 13 Rattazzi, M.C., Scandalios, J.G. & Whitt, G.S. eds), pp. 1–59. Alan R. Liss, New York.

Subject index

Author index

Index of Latin Names

Printed and bound in the UK by
CPI Antony Rowe, Eastbourne

Printed and bound by CPI Group (UK) Ltd, Croydon, CR0 4YY
16/09/2026
14963385-0006